Food Chemistry and Bioactive Compounds in Relation to Health

Food Chemistry and Bioactive Compounds in Relation to Health

Guest Editors

Leontina Grigore-Gurgu
Elena Enachi
Iuliana Aprodu

MDPI

Basel • Beijing • Wuhan • Barcelona • Belgrade • Novi Sad • Cluj • Manchester

Guest Editors
Leontina Grigore-Gurgu
Faculty of Food Science
and Engineering
Dunarea de Jos University
of Galati
Galati
Romania

Elena Enachi
Faculty of Food Science
and Engineering
Dunarea de Jos University
of Galati
Galati
Romania

Iuliana Aprodu
Faculty of Food Science
and Engineering
Dunarea de Jos University
of Galati
Galati
Romania

Editorial Office
MDPI AG
Grosspeteranlage 5
4052 Basel, Switzerland

This is a reprint of the Special Issue, published open access by the journal *Molecules* (ISSN 1420-3049), freely accessible at: https://www.mdpi.com/journal/molecules/special_issues/89GEWIPG62.

For citation purposes, cite each article independently as indicated on the article page online and as indicated below:

Lastname, A.A.; Lastname, B.B. Article Title. *Journal Name* **Year**, *Volume Number*, Page Range.

ISBN 978-3-7258-5885-9 (Hbk)
ISBN 978-3-7258-5886-6 (PDF)
https://doi.org/10.3390/books978-3-7258-5886-6

Contents

About the Editors

Leontina Grigore-Gurgu

Leontina Grigore-Gurgu is a lecturer at the Faculty of Food Science and Engineering, Dunarea de Jos University of Galati, Romania. She graduated from the Faculty of Biology, Bucharest University in 2001 and completed her PhD in Industrial Engineering in 2011. Her academic portfolio includes courses in microbiology, biotechnology, and hygiene in food industry, where she combines theoretical knowledge with practical insights. Her research focuses on microbial genetics, foodborne pathogens, and functional food components analysis, with publications in applied science and molecular biology fields.

Elena Enachi

Elena Enachi is a lecturer and a researcher at the Faculty of Medicine and Pharmacy and at Faculty of Food Science and Engineering—the Integrated Center of Research, Expertise and Technological Transfer in the Food Industry (BioAliment—TehnIA), "Dunarea de Jos" University of Galati, Romania. Her doctoral research focused on studying and assessing the kinetic and molecular behavior of oxidative enzymes in terms of a process–structure–function relationship, as part of her PhD in Biotechnology. Her academic work and expertise explore the field of Instrumental Analysis with a particular focus on analytical techniques such as chromatographic and spectroscopy techniques and molecular biology techniques such as confocal microscopy. Her main research interests include both the study of various bioactive molecules from different sources that have the potential to induce numerous health benefits and the study of pharmaceutical compounds in terms of their pharmacological effects, formulation strategies, and therapeutic potential.

Iuliana Aprodu

Iuliana Aprodu is a Professor at the Faculty of Food Science and Engineering at Dunarea de Jos University of Galați, Romania, with extensive expertise in food biochemistry, protein characterization, and innovative food processing technologies. Her research focuses on the structural and functional properties of food proteins, enzymatic hydrolysis, and bioactive compounds, contributing to advancements in food quality, safety, and sustainability. Through her work, she integrates cutting-edge analytical techniques and computational modeling to optimize food formulations and develop novel ingredients with enhanced functional properties. She has been actively involved in national and international research projects, fostering interdisciplinary collaborations in bioprocessing and food technology.

Preface

This Special Issue reprint is focused on the chemical composition of foods and, in particular, on the bioactive compounds they contain, and how these elements influence human health. It encompasses studies regarding the identification, characterization, and quantification of bioactive molecules in food matrices.

The primary aim was to provide a comprehensive overview of current research on the health-related benefits of the bioactive compounds, linking important food chemistry knowledge with well-being outcomes.

By compiling diverse studies, this reprint seeks to highlight the role of bioactive compounds in disease prevention and health promotion, and to encourage the development of functional foods with enhanced health benefits, while discussing the implications of food processing on the bioavailability and efficacy of bioactive molecules.

With the rise in chronic diseases and lifestyle-related health issues, the need to understand how dietary components affect human physiology became urgent. Therefore, the motivation behind this Special Issue reprint stems from the growing recognition of the diet's impact on health. The studies included in this Special Issue reprint contribute to the following:

- providing evidence-based insights into the health benefits of bioactive compounds;
- supporting the development of dietary guidelines and public health strategies;
- promoting interdisciplinary research between food science, chemistry, and health disciplines.

The purpose of this reprint is to inform and inspire on the potential of food-based interventions in public health. This Special Issue reprint is addressed to the following:

- researchers and academics in food science, chemistry, and nutrition;
- healthcare professionals interested in dietary interventions;
- policy makers involved in public health and nutrition;
- industry professionals in the food and nutraceutical sectors.

By engaging this diverse audience, new collaborations and innovations in the field of food chemistry and health might be fostered.

Leontina Grigore-Gurgu, Elena Enachi, and Iuliana Aprodu
Guest Editors

molecules

MDPI

Editorial

Food Chemistry and Bioactive Compounds in Relation to Health

Leontina Grigore-Gurgu [1,*], Elena Enachi [1,2,*] and Iuliana Aprodu [1,*]

[1] Faculty of Food Science and Engineering, Dunarea de Jos University of Galati, 111 Domneasca Street, 800201 Galati, Romania

[2] Faculty of Medicine and Pharmacy, Dunarea de Jos University of Galati, 47 Domneasca Street, 800008 Galati, Romania

* Correspondence: leontina.gurgu@ugal.ro (L.G.-G.); elena.enachi@ugal.ro (E.E.); iuliana.aprodu@ugal.ro (I.A.)

Bioactive compounds present in food are essential factors in maintaining optimal health, through their ability to modulate physiological processes, reduce oxidative stress, and contribute to the prevention of chronic diseases related to aging. This Special Issue of *Molecules*, entitled "Food Chemistry and Bioactive Compounds in Relation to Health", includes ten original research articles and one review paper, manuscripts that address several advanced methods for extraction, purification, and structural characterization of bioactive compounds from food and natural sources, as well as the functional evaluation of their biological activities.

The studies included in this Special Issue can be grouped into six categories: (i) the antioxidant and anti-inflammatory potential of polyphenols, illustrated by the inhibition activity of metalloproteinases, the evaluation of the antioxidant properties of *Echinacea purpurea*, and the newest insights into the phenolic compounds from *Prunus spinosa*; (ii) the protective mechanisms against metabolic stress and cellular apoptosis induced by dicarbonyls; (iii) bioactive peptides and enzymatic hydrolysates with specific activity, including peptides that stimulate alcohol dehydrogenase activity and functional modulators obtained from brewer's yeast; (iv) antimicrobial activities and efflux pump inhibition by sesquiterpenes (nerolidol, farnesol, α-bisabolol) formulated in nano-liposomes; (v) compositional and sensory analyses of cold-pressed oils and functional plant extracts, with risk assessment of polycyclic aromatic hydrocarbons in Chinese medicinal plants and characterization of cold-pressed plant oils; and (vi) metabolomic and prebiotic evaluations of fruit-derived compounds through the bioactive potential of micronized raspberry pomace powders and the bioaccessibility and prebiotic potential of polyphenols from black elderberry (*Sambucus nigra*). In order to motivate readers to discover the papers of this Special Issue, a short and comprehensive description is provided in the following paragraphs.

Latronico et al. [Contribution 1] investigated the effects of four dietary antioxidants, green tea extract (GTE), resveratrol (RSV), curcumin (CRC), and olive fruit extract (OLI), on oxidative stress and the activity of MMP-2 and MMP-9 metalloproteinases. The targeted enzymes are known to be involved in breast cancer progression. All tested compounds and extracts demonstrated in vitro antioxidant activity, the most effective being the GTE. On the THP-1 macrophages activated with LPS (lipopolysaccharide), GTE, RSV, and CRC significantly reduced reactive oxygen species (ROS) production and the levels of MMP-2/MMP-9, while OLI showed minimal effects. In-gel activity analysis of patient sera confirmed a dose-dependent inhibition, with GTE being the most potent overall inhibitor, and CRC and OLI exhibiting selective effects on the MMP-2 gelatinase. The results suggested that polyphenols, through several different mechanisms including oxidative stress reduction and metal ion chelation, may represent promising agents for the prevention and adjuvant therapy of breast cancer. However, to maximize the in vivo efficacy, strategies to increase bioavailability, such as the use of nanomaterial-based delivery systems, are still needed. These results were consistent with previous reports showing that the green tea catechins inhibit MMP activity, both through direct binding and modulation of ROS [1,2], and RSV reduces MMP-9 expression through the NF-κB and AP-1 pathways [1,2].

Liccardo et al. [Contribution 2] investigated the effect of genistein, an isoflavone from soy, on methylglyoxal (MG)-induced cytotoxicity in endothelial cells, a process associated with vascular complications of diabetes. The methylglyoxal compound has been identified as a promoter of oxidative stress and apoptosis, through the generation of reactive oxygen species (ROS) and activation of the mitogen-activated protein kinases (MAPKs), extracellular signal-regulated kinase (ERK/p38), and caspase-3 pathways. The results showed that genistein significantly reduced ROS production, inhibited nuclear erythroid 2-related factor (Nrf2) translocation, and blocked MAPK and caspase-3 activation, thereby preventing apoptosis. These findings suggest that genistein could be used as a therapeutic supplement to prevent endothelial dysfunction and vascular complications associated with diabetes. The obtained results are similar to those reported by Pang et al. [3], who showed that polydatin protects human endothelial cells against MG-induced apoptosis by reducing oxidative stress and maintaining mitochondrial potential [3]. Do et al. [4] also showed that genistein-rich plant extracts reduce ROS formation and apoptosis under glucotoxicity conditions.

Wang et al. [Contribution 3] investigated several corn gluten-derived peptides with alcohol dehydrogenase (ADH) activation potential, after simulated gastrointestinal digestion and absorption through an in vitro model of the human intestine with Caco-2 cell monolayers. While the hydrolysate obtained with alcalase showed the highest ADH activation activity (≈55%), the peptides fraction with a molecular weight lower than 1 kDa was the most active. Three peptides, namely SSNCQPF, TGCPVLQ, and QPQQPW, were identified in this fraction. The identified peptides exhibited potent in vitro ADH activation activity, with EC_{50} values of 1.35 mM, 2.26 mM, and 2.73 mM, respectively. Molecular docking analyses showed that these peptides bind with the active site of ADH through hydrogen bonds and hydrophobic interactions and are able to create a stable complex. Compared to chickpea peptides (EC_{50} > 3 mM) [5], those from corn showed stronger activity. When compared to the tripeptide KPC from chicken breast (binding energy −6.6 kcal/mol) [6], the SSNCQPF and TGCPVLQ peptides had lower binding energies (−8.69 and −8.20 kcal/mol), indicating a more stable interaction with ADH.

Cao et al. [Contribution 4] assayed the concentrations, sources, and health risks associated with polycyclic aromatic hydrocarbons (PAHs) in seven Chinese herbal medicines using a rapid extraction and purification method, followed by gas chromatography–mass spectrometry analysis (GC-MS). The total PAH content ranged from approximately 177 to 1414 μg/kg, with phenanthrene being the main compound. The characteristic ratios analysis indicated that the main sources of contamination were from oil, coal, and biomass combustion. Through the incremental lifetime cancer risk assessment, it was shown that honeysuckle ingestion posed a potential risk, while the *Lycium chinense* fruit posed a much lower risk. The other five investigated plants, namely ginseng, glycyrrhizae, *Coix lacryma*, and seeds of lotus and *Sterculia lychnophora*, were found to be within acceptable limits. The sensitivity analysis, carried out using Monte Carlo simulations, highlighted that PAH concentration is considered to be the main risk determinant. The study confirmed the observations of Ishizaki et al. [7] regarding the predominance of 2–3-ring PAHs in medicinal plants, suggesting the recent contamination and increased mobility throughout the environment. Also, the level of benzo[a]pyrene (BaP) detected in the honeysuckle samples exceeded the safety limit set by European regulations, which is very similar to the results obtained by Yu et al. [8] in the analysis of PAHs in medicinal plants used as food additives.

Adebimpe Ojo et al. [Contribution 5] evaluated the antioxidant properties of flowers and leafy stems of *Echinacea purpurea* grown with synthetic mulches of different colors (black, green, brown) and thicknesses (80 and 100 g/m^2). The panel of analyses included chemical composition (protein, fat, ash), total polyphenol content, and antioxidant activity (DPPH, ABTS, FRAP), as well as polyphenol profile by HPLC. The results showed that the flowers had significantly higher protein, ash, and polyphenol contents and antioxidant activity compared to the stems and leaves. The main phenolic compounds identified were p-coumaric acid, chlorogenic acid, and rutin. The mulch did not significantly influence the overall chemical composition but instead caused variations in the concentration of some polyphenols, thus confirming the potential of *E. purpurea* flowers as a source of natural antioxidants for functional food and cosmetic and pharmaceutical products. The total polyphenol concentrations in the flowers were within the range reported by the study of Tsai et al. [9], hence confirming the high antioxidant potential of this species. Moreover, rutin was identified as the major flavonoid in all the flower samples, complying with the observations of Skrzypczak-Pietraszek et al. [10], who highlighted the essential role it plays in the antioxidant activity of plant extracts.

Santana et al. [Contribution 6] compared the antibacterial activity and efflux pump (NorA, Tet(K), MsrA, MepA) inhibition capacity of the nerolidol, farnesol, and α-bisabolol sesquiterpenes, used individually and as liposomal nanoformulations, against multidrug-resistant strains of *Staphylococcus aureus*. In vitro tests showed that the isolated sesquiterpenes exhibited direct antibacterial activity and reduced the minimum inhibitory concentrations (MIC) of the antibiotics and ethidium bromide (EtBr), indicating the inhibition of efflux pumps. Nerolidol reduced the MIC of EtBr to 5 μg/mL for the 1199B strain, exceeding the standard carbonyl cyanide 3-chlorophenyl hydrazone inhibitor, and farnesol exhibited the lowest MIC (16 μg/mL) against the strain RN4220. Liposomal nanoformulations did not significantly improve the activity, except for farnesol liposomes, which reduced the MIC of EtBr to 40.3 μg/mL. The study results align with the observations of de Moura et al. [11], who demonstrated the efficacy of nerolidol against multidrug-resistant bacteria, including by inhibiting the formation of biofilm. Also, da Cruz et al. [12] highlighted the ability of α-bisabolol to inhibit Tet(K) and NorA pumps, and Oliveira et al. [13] showed that farnesol can potentiate the effect of conventional antibiotics, such as fusidic acid, against *S. aureus* strains expressing the MrsA mechanism.

Rabiej-Kozioł et al. [Contribution 7] presented a comprehensive characterization of the chemical composition, antioxidant activity, oxidative stability, and sensory attributes of six cold-pressed oils, namely flax (*Linum usitatissimum*), pumpkin (*Cucurbita pepo*), milk thistle (*Silybum marianum*), rapeseed (*Brassica napus*), camelina (*Camelina sativa*), and sunflower (*Helianthus annuus*), commercialized on the Polish market. The analyses included the determination of the fatty acid profile, the quantification of the content of tocopherols (44.04–76.98 mg/100 g), sterols (300–684 mg/100 g). and polyphenols (2.93–8.32 mg GA/100 g), as well as the evaluation of antioxidant activity by DPPH, ABTS, and FRAP methods, with values ranging between 185.36 and 396.63 μmol TE/100 g, 958.59 and 1638.58 μmol TE/100 g, and 61.93 and 119.21 μmol TE/100 g, respectively. The calculated nutritional indices (AI, TI, HH) revealed a beneficial potential for cardiovascular health, most pronounced in the case of rapeseed oil (AI = 0.02; HH = 19.53). The oxidative stability, expressed by the induction period (IP), ranged between 4.87 h (linseed oil) and 12.93 h (rapeseed oil), being correlated with the proportion of polyunsaturated fatty acids (PUFAs) and the level of antioxidant biocompounds. All the samples respected the legal limits concerning the content of polycyclic aromatic hydrocarbons (PAH $\leq$ 8.76 μg/kg).

The descriptive sensory analysis (QDA) and the hedonic tests indicated the highest acceptability for pumpkin oil, attributed to the sweet and roasted aromatic note, while the bitter taste and astringency decreased the scores for linseed oil. The results are similar to the studies of Grajzer et al. [14], who assessed the role of tocopherols in oxidative stability, and to those of Symoniuk et al. [15,16], who confirmed the variability in the composition and stability of cold-pressed oils, depending on the vegetable source and processing. The study also expands the results of Bou Fakhreddine and Sánchez [17], who showed that the sensory perception significantly influenced the purchase intention, even in the case of products with well-known nutritional benefits.

Różyło et al. [Contribution 8] investigated the effects of micronization on the bioactive composition and antioxidant properties of raspberry pomace powder (*Rubus idaeus* L.). Ball mill micronization reduced the average particle size from 225 μm (control) to 25 μm (after 10 min of micronization) and 10.5 μm (after 20 min of micronization), with minor changes in their color. They also performed an FTIR analysis that revealed changes in characteristic bands (~1720, 1635, and 1326 cm^{-1}), indicating the breaking of the intramolecular hydrogen bonds in polysaccharides, and the increase in the proportion of simple sugars. The HPLC confirmed higher levels of glucose and fructose in the micronized samples. Nine types of phenolic compounds (including ellagic acid derivatives, anthocyanins, and rutin) were identified, and their concentration increased significantly after micronization. Also, the antioxidant activity (ABTS, FRAP) and total phenolic content (up to 23.94 mg GAE/g) were both higher in the micronized samples, while the DPPH-based method revealed slightly decreased activity. The results were in agreement with the studies of Sadowska et al. [18], who concluded that fluidized-jet micronization increased both the anthocyanin content and antioxidant activity in raspberry powder. The results were also validated by the observations of Różyło et al. [19] regarding the effects of micronization on phenolic composition and antioxidant activity of spinach leaves.

Dumitrașcu et al. [Contribution 9] determined the effect of enzymatic hydrolysis on the antioxidant activity and functionality of proteins from spent brewer's yeast (SY). Three proteases (bromelain, neutrase, and trypsin) were used for SY hydrolysis for a period of time between 4 and 67 h, and the obtained hydrolysates were characterized from physicochemical, antioxidant, and techno-functional points of view. Neutrase generated the highest degree of hydrolysis and the highest levels of soluble proteins and antioxidant activity ($ABTS^+$ and DPPH), while bromelain produced the <3 kDa fraction with the highest DPPH antioxidant activity. Trypsin led to the best foaming properties, and bromelain provided emulsions with the highest structural rigidity. The SDS-PAGE analyses confirmed the predominance of peptides with molecular weights < 10 kDa in all hydrolyzed samples. From a functional point of view, trypsin hydrolysates had the best foaming capacity and stability (up to 132.5% and 98%), and bromelain-based hydrolysate emulsions showed the highest structural rigidity. The results reported by Dumitrașcu et al. (Contribution 9 of this Special Issue) are in agreement with previous studies available in the literature. Marson et al. [20] reported a significant increase in the antioxidant activity of yeast hydrolysates obtained with Protamex™ (Novozymes A/S, Bagsværd, Denmark) and Brauzyn® (Prozyn, São Paulo, Brazil), whereas Vieira et al. [21] confirmed the efficiency of neutrase in obtaining fractions with high antioxidant potential after ultrafiltration. Also, Mirzaei et al. [22] determined the antioxidant and ACE inhibitory activity of peptides with a molecular weight lower than 3 kDa, obtained from Saccharomyces cerevisiae with trypsin, hence further supporting the results of the abovementioned study regarding the bioactivity of small peptides.

Haş et al. [Contribution 10] investigated the bioactive potential of black elderberry (*Sambucus nigra* L.) from the spontaneous Romanian flora, focusing on the antioxidant and antimicrobial activity and bioaccessibility of phenolic compounds and their prebiotic potential. The methanolic extract of the freeze-dried elderberry powder displayed high antioxidant activity, with maximum FRAP values of 185 μmol Fe^{2+}/g DW, and significant antimicrobial activity, especially against *Staphylococcus aureus* and *Candida parapsilosis* (MIC = 1.95 mg/mL). The HPLC-DAD-ESI-MS analysis identified 12 phenolic compounds, with anthocyanins being predominant (41.8%), especially cyanidin–glucosides and cyanidin–sambubiosides. Following the simulated digestion, a bioaccessibility of 74.54% was observed, with a remarkable increase in the hydroxybenzoic acid derivatives (224.64%) and thus explaining the transformation of anthocyanins into more stable compounds. The study demonstrated for the first time the prebiotic potential of the extract on five probiotic strains, with the highest growth index (GI = 152.44%) being observed for *Lactobacillus casei* at 1.5%. The results showed many similarities with the results of Imenšek et al. [23], scientists who reported similar values for DPPH and FRAP in *S. nigra* fruits. These authors reported significantly higher ABTS values, suggesting variations depending on the degree of ripening and the plant's geographical origin. Furthermore, Reider et al. [24] assessed the prebiotic effect of a purified elderberry extract on microbial diversity, especially on *Akkermansia* spp. In addition, the studies of Mohammadsadeghi et al. [25] confirmed the antifungal activity of an elderberry extract on *Candida albicans*, although without any previous data on *C. parapsilosis*, which underlines the novelty of the current contribution.

Negrean et al. [Contribution 11] conducted a comprehensive analysis of the bioactive compounds from blackthorn fruits (*Prunus spinosa* L.), thus highlighting their antioxidant and antimicrobial potential and applications in the food, pharmaceutical, and cosmetic industries. The fruits contained several antioxidants such as flavonoids (quercetin, rutin, catechins), phenolic acids (chlorogenic, caffeic), anthocyanins (cyanidin-3-glucoside), tannins, vitamins (C), and minerals. These compounds exhibit antioxidant, anti-inflammatory, an tidiabetic, antihypertensive, and antitumor effects. Other studies have demonstrated strong correlations between polyphenol content and antioxidant activity (DPPH, ABTS, FRAP), as well as antimicrobial effects against Gram-positive and Gram-negative bacteria. The extracts showed antidiabetic potential by inhibiting α-glucosidase and anti-inflammatory effects by stimulating IL-10 secretion. Antiproliferative effects on various tumor cell lines (MCF-7, HepG2) and protection against in vivo oxidative stress have also been reported. Modern extraction techniques allow for the yield of polyphenols to be increased, and the extracts can be used as natural colorants, preservatives, ingredients for supplements, and cosmetics with a sunscreen factor. The conclusions emphasized the need for further studies to clarify the bioactive mechanisms and develop functional products and smart packaging based on blackthorn extracts. The study of Marcetić et al. [26] confirmed the phenolic profile and prebiotic activity of blackthorn fruits, and the study of Magiera et al. [27] highlighted the anti-inflammatory effects ex vivo on human immune cells. Also, Condello et al. [28] demonstrated the effectiveness of a blackthorn extract in inhibiting the in vivo growth of colorectal tumors, supporting the oncological potential of these compounds.

Author Contributions: L.G.-G., E.E., and I.A.; methodology, validation, formal analysis, investigation, data curation, writing—original draft preparation, writing—review and editing. All authors have read and agreed to the published version of the manuscript.

Funding: This research received no external funding.

Conflicts of Interest: The authors declare no conflicts of interest.

List of Contributions:

1. Latronico, T.; Petraglia, T.; Sileo, C.; Bilancia, D.; Rossano, R.; Liuzzi, G.M. Inhibition of MMP-2 and MMP-9 by Dietary Antioxidants in THP-1 Macrophages and Sera from Patients with Breast Cancer. *Molecules* **2024**, *29*, 1718.
2. Liccardo, M.; Sapio, L.; Perrella, S.; Sirangelo, I.; Iannuzzi, C. Genistein Prevents Apoptosis and Oxidative Stress Induced by Methylglyoxal in Endothelial Cells. *Molecules* **2024**, *29*, 1712.
3. Wang, Z.; Li, G.; Liu, X. Identification of Corn Peptides with Alcohol Dehydrogenase Activating Activity Absorbed by Caco-2 Cell Monolayers. *Molecules* **2024**, *29*, 1523.
4. Cao, D.; Zhu, Z.; Zhao, S.; Zhang, X.; Lin, J.; Wang, J.; Zeng, Q.; Zhu, M. Concentrations, Sources and Health Risk Assessment of Polycyclic Aromatic Hydrocarbons in Chinese Herbal Medicines. *Molecules* **2024**, *29*, 972.
5. Adebimpe Ojo, C.; Dziadek, K.; Sadowska, U.; Skoczylas, J.; Kopeć, A. Analytical Assessment of the Antioxidant Properties of the Coneflower (*Echinacea purpurea* L. Moench) Grown with Various Mulch Materials. *Molecules* **2024**, *29*, 971.
6. Santana, J.E.G.; Oliveira-Tintino, C.D.d.M.; Gonçalves Alencar, G.; Siqueira, G.M.; Sampaio Alves, D.; Moura, T.F.; Tintino, S.R.; de Menezes, I.R.A.; Rodrigues, J.P.V.; Gonçalves, V.B.P.; et al. Comparative Antibacterial and Efflux Pump Inhibitory Activity of Isolated Nerolidol, Farnesol, and α-Bisabolol Sesquiterpenes and Their Liposomal Nanoformulations. *Molecules* **2023**, *28*, 7649.
7. Rabiej-Kozioł, D.; Momot-Ruppert, M.; Stawicka, B.; Szydłowska-Czerniak, A. Health Benefits, Antioxidant Activity, and Sensory Attributes of Selected Cold-Pressed Oils. *Molecules* **2023**, *28*, 5484.
8. Różyło, R.; Amarowicz, R.; Janiak, M.A.; Domin, M.; Gawłowski, S.; Kulig, R.; Łysiak, G.; Rząd, K.; Matwijczuk, A. Micronized Powder of Raspberry Pomace as a Source of Bioactive Compounds. *Molecules* **2023**, *28*, 4871.
9. Dumitrașcu, L.; Lanciu Dorofte, A.; Grigore-Gurgu, L.; Aprodu, I. Proteases as Tools for Modulating the Antioxidant Activity and Functionality of the Spent Brewer's Yeast Proteins. *Molecules* **2023**, *28*, 3763.
10. Haș, I.M.; Teleky, B.-E.; Szabo, K.; Simon, E.; Ranga, F.; Diaconeasa, Z.M.; Purza, A.L.; Vodnar, D.-C.; Tit, D.M.; Nițescu, M. Bioactive Potential of Elderberry (*Sambucus nigra* L.): Antioxidant, Antimicrobial Activity, Bioaccessibility and Prebiotic Potential. *Molecules* **2023**, *28*, 3099.
11. Negrean, O.-R.; Farcas, A.C.; Pop, O.L.; Socaci, S.A. Blackthorn—A Valuable Source of Phenolic Antioxidants with Potential Health Benefits. *Molecules* **2023**, *28*, 3456.

References

1. Lee, S.J.; Kim, M.M. Resveratrol with antioxidant activity inhibits matrix metalloproteinase via modulation of SIRT1 in human fibrosarcoma cells. *Life Sci.* **2011**, *88*, 465–472. [CrossRef]
2. Demeule, M.; Brossard, M.; Page, M.; Gingras, D.; Beliveau, R. Matrix metalloproteinase inhibition by green tea catechins. *BBA* **2000**, *1478*, 51–60. [CrossRef] [PubMed]
3. Pang, N.; Chen, T.; Deng, X.; Chen, N.; Li, R.; Ren, M.; Li, Y.; Luo, M.; Hao, H.; Wu, J.; et al. Polydatin Prevents Methylglyoxal-Induced Apoptosis through Reducing Oxidative Stress and Improving Mitochondrial Function in Human Umbilical Vein Endothelial Cells. *Oxid. Med. Cell. Longev.* **2017**, *2017*, 7180943. [CrossRef]
4. Do, M.; Lee, J.H.; Wahedi, H.M.; Pak, C.; Lee, C.H.; Yeo, E.J.; Lim, Y.; Ha, S.K.; Choi, I.; Kim, S.Y. Lespedeza bicolor ameliorates endothelial dysfunction induced by methylglyoxal glucotoxicity. *Phytomedicine* **2017**, *36*, 26–36. [CrossRef] [PubMed]
5. Zan, R.; Zhu, L.; Wu, G.; Zhang, H. Identification of Novel Peptides with Alcohol Dehydrogenase (ADH) Activating Ability in Chickpea Protein Hydrolysates. *Foods* **2023**, *12*, 1574. [CrossRef] [PubMed]
6. Xiao, C.; Toldrá, F.; Zhou, F.; Mora, L.; Luo, L.; Zheng, L.; Luo, D.; Zhao, M. Chicken-derived tripeptide KPC (Lys-Pro-Cys) stabilizes alcohol dehydrogenase (ADH) through peptide–enzyme interaction. *LWT* **2022**, *161*, 113376. [CrossRef]
7. Ishizaki, A.; Sito, K.; Kataoka, H. Analysis of contaminant polycyclic aromatic hydrocarbons in tea products and crude drugs. *Anal. Methods* **2011**, *3*, 299–305. [CrossRef]
8. Yu, L.; Cao, Y.; Zhang, J.; Cui, Z.; Sun, H. Isotope dilution-GC-MS/MS analysis of 16 polycyclic aromatic hydrocarbons in selected medicinal herbs used as health food additives. *Food Addit. Contam. Part A* **2012**, *29*, 1800–1809. [CrossRef]
9. Tsai, Y.L.; Chiou, S.Y.; Chan, K.C.; Sung, J.M.; Lin, S.D. Caffeic acid derivatives, total phenols, antioxidant and antimutagenic activities of *Echinacea purpurea* flower extracts. *LWT* **2012**, *46*, 169–176. [CrossRef]

10. Skrzypczak-Pietraszek, E.; Piska, K.; Pietraszek, J. Enhanced production of the pharmaceutically important polyphenolic compounds in *Vitex agnus castus* L. shoot cultures by precursor feeding strategy. *Eng. Life Sci.* **2018**, *18*, 287–297. [CrossRef]
11. de Moura, D.F.; Rocha, T.A.; de Melo Barros, D.; da Silva, M.M.; dos Santos Santana, M.; Neta, B.M.; Cavalcanti, I.M.F.; Martins, R.D.; da Silva, M.V. Evaluation of the Antioxidant, Antibacterial, and Antibiofilm Activity of the Sesquiterpene Nerolidol. *Arch. Microbiol.* **2021**, *203*, 4303–4311. [CrossRef]
12. da Cruz, R.P.; de Freitas, T.S.; Costa, M.D.S.; Dos Santos, A.T.L.; Campina, F.F.; Pereira, R.L.S.; Bezerra, J.W.A.; Quintans-Júnior, L.J.; Araújo, A.A.D.S.; De Siqueira Júnior, J.P.; et al. Effect of α-Bisabolol and Its β-Cyclodextrin Complex as TetK and NorA Efflux Pump Inhibitors in *Staphylococcus aureus* Strains. *Antibiotics* **2020**, *9*, 28.
13. Oliveira, D.; Borges, A.; Saavedra, M.J.; Borges, F.; Simões, M. Screening of Natural Molecules as Adjuvants to Topical Antibiotics to Treat *Staphylococcus aureus* from Diabetic Foot Ulcer Infections. *Antibiotics* **2022**, *11*, 620. [CrossRef]
14. Grajzer, M.; Szmalcel, K.; Kuźmiński, Ł.; Witkowski, M.; Kulma, A.; Prescha, A. Characteristics and Antioxidant Potential of Cold-Pressed Oils—Possible Strategies to Improve Oil Stability. *Foods* **2020**, *9*, 1630. [CrossRef]
15. Symoniuk, E.; Ratusz, K.; Ostrowska-Ligęza, E.; Krygier, K. Impact of Selected Chemical Characteristics of Cold-Pressed Oils on Their Oxidative Stability. *Food Anal. Methods* **2018**, *11*, 1095–1104. [CrossRef]
16. Symoniuk, E.; Wroniak, M.; Napiórkowska, K.; Brzezinska, R.; Ratusz, K. Oxidative Stability and Antioxidant Activity of Selected Cold-Pressed Oils and Oil Mixtures. *Foods* **2022**, *11*, 1597. [CrossRef] [PubMed]
17. Bou Fakhreddine, L.; Sánchez, M. The Interplay between Health Claims and Sensory Attributes in Determining Consumers' Purchase Intentions for Extra Virgin Olive Oil. *Food Qual. Prefer.* **2023**, *106*, 104819. [CrossRef]
18. Sadowska, A.; Świderski, F.; Hallmann, E. Properties of raspberry powder obtained by a new method of fluidized-bed jet milling and drying compared to other drying methods. *J. Sci. Food Agric.* **2020**, *100*, 4303–4309. [CrossRef] [PubMed]
19. Rózyło, R.; Piekut, J.; Dziki, D.; Smolewska, M.; Gawłowski, S.; Wójtowicz, A.; Gawlik-Dziki, U. Effects of Wet and Dry Micronization on the GC-MS Identification of the Phenolic Compounds and Antioxidant Properties of Freeze-Dried Spinach Leaves and Stems. *Molecules* **2022**, *27*, 8174. [CrossRef]
20. Marson, G.V.; de Castro, R.J.S.; Machado, M.T.D.C.; da Silva Zandonadi, F.; Barros, H.D.D.F.Q.; Maróstica Júnior, M.R.; Sussulini, A.; Hubinger, M.D. Proteolytic enzymes positively modulated the physicochemical and antioxidant properties of spent yeast protein hydrolysates. *Process Biochem.* **2020**, *91*, 34–45. [CrossRef]
21. Vieira, E.F.; Dias, D.; Carmo, H.; Ferreira, I. Protective ability against oxidative stress of brewers' spent grain protein hydrolysates. *Food Chem.* **2017**, *228*, 602–609. [CrossRef]
22. Mirzaei, M.; Mirdamadi, S.; Ehsani, M.R.; Aminlari, M.; Hosseini, E. Purification and identification of antioxidant and ACE-inhibitory peptide from *Saccharomyces cerevisiae* protein hydrolysate. *J. Funct. Foods* **2015**, *19*, 259–268. [CrossRef]
23. Imenšek, N.; Kristl, J.; Šumenjak, T.K.; Ivančič, A. Antioxidant activity of elderberry fruits during maturation. *Agriculture* **2021**, *11*, 555. [CrossRef]
24. Reider, S.; Watschinger, C.; Längle, J.; Pachmann, U.; Przysiecki, N.; Pfister, A.; Zollner, A.; Tilg, H.; Plattner, S.; Moschen, A.R. Short- and Long-Term Effects of a Prebiotic Intervention with Polyphenols Extracted from European Black Elderberry—Sustained Expansion of *Akkermansia* spp. *J. Pers. Med.* **2022**, *12*, 1479. [CrossRef]
25. Mohammadsadeghi, S.; Malekpour, A.; Zahedi, S.; Eskandari, F. The antimicrobial activity of elderberry (*Sambucus nigra* L.) extract against gram positive bacteria, gram negative bacteria and yeast. *Res. J. Appl. Sci.* **2013**, *8*, 240–243.
26. Marcetić, M.; Samardžić, S.; Ilić, T.; Božić, D.D.; Vidović, B. Phenolic Composition, Antioxidant, Anti-Enzymatic, Antimicrobial and Prebiotic Properties of *Prunus spinosa* L. Fruits. *Foods* **2022**, *11*, 3289. [CrossRef]
27. Magiera, A.; Czerwinska, M.E.; Owczarek, A.; Marchelak, A.; Granica, S.; Olszewska, M.A. Polyphenol-Enriched Extracts of Prunus spinosa Fruits: Anti-Inflammatory and Antioxidant Effects in Human Immune Cells Ex Vivo in Relation to Phytochemical Profile. *Molecules* **2022**, *27*, 1691. [CrossRef]
28. Condello, M.; Pellegrini, E.; Spugnini, E.P.; Baldi, A.; Amadio, B.; Vincenzi, B.; Occhionero, G.; Delfine, S.; Mastrodonato, F.; Meschini, S. Anticancer activity of "Trigno M", extract of Prunus spinosa drupes, against in vitro 3D and in vivo colon cancer models. *Biomed. Pharmacother.* **2019**, *118*, 109281. [CrossRef]

molecules

MDPI

Article

Inhibition of MMP-2 and MMP-9 by Dietary Antioxidants in THP-1 Macrophages and Sera from Patients with Breast Cancer

Tiziana Latronico [1], Tania Petraglia [2], Carmela Sileo [2], Domenico Bilancia [3], Rocco Rossano [2,*] and Grazia Maria Liuzzi [1]

[1] Department of Biosciences, Biotechnologies and Environment, University of Bari "Aldo Moro", 70126 Bari, Italy; tiziana.latronico@uniba.it (T.L.); graziamaria.liuzzi@uniba.it (G.M.L.)
[2] Department of Sciences, University of Basilicata, 85100 Potenza, Italy; tania.petraglia@unibas.it (T.P.); carmelasileo29@gmail.com (C.S.)
[3] Operating Unit, Medical Oncology, Hospital "Azienda Ospedaliera S. Carlo", 85100 Potenza, Italy; domenicobilancia@gmail.com
* Correspondence: rocco.rossano@unibas.it

Abstract: Polyphenols, the main antioxidants of diet, have shown anti-inflammatory, antioxidant and anticarcinogenic activities. Here, we compared the effects of four polyphenolic compounds on ROS production and on the levels of matrix metalloproteinase (MMP)-2 and -9, which represent important pathogenetic factors of breast cancer. THP-1 differentiated macrophages were activated by LPS and simultaneously treated with different doses of a green tea extract (GTE), resveratrol (RSV), curcumin (CRC) and an olive fruit extract (oliplus). By using the 2,2-Diphenyl-1-picrylhydrazyl (DPPH) radical scavenging assay, we found that all of the tested compounds showed antioxidant activity in vitro. In addition, GTE, RSV and CRC were able to counteract ROS production induced by H_2O_2 in THP-1 cells. As assessed by a zymographic analysis of THP-1 supernatants and by an "in-gel zymography" of a pool of sera from patients with breast cancer, the antioxidant compounds used in this study inhibited both the activity and expression of MMP-2 and MMP-9 through different mechanisms related to their structures and to their ability to scavenge ROS. The results of this study suggest that the used antioxidants could be promising agents for the prevention and complementary treatment of breast cancer and other diseases in which MMPs play a pivotal role.

Keywords: dietary antioxidants; matrix metalloproteinases; breast cancer; THP-1 macrophages; oxidative stress

1. Introduction

Breast cancer stands as one of the most prevalent and life-threatening malignancies affecting women worldwide [1]. This multifaceted disease arises from the uncontrolled growth of abnormal breast cells and can manifest in various forms, making it a formidable challenge in the realm of oncology. Although it is a multifactorial pathology, it is now established that lifestyle can affect the onset and development of breast cancer. In this regard, it can be noted that with globalization, in recent years, the incidence of breast cancer has increased in developing countries where the habits of Western countries are taking over. It has been highlighted, in fact, that a sedentary lifestyle and a diet rich in saturated fatty acids and poor in fibers and vegetables affects the onset and relapses of breast cancer [2,3].

Understanding the intricate molecular mechanisms underlying breast cancer is crucial for developing effective diagnostic tools and therapeutic strategies.

It is now known that inflammation plays a key role in tumor development and progression and that these processes are driven by a complex interplay between malignant tumor cells and the surrounding nonmalignant stroma, comprising the extracellular matrix (ECM); stromal cells such as endothelial cells (ECs); fibroblasts; and infiltrating immune cells. Among infiltrated immune cells, tumor-associated macrophages (TAMs) represent the primary resident cells of the breast tumor microenvironment (TME) and are the main cells implicated in inflammatory processes [4] via their ability to secrete inflammatory factors such as pro-inflammatory cytokines and matrix metalloproteinases (MMPs) [5,6].

MMPs are part of a family of proteolytic enzymes responsible for remodeling the ECM, a dynamic network of proteins that provides structural support to tissues. MMPs are essential for normal physiological processes, like tissue repair and organ development, but their dysregulated activity has been intricately linked to cancer progression. In the context of breast cancer, MMPs have garnered significant attention due to their role in several key aspects of cancer progression, including invasion, metastasis and angiogenesis [7]. Elevated levels of specific MMPs, such as gelatinases MMP-2 and MMP-9, have been found in breast cancer tissues and are associated with a more aggressive tumor phenotype [8]. MMPs can promote cancer cell invasion into surrounding tissues and facilitate the establishment of metastases [9]. Additionally, MMPs can indirectly affect the tumor microenvironment by altering the ECM composition, thus contributing to the remodeling process.

Although it was initially assumed that MMPs present in the tumor microenvironment were produced by tumor cells, in situ hybridization techniques have demonstrated that MMPs are predominantly produced by adjacent host stromal and inflammatory cells in response to factors released by tumors [10].

An important role in breast cancer initiation and progression is also played by oxidative stress. Oxidative stress, generated by an imbalance between reactive oxygen species (ROS) and endogenous antioxidants, can contribute to DNA damage, genetic mutations and alterations to cellular signaling pathways, which can lead to the transformation of normal breast cells into cancerous cells. Additionally, oxidative stress may promote the survival and growth of breast cancer cells and contribute to resistance to chemotherapy and radiation therapy [11].

It is now common knowledge that there is an interplay between oxidative stress and MMPs. Reactive oxygen species have been implicated in MMP-9 zymogen activation [12]. In addition, oxidative stress can lead to the activation of signaling pathways that regulate MMP expression. For example, the nuclear factor-kappa B (NF-κB) pathway, which is activated in response to oxidative stress, can stimulate the expression of certain MMPs [11].

Based on these considerations, MMPs represent attractive therapeutic targets that are susceptible to antioxidant modulation. Therefore, compounds capable of inhibiting MMPs and, at the same time, neutralizing ROS, can act as powerful bullets in the fight against breast cancer.

Polyphenols are ubiquitous organic molecules present in vegetables, fruits, coffee, tea, cereals and wine. Together with vitamins, they represent the main antioxidants of diet. Polyphenols are characterized by the presence of hydroxylated phenolic groups organized in more or less complex structures. Based on the structural elements that bind these phenolic rings, polyphenols are divided into two categories, namely flavonoids (flavones, flavanones, flavanols, flavonols, isoflavones and anthocyanins) and non-flavonoids (phenolic acids, stilbenes, coumarins, curcumin, tannins and lignans) [13].

The aim of this study was to evaluate the antioxidant activity of four different dietary polyphenolic compounds and their ability to inhibit ROS production in THP-1 macrophages and the activity of MMP-2 and MMP-9 in both THP-1 macrophages and in sera from patients with breast cancer. For our experiments, as a member of a flavonoids group, we chose a green tea extract (GTE) containing high concentrations of the flavanols catechin, epicathechin, epigallocatechin, epicatechin gallate and epigallocatechin gallate, whereas the other antioxidants were represented by three non-flavonoids: resveratrol (a stilbene), curcumin and tyrosol/hydroxytyrosol. The latter are the main antioxidants present in an olive fruit extract (oliplus) (Figure 1).

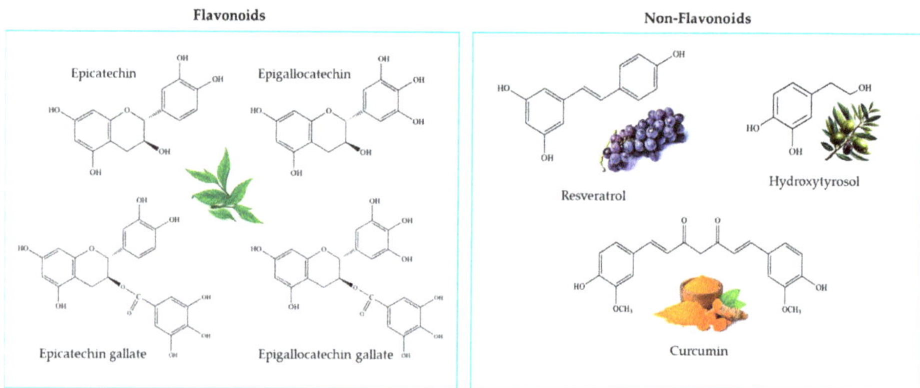

Figure 1. The structures of the different polyphenols used in this study showing the nature of their chemical groups.

The results suggest that the polyphenols used in this study exert antioxidant activity at low concentrations and inhibit MMP-2 and MMP-9 activity with different mechanisms of action possibly related to their different structures. On these grounds, a diet rich in polyphenols characterized by different chemical structures may represent a winning strategy for the inhibition of key factors, such as MMPs, involved in breast cancer.

2. Results

2.1. *Antioxidant Activity In Vitro*

Figure 2 shows the free radical scavenging activity of the four dietary antioxidant compounds used, as detected by the 2,2-Diphenyl-1-picrylhydrazyl (DPPH) radical scavenging assay. Among the studied compounds, the highest antioxidant activity was observed for the green tea extract (GTE) (1.80 ± 0.11 μg/mL), followed by resveratrol (RSV) (7.54 ± 0.87 μg/mL) and curcumin (CRC) (9.71 ± 0.33 μg/mL), whereas the lowest antioxidant activity (15.23 ± 1.66 μg/mL) was measured in the extract derived from olive fruits (OLI).

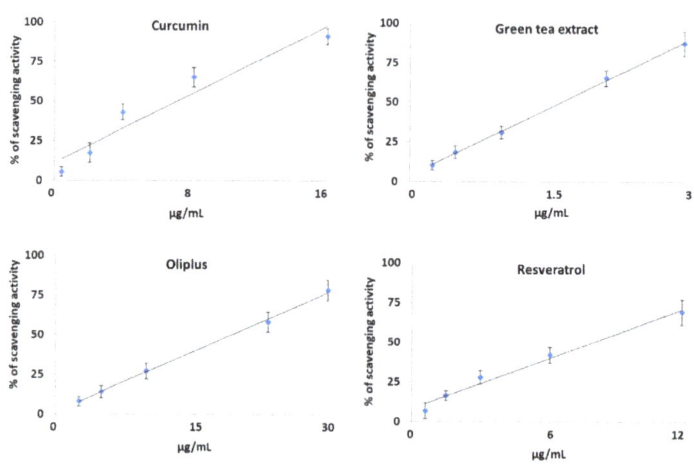

Antioxidant	IC_{50} (μg/mL)
Curcumin	9.71 ± 0.33
Green tea extract	1.80 ± 0.11
Oliplus	15.23 ± 1.66
Resveratrol	7.54 ± 0.87

Figure 2. The scavenging activities of the dietary antioxidants. The graphs represent the scavenging activities of the antioxidants against 2,2-Diphenyl-1-picrylhydrazyl (DPPH). The results are expressed as the IC_{50} values that represent the concentration of antioxidant necessary to scavenge 50% of free radicals. The data are shown as the mean values ± S.D. of three replicates.

2.2. Effect of Antioxidant Compounds on THP-1 Macrophage Viability

Preliminary experiments were performed to identify the concentrations of the antioxidant compounds that were not toxic for cells, a basic assumption to exclude that the effects observed were the consequence of cellular suffering. To this purpose, differentiated THP-1 macrophages were treated for 20 h with the antioxidants at concentrations ranging between 10 and 250 µg/mL, as described in the Materials and Methods Section. As shown in Figure 3, among the investigated compounds, GTE and OLI were the least toxic since they did not show any toxicity at any of the concentrations tested. By contrast, CRC and RSV were toxic to cells at concentrations higher than 25 µg/mL. No difference in cell viability was observed in the cells activated with LPS and treated with the antioxidants.

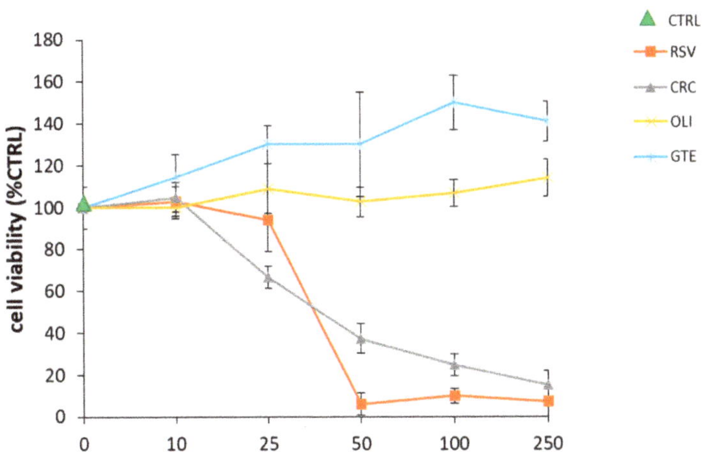

Figure 3. The cell viability of the THP-1 macrophages treated with the antioxidant compounds. The THP-1 macrophages were treated for 20 h with resveratrol (RSV), oliplus (OLI), green tea extract (GTE) or curcumin (CRC) at the indicated concentrations and then subjected to the MTT assay. The control (CTRL) was represented from untreated cells in serum-free RPMI. The graphs represent the dose–response curves of cell viability, expressed as the percentage of cell survival in comparison to the control, which was set at 100%. The horizontal dashed line, set at 60%, indicates the threshold of cell viability. Concentrations of antioxidant compounds that yielded cell viability values <60% of the control were considered toxic doses. The values are shown as the mean ± SD of n = 3 experiments performed on different cell populations.

2.3. Effect of Antioxidant Compounds on MMP-2 and MMP-9 Released from LPS-Activated THP-1 Macrophages

As shown in the representative zymograms reported in Figure 4a–c, the treatment of THP-1 macrophages with LPS increased the levels of both MMP-2 and MMP-9. By contrast, the co-treatment with GTE, RSV or CRC induced a dose-dependent inhibition of MMP-2 and MMP-9. Differently, OLI had no inhibitory effects on the two proteinases. The statistical analysis of data (Figure 4d–f) evidenced that CRC was the most potent antioxidant among those used, since at the concentration of 25 µg/mL, it was able to achieve the highest inhibition of both the MMP-2 and MMP-9 levels (45 and 80%, respectively) compared to the other antioxidants at the same concentration. However, RSV at the lowest concentration used (10 µg/mL) was the only antioxidant capable of inducing a statistically significant inhibition of MMP-9 (approximately 50%). At the maximum non-toxic concentration of 250 µg/mL, GTE completely inhibited the activity of both MMP-2 and MMP-9.

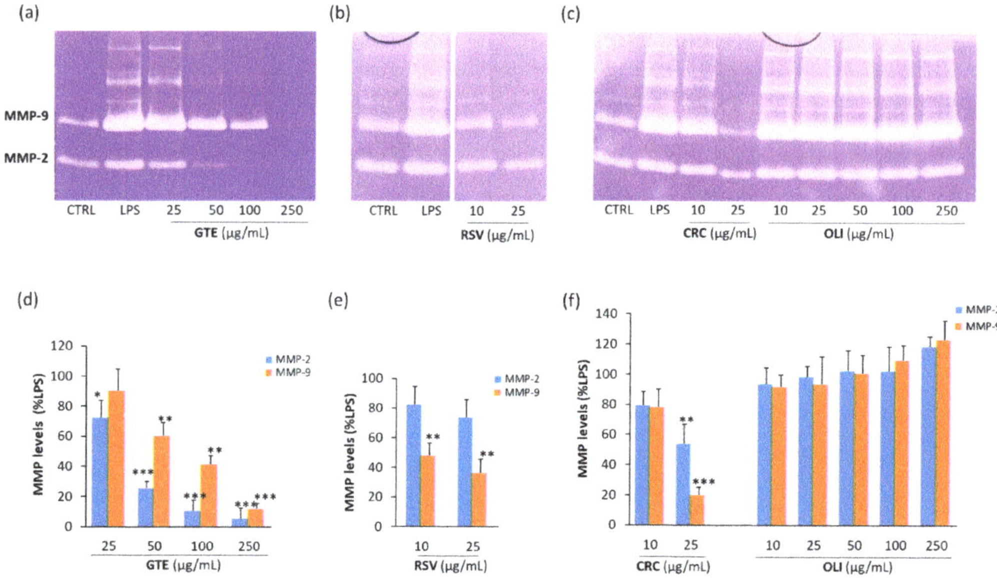

Figure 4. Effect of green tea extract (GTE), resveratrol (RSV), curcumin (CRC) and oliplus (OLI) on MMP-2 and MMP-9 levels in LPS-activated THP-1 macrophages. (**a**–**c**) Representative zymographic gels in (**a**–**c**) show MMP-2 and MMP-9 levels in culture supernatants from THP-1 macrophages activated with LPS (10 µg/mL) and simultaneously treated for 20 h with GTE, RSV, CRC or OLI at indicated concentrations. Negative and positive controls were obtained from unstimulated and untreated cells in serum-free medium (CTRL) and LPS-activated cells (LPS), respectively. Histograms in (**d**–**f**) represent mean ± SD of $n = 3$ experiments performed on different cell populations. Results are expressed as percentage of MMP levels in comparison to LPS, calculated after scanning densitometry and computerized analysis of gels. Asterisks represent values statistically different from positive control (one-way ANOVA followed by Dunnet's post hoc test; * $p < 0.05$, ** $p < 0.01$, *** $p < 0.001$).

2.4. Effect of Antioxidant Compounds on ROS Production in THP-1 Macrophages Treated with Hydrogen Peroxide

To evaluate the potential antioxidant activity of the studied compounds, differentiated THP-1 macrophages were pretreated with GTE, RSV, OLI or CRC at concentrations ranging between 5 and 25 µg/mL, and then activated with hydrogen peroxide (H_2O_2) as described in the Materials and Methods Section.

As shown in Figure 5 GTE, RSV and CRC showed a statistically significant ability to counteract the H_2O_2-induced ROS production. Among the compounds tested, CRC and RSV were the most effective; at the dose of 5 µg/mL, they were able to induce reductions of about 65 and 50%, respectively, in the ROS levels. However, at the concentration of 25 µg/mL, CRC and GTE were the most efficacious compounds in counteracting H_2O_2-induced ROS production (75%). No statistically significant inhibition of ROS was observed by OLI at any of the doses used.

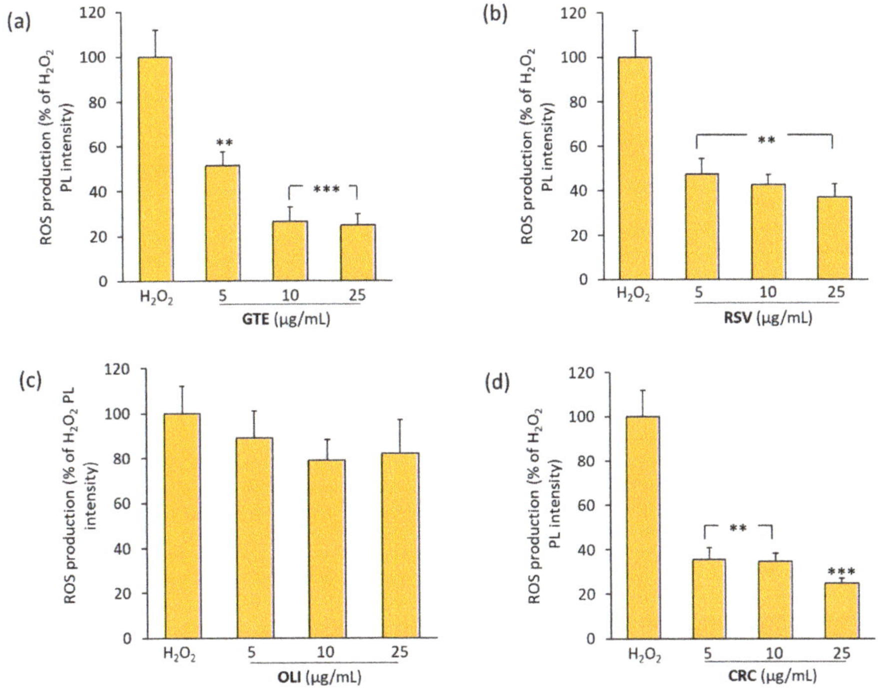

Figure 5. The effects of green tea extract (GTE), resveratrol (RSV), oliplus (OLI) and curcumin (CRC) on ROS in THP-1 macrophages. ROS production, assayed by measuring the changes in the fluorescent signal of 2′,7′-dichlorofluorescein (DCFA), was expressed as the percentage (%) of photoluminescence (PL) intensity in comparison to the positive control (100%), represented by cells treated with 100 μM H_2O_2. The histograms (**a**–**d**) represent the mean values ± SD of $n = 3$ experiments performed on different cell populations. A statistically significant decrease in comparison to H_2O_2 is indicated by asterisks (one-way ANOVA followed by Dunnet's post hoc test; ** $p < 0.01$ and *** $p < 0.001$).

2.5. "In-Gel" Inhibition of MMP-2 and MMP-9 Activity by Antioxidant Compounds

Figure 6 shows the results on the ability of antioxidants to inhibit the "in-gel" activity of MMP-2 and MMP-9 in a pool of sera obtained from patients affected by breast cancer. In Figure 6a–d, in the lanes corresponding to the control (CTRL, lanes of sera incubated in the absence of antioxidants), two main bands of digestion corresponding to MMP-9 and MMP-2 are observed. The incubation with 1,10 phenanthroline (1,10-PA), a specific inhibitor of metalloproteinases, completely inhibited the activity of both MMP-2 and MMP-9 (gel a). Treatment with GTE (gel a) determined an inhibition for both gelatinases in a concentration-dependent manner. As shown in gel b, RSV induced a slight inhibition of MMP-2 activity, whereas it did not exert any inhibitory activity on MMP-9 at any of the concentrations tested. In the lanes incubated with OLI (gel c), a dose-dependent reduction in MMP-2 activity was evident, whereas MMP-9 was inhibited at a low extent only at the highest concentrations of 50 and 100 μg/mL. Finally, CRC (gel d) induced a strong inhibition of MMP-2 at the highest concentration of 100 μg/mL and did not inhibit MMP-9.

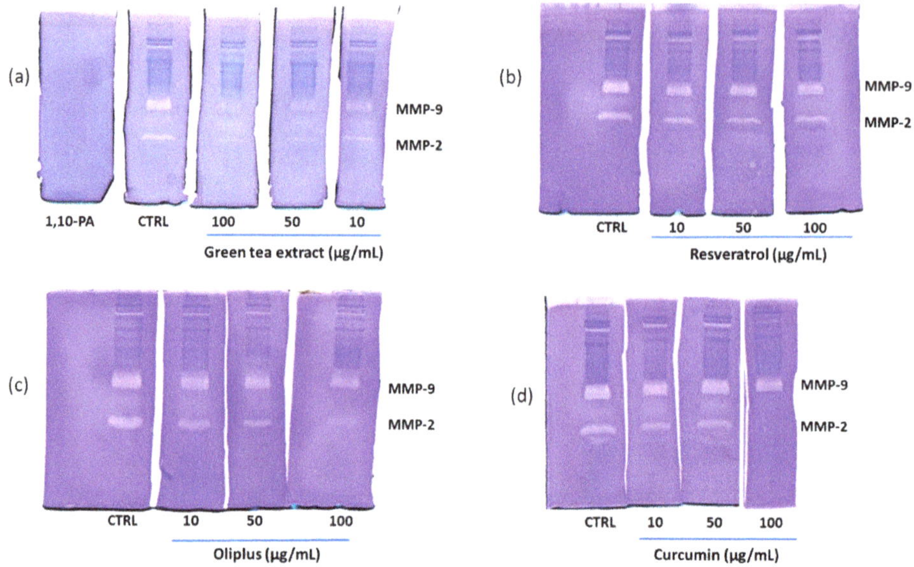

Figure 6. The "in-gel" inhibition of MMP-2 and MMP-9 activity by the antioxidant compounds. A pool of sera obtained from patients affected by breast cancer was applied to 1D gelatin zymography. After electrophoresis, gels were cut in lanes, and each lane was incubated individually in the presence of the following single antioxidants at different concentrations: green tea extract (**a**), resveratrol (**b**), oliplus (**c**) and curcumin (**d**), respectively.

The percentages of inhibition of MMP-2 and MMP-9, calculated in comparison with CTRL, after the densitometric analysis of gels, are shown in Table 1.

Table 1. In-gel inhibition of MMP-2 and MMP-9 activity by antioxidants.

	Inhibition (%)	
	MMP-2	**MMP-9**
1,10 PA	100	100
CTRL	0	0
Antioxidant (μg/mL)		
Green tea extract		
10	50.4 ± 3.7	56.4 ± 2.1
50	65.7 ± 2.5	53.6 ± 5.5
100	85.8 ± 6.1	89.9 ± 2.8
Resveratrol		
10	20.1 ± 1.2	3.2 ± 0.2
50	22.6 ± 1.9	2.6 ± 0.5
100	21.4 ± 1.1	2.1 ± 0.4
Oliplus		
10	34.9 ± 2.9	10.3 ± 2.3
50	59.0 ± 8.3	15.4 ± 3.3
100	74.5 ± 3.4	28.4 ± 4.1
Curcumin		
10	20.2 ± 1.3	1.3 ± 0.2
50	18.6 ± 6.3	2.2 ± 0.8
100	100	18.2 ± 3.7

CTRL: control (MMPs incubated in the absence of antioxidants); 1,10 PA: 1,10 phenanthroline, specific inhibitor of MMPs. Values represent percentage of inhibition in comparison to CTRL, calculated as mean ± SD of three independent experiments.

3. Discussion

In the past two decades, numerous studies have been directed towards the functional role of diet for the prevention and complementary treatment of numerous pathologies including tumors and cardiovascular and neurodegenerative diseases [14]. A common hallmark of all of these multifactorial diseases is represented by inflammation. To understand how nutrition can improve the course of these pathologies, it is important to identify the molecular mechanisms through which food molecules are able to dampen inflammatory processes by intervening on cellular targets.

Among the molecules introduced by diet, polyphenols are the most studied compounds for their multiple biological functions. Their biological effects are mainly due to their anti-inflammatory, immunomodulatory, anti-angiogenic, neuroprotective, antioxidant and radical scavenging properties that might be beneficial for the prevention and treatment of different diseases, including breast cancer [15–17].

In this paper, we compared the effects of two purified polyphenols (CRC and RSV) and two extracts rich in polyphenols (GTE and OLI), which are present in numerous vegetables that are commonly consumed, on the production of ROS and on the levels of MMP-2 and MMP-9 in an experimental model represented by THP-1 cells differentiated in macrophages and activated with LPS.

Macrophages, key components of the immune system, play a multifaceted role in the context of cancer, particularly breast cancer [18]. In the tumor microenvironment, tumor-associated macrophages (TAMs) often adopt distinct phenotypes, ranging from pro-inflammatory (M1) to anti-inflammatory (M2), influenced by signals from the surrounding milieu. The crosstalk between breast cancer cells and inflammatory macrophages not only influence the tumor microenvironment but also impact various aspects of cancer biology, such as proliferation, invasion, angiogenesis and metastasis.

TAMs are implicated in the generation of ROS, contributing to oxidative stress within the tumor microenvironment, which, in turn, can promote tumor progression through various mechanisms, including genomic instability, DNA damage and the activation of signaling pathways promoting the release of pro-inflammatory molecules [19].

THP-1 macrophages represent a good model of study and are already used by other authors because, through their activation with stimuli such as LPS, they can be differentiated into a pro-inflammatory M1-like phenotype that is capable of secreting pro-inflammatory factors including cytokines, ROS and enzymes such as MMPs [4].

In this study, we chose MMPs as targets of the effects of the studied polyphenols since these enzymes play a key role in cancer cell invasion by digesting the ECM, supporting cancer cell growth and tumor metastasis [20,21]. In this contest, numerous attempts have been made to generate drugs that are capable of inhibiting MMPs, mainly with the aim of blocking the invasion and metastasis of tumor cells, but the results of clinical trials have proved disappointing, with unwanted side effects and a lack of effectiveness [22]. To date, there are no MMP inhibitors available at the clinical level for the treatment of cancer, or those available are not therapeutically useful due to the lack of specificity [23]. As reported by Vandooren et al. [24], there is a persistent lack of knowledge on the complexity of MMP biology, which hinders the development of safe and effective MMP-targeted drugs. On these bases, the use of natural compounds as dietary antioxidants that are able to inhibit MMPs with different mechanisms of action may represent an interesting challenge for research, considering the role that ROS play in the activation of MMPs [25].

Several phytochemicals have been found to act as direct inhibitors of MMP activation or as modulators of signaling pathways associated with MMP expression [20,26,27]. In this regard, numerous studies have demonstrated the ability of polyphenols to inhibit MMP activity and expression [28–32].

By using the DPPH radical scavenging assay, we measured the in vitro antioxidant activity of RSV, CRC, OLI and GTE and found that all of the tested compounds are good antioxidants which act at low concentrations. In accordance with the data reported in the literature, we found different antioxidant capacities for the studied compounds, which were comparable to those found by other authors using the same DPPH assay [33–36]. As already reported, the different antioxidant capacities of the studied polyphenols could be attributed to their different molecular structures [30]. Indeed, the antioxidant power of polyphenols may depend on the number of phenolic rings, the number and positions of hydroxyl groups and the double bonds present in the molecule.

The results of this study indicate that the used polyphenols are able to inhibit MMP-2 and MMP-9 in relation to their antioxidant power. Among the antioxidants used in this study, RSV, CRC and GTE were able to counteract the increases in the MMP-9 and MMP-2 levels with a different inhibitory capacity, whereas OLI, which was the compound with the lowest antioxidant power, was not able to significantly inhibit the levels of MMP-2 and MMP-9 released in THP-1 supernatants. The lack of inhibition of MMPs by OLI could depend on the doses used, the composition of the extract and the cell population tested. In fact, in another work in which OLI was tested on a different cell population and at a higher concentration, this antioxidant significantly inhibited MMP-2 and MMP-9 levels [30]. Similarly, an inhibitory effect on MMP-9 was observed by other authors using a different extract of olive oil or hydroxytyrosol [32,37].

The cytotoxicity test performed on cells treated with the antioxidants indicated that the used compounds were not toxic for cells at the inhibitory doses, and therefore, the observed changes in the levels of MMP-9 and MMP-2 in LPS-activated THP-1 supernatants were due to a real inhibitory effect on gelatinases.

One important mechanism by which polyphenols exert their anti-MMP activities is likely to be via the downregulation of ROS. Numerous studies have demonstrated a strict interplay between MMP levels and ROS production [38]. ROS activate signaling pathways that are involved in the transcriptional activation of the gene that promotes MMP expression (Figure 7) [39].

In this work, we tested the ability of the antioxidants used to block the production of ROS in THP-1 cells activated with a non-specific pro-oxidant stimulus represented by hydrogen peroxide. We tested concentrations between 5 and 25 μg/mL to compare the ROS scavenging capacity between all of the antioxidants used. The results obtained evidence that GTE, RSV and CRC showed a protective effect in counteracting the oxidative stress induced in THP-1 cells by hydrogen peroxide. The antioxidant that was less effective in blocking the production of ROS was OLI. This result seems to confirm our hypothesis, according to which the ability to block the production of ROS, resulting in the inhibition of MMP-2 and MMP-9 levels in THP-1 cells, could be related to the antioxidant power of the compounds tested.

According to the results obtained from the analysis of the THP-1 supernatants and from the "in-gel" zymography of a pool of sera from patients with breast cancer, we can hypothesize that, as suggested by other authors, the antioxidant compounds used in this study can inhibit MMPs through different mechanisms, involving the inhibition of both MMP activity and expression, with the latter occurring through the downregulation of different signaling pathways involved in MMP-2 and MMP-9 gene transcription (Figure 7). At the molecular level, polyphenols may downregulate the ROS-induced activation of ERK 1/2, p38 and JNK, as well as PI3K/AKT, preventing the nuclear translocation of NF-kB and the downstream activation of the transcription factor AP-1. The consequent downregulation of MMP gene transcription results in the inhibition of tumor cell growth, metastasis, angiogenesis and inflammation [22,40,41]. As a result of our experiments, RSV inhibited MMP-9 in THP-1 supernatants but not in "in-gel" zymography, suggesting that this antioxidant only exerted its inhibitory action on the expression of the enzyme. This result is consistent with previous studies showing the ability of RSV to downregulate MMP-9 expression by inhibiting the activation of the transcription factors NF-kB and AP-1 [30,40].

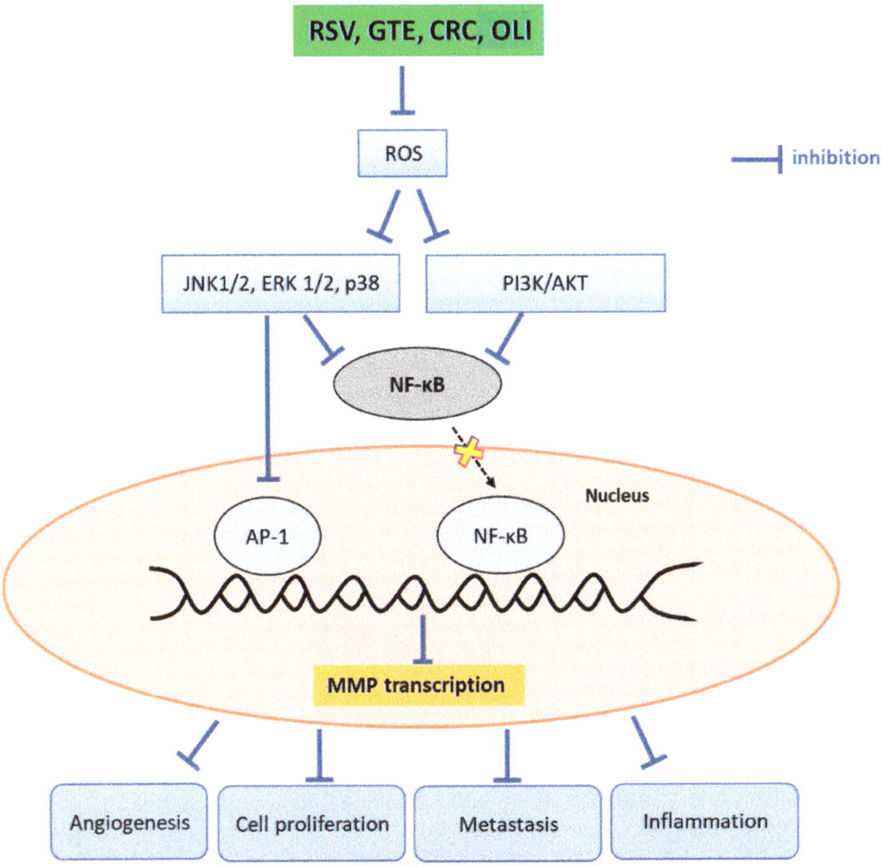

Figure 7. Antioxidant-mediated downregulation of MMP gene expression via ROS-triggered signal transduction pathways.

Differently, we found that GTE was able to inhibit, in a dose-dependent manner, both the activity and expression of MMP-2 and MMP-9 as assessed by the "in-gel" zymography and by the analysis of the THP-1 supernatants, respectively.

By contrast, CRC inhibited MMP-2 and MMP-9 with different mechanisms. In fact, the inhibition of MMP-9 seems to be directed exclusively on the expression of the enzyme, as demonstrated by the analysis of the THP-1 supernatants. Conversely, MMP-2 expression and activity levels were both inhibited. In the latter case, CRC, at the highest concentration used, inhibited MMP-2 with an efficiency comparable to that of 1,10 phenanthroline, a specific inhibitor of MMPs. However, it must be considered that, due to the cytotoxicity of CRC on THP-1 cells, it was not possible to compare its inhibitory effect at concentrations higher than 25 μg/mL in the two systems used.

One of the possible mechanisms to explain the "in-gel" inhibition of MMPs by polyphenols could be their ability to chelate metals [42], including zinc [43,44], the indispensable cofactor for the activity of MMP-2 and MMP-9. Polyphenols are excellent metal chelators; their chelating capacities are closely linked to the presence of catechol groups and the combination of hydroxyl and carbonyl groups from which the metal binding sites originate. Polyphenols with galloyl or catechol groups are more potent metal chelators than those without these groups [45]. Lakey-Beitia et al. [46] proposed that polyphenols could be classified into three groups: a group with only one binding site for metals that includes most polyphenols, such as curcuminoids, lignans, stillbenes, isoflavonoids, flavanols and anthocyanins; a group with two binding sites, including flavones and flavanones; and finally, a group with three binding sites, which includes flavonols, flavanols and tannins. All of these considerations could explain the reason why, in our experiments, GTE is the most active "in-gel" inhibitor of MMPs even at the lowest concentrations. However, as reported by Suzuki et al. [40], another important mechanism to explain the inhibitory activity of polyphenols towards MMPs might be their direct binding to MMPs, as demonstrated by several molecular docking studies [47–50].

On these bases, the natural MMP inhibitors tested in this study could be used to plan alternative therapeutic strategies for the treatment of cancer in synergy with conventional therapies.

4. Materials and Methods

4.1. Chemicals and Reagents

Gelatin, DNase, Roswell Park Memorial Institute (RPMI) 1640 medium, Roswell Park Memorial Institute (RPMI) 1640 medium without Phenol Red, fetal bovine serum (FBS), penicillin and streptomycin were provided by GIBCO (Paisley, Scotland). Trypan Blue, 3-(4,5-dimethylthiazol-2-yl)-2.5-diphenyltetrazolium bromide (MTT), 1,10 phenanthroline (PA), phorbol 12-myristate 13-acetate (PMA), lipopolysaccharide (LPS), hydrogen peroxide (H_2O_2) and 2,2-diphenyl-1-picrylhydraziyl radical (DPPH) were provided by Sigma (St. Louis, MO, USA). 2′,7′-dichlorofluorescein diacetate (DCFH-DA) was purchased from Calbiochem (Milano, Italy). Standard proteins and R-250 Coomassie Brilliant Blue were from Bio-Rad (Hercules, CA, USA). Human monocytic leukemia cell line (THP-1 ATCC® TIB-202™) was obtained from American Type Culture Collection (ATCC), Manassas, VI, USA.

4.2. Antioxidants

The antioxidant samples, namely resveratrol (99.7%); oliplus (total polyphenols: 45.5%); green tea extract (total polyphenols: 50%) and curcumin (96.6%), were purchased from Nutraceutica s.r.l. (Monterenzio, (BO), Italy). All of the antioxidants, except for green tea extract (prepared in 40% ethanol in water solution), were solubilized with 80% ethanol in water solution.

4.3. DPPH Radical Scavenging Activity

2,2-Diphenyl-1-picrylhydraziyl radical (DPPH) is commonly employed to measure the capability of antioxidant compounds to scavenge free radicals or serve as hydrogen providers. DPPH is a deep purple nitrogenous organic radical that absorbs at 517 nm. When an antioxidant compound reacts with the DPPH radical, the reduced molecular form is generated; thus, the absorbance decreases proportionally and can be spectrophotometrically monitored over time at 517 nm. The capacity of antioxidants to scavenge DPPH was evaluated as reported by Petraglia et al. [51]. Briefly, 0.2 mL of antioxidant samples at different concentrations was added to 0.8 mL of 0.2 mM DPPH in ethanol. After 30 min of incubation at room temperature in the dark, absorbance was measured at 517 nm. Results were expressed as percentage of DPPH radical scavenging using the following equation: (%) = [(Abs control − Abs sample)/Abs control] × 100. Antioxidant activity was expressed as IC_{50}, representing the sample concentration (μg/mL) required to scavenge 50% of DPPH free radicals.

4.4. THP-1 Differentiation

THP-1 monocytes were maintained in T-75 flasks (75 cm^2) in Roswell Park Memorial Institute (RPMI) 1640 Medium supplemented with 100 U/mL penicillin, 100 μg/mL streptomycin, 10% FBS at 37 °C and 5% CO_2. For differentiation into macrophages, THP-1 monocytes were plated at a density of 1.5×10^5 in 96-well plates and incubated for 24 h with 1 μM phorbol 12-myristate 13-acetate (PMA). After incubation for another 24 h in RPMI medium, THP-1, differentiated in macrophages, was used for the different assays.

4.5. Cell Viability

To assess the biocompatibility of antioxidant compounds, THP-1 macrophages plated in 96-well plates in serum-free medium were treated with resveratrol (RSV), oliplus (OLI), green tea extract (GTE) or curcumin (CRC) at the final concentrations of 10, 25, 50, 100 and 250 μg/mL. After incubation for 20 h at 37 °C and 5% CO_2, the culture medium was removed, and the cell viability was evaluated by MTT [3-(4,5-dimethylthiazol-2-yl)-2,5-diphenyl tetrazolium bromide] assay as previously reported [52].

4.6. Treatment of LPS-Activated THP-1 Macrophages with Antioxidant Compounds

Differentiated THP-1 macrophages, plated in 96-well microplates in serum-free medium, were stimulated with 10 μg/mL of LPS and simultaneously treated with different concentrations of RSV, OLI, GTE and CRC. THP-1 macrophages in serum-free RPMI 1640 medium and LPS-activated THP-1 macrophages represented negative and positive controls, respectively. After 20 h of incubation at 37 °C and 5% CO_2, the culture medium was collected and stored at −80 °C until analysis, whereas the cells were subjected to MTT assay to assess cell viability.

4.7. Reactive Oxygen Species Detection

To evaluate the potential antioxidant activity of the studied compounds, differentiated THP-1 macrophages were plated in 96-well microplates and pretreated with GTE, OLI, RSV and CRC at concentrations in the range of 5–25 μg/mL in RPMI without phenol R = red. After 1 h of incubation, cells were stimulated for 1 h with 100 μM of H_2O_2 in the presence of antioxidant compounds, and the detection of ROS was performed by loading cells with 10 μM of 2′,7′-dichlorofluorescein diacetate (DCFH-DA) in phenol-red-free RPMI, as previously reported [53]. After incubation for 30 min at 37 °C, the culture medium was removed, and cells were rinsed twice with PBS. Cells were resuspended in phenol-red-free RPMI, and a spectrofluorometric analysis was performed at 485 nm excitation/525 nm emission using a multi plate reader (Cytation 3, BioTek, Winooski, VT, USA). Negative control was represented by cells treated only with DCFH-DA in the same experimental conditions. H_2O_2-stimulated THP-1 macrophages represented the positive control (H_2O_2). The ability of the antioxidant compounds to inhibit ROS production was expressed as relative percentage of photoluminescence intensity (PLI) in comparison to the positive control.

4.8. Detection of MMP-2 and MMP-9 by Zymography

The MMP-2 and MMP-9 levels in cell supernatants were detected by zymographic analysis as reported by Latronico et al. [52]. Briefly, 50 μL of supernatants, containing about 10 μg of total proteins, was precipitated with 1 mL of ice-cold acetone. After incubation for 1 h at −20 °C and centrifugation at 13,000× *g* at 4 °C, dry pellets were solubilized with 15 μL of Laemmli sample buffer without β-mercaptoethanol. Samples were run in a 7.5% polyacrylamide gel copolymerized with 0.1% (*wt*/*v*) gelatin. After the electrophoretic run at 120 V, gels were rinsed twice with 2.5% Triton X-100/10 mM $CaCl_2$ in 50 mM Tris–HCl, pH 7.4 (washing buffer) and incubated for 24 h at 37 °C in 1% Triton X-100/50 mM Tris–HCl/10 mM $CaCl_2$, pH 7.4 (incubation buffer). After staining and destaining of gels, MMP-2 and MMP-9 levels were visualized as a clear band of digestion on a blue background of the gel and were quantified by computerized densitometric image analysis using Image LabTM Software, Version 5.2 (Bio-Rad Laboratories, Hercules, CA, USA). Gelatinase levels were expressed as percentage of positive control (LPS-activated cells).

4.9. Determination of Inhibitory Capacity (In-Gel Inhibition) of Dietary Antioxidant on Gelatinases Present in Pool of Sera from Patients with Breast Cancer

The effects of natural antioxidants on the activity of gelatinases present in a pool of sera from patients with breast cancer were evaluated by one-dimensional zymography on polyacrylamide gel copolymerized with gelatin as already reported [54]. The serum samples were provided by the Medical Oncology Unit of the S. Carlo Hospital in Potenza (ethical committee authorization no. 74/2021). To perform the inhibition tests, 1.5 µL of aliquots of the serum pool was analyzed by "in-gel" zymography. After reactivation of the proteases, the various lanes were cut from the gels and incubated individually (for 16 h at 37 °C) in incubation buffer containing various antioxidants at the concentrations of 10, 50 and 100 µg/mL. 1,10-phenanthroline (1,10 PA) was used as a positive control. After staining and destaining of gels, MMP-2 and MMP-9 activity was detected as reported in Section 4.8.

5. Conclusions

The results obtained in this study confirmed the role of some polyphenols extracted from natural matrices as precious allies for our health, since they act as inhibitors of important enzymes involved in pathological events, such as MMPs. It is therefore reasonable to consider polyphenols as potential reservoirs of innovative therapeutic solutions for human health. However, the changes in their structure occurring during metabolic processes could influence their inhibitory action in vivo.

To maximize the anticancer effects of polyphenols, different drug delivery nanosystems made with biocompatible materials could be used to increase their bioavailability and bioaccessibility. The main advantages derived from the encapsulation of polyphenols in biocompatible nanomaterials are represented by the increases in their stability and solubility, protection from degradation and the possibility of targeting them towards specific cells or tissues.

In conclusion, the results of this study represent a good starting point for more in-depth future studies aimed at understanding the mechanisms underlying inhibition by polyphenols, as well as the biotransformations that these compounds undergo in vivo.

Author Contributions: Conceptualization, R.R. and G.M.L.; investigation, T.L., T.P. and C.S.; resources, D.B.; writing—original draft preparation, R.R. and G.M.L.; writing—review and editing, R.R. and G.M.L. All authors have read and agreed to the published version of the manuscript.

Funding: This research received no external funding.

Institutional Review Board Statement: The part of this study involving the use of sera from patients with breast cancer was conducted in accordance with the Declaration of Helsinki and approved by the Ethics Committee of S. Carlo Hospital, Potenza, Italy (authorization no. 74/2021).

Informed Consent Statement: Informed consent was obtained from all subjects involved in this study.

Data Availability Statement: Data are contained within the article.

Conflicts of Interest: The authors declare no conflicts of interest.

References

1. Arnold, M.; Morgan, E.; Rumgay, H.; Mafra, A.; Singh, D.; Laversanne, M.; Vignat, J.; Gralow, J.R.; Cardoso, F.; Siesling, S.; et al. Current and future burden of breast cancer: Global statistics for 2020 and 2040. *Breast* **2022**, *66*, 15–23. [CrossRef] [PubMed]
2. De Cicco, P.; Catani, M.V.; Gasperi, V.; Sibilano, M.; Quaglietta, M.; Savini, I. Nutrition and Breast Cancer: A Literature Review on Prevention, Treatment and Recurrence. *J. Nutr.* **2019**, *11*, 1514. [CrossRef] [PubMed]
3. Cava, E.; Marzullo, P.; Farinelli, D.; Gennari, A.; Saggia, C.; Riso, S.; Prodam, F. Breast Cancer Diet "BCD": A Review of Healthy Dietary Patterns to Prevent Breast Cancer Recurrence and Reduce Mortality. *J. Nutr.* **2022**, *14*, 476. [CrossRef] [PubMed]
4. Bednarczyk, R.B.; Tuli, N.Y.; Hanly, E.K.; Rahoma, G.B.; Maniyar, R.; Mittelman, A.; Geliebter, J.; Tiwari, R.K. Macrophage inflammatory factors promote epithelial-mesenchymal transition in breast cancer. *Oncotarget* **2018**, *9*, 24272–24282. [CrossRef] [PubMed]
5. Basak, U.; Sarkar, T.; Mukherjee, S.; Chakraborty, S.; Dutta, A.; Dutta, S.; Nayak, D.; Kaushik, S.; Das, T.; Sa, G. Tumor-associated macrophages: An effective player of the tumor microenvironment. *Front. Immunol.* **2023**, *14*, 1295257. [CrossRef]
6. Li, M.; He, L.; Zhu, J.; Zhang, P.; Liang, S. Targeting tumor-associated macrophages for cancer treatment. *Cell Biosci.* **2022**, *12*, 85. [CrossRef]
7. Kessenbrock, K.; Plaks, V.; Werb, Z. Matrix metalloproteinases: Regulators of the tumor microenvironment. *Cell* **2010**, *141*, 52–67. [CrossRef]
8. Kwon, M.J. Matrix metalloproteinases as therapeutic targets in breast cancer. *Front. Oncol.* **2023**, *12*, 1108695. [CrossRef]

9. Gonzalez-Avila, G.; Sommer, B.; Mendoza-Posada, D.A.; Ramos, C.; Garcia-Hernandez, A.A.; Falfan-Valencia, R. Matrix metalloproteinases participation in the metastatic process and their diagnostic and therapeutic applications in cancer. *Crit. Rev. Oncol./Hematol.* **2019**, *137*, 57–83. [CrossRef]
10. Nelson, A.R.; Fingleton, B.; Rothenberg, M.L.; Matrisian, L.M. Matrix Metalloproteinases: Biologic Activity and Clinical Implications. *J. Clin. Oncol.* **2000**, *18*, 1135. [CrossRef]
11. Arfin, S.; Jha, N.K.; Jha, S.K.; Kesari, K.K.; Ruokolainen, J.; Roychoudhury, S.; Rathi, B.; Kumar, D. Oxidative Stress in Cancer Cell Metabolism. *Antioxidants* **2021**, *10*, 642. [CrossRef] [PubMed]
12. Ra, H.J.; Parks, W.C. Control of matrix metalloproteinase catalytic activity. *Matrix Biol.* **2007**, *26*, 587–596. [CrossRef]
13. Santhiravel, S.; Bekhit, A.E.-D.A.; Mendis, E.; Jacobs, J.L.; Dunshea, F.R.; Rajapakse, N.; Ponnampalam, E.N. The impact of plant phytochemicals on the gut microbiota of humans for a balanced life. *Int. J. Mol. Sci.* **2022**, *23*, 8124. [CrossRef] [PubMed]
14. Gropper, S.S. The Role of Nutrition in Chronic Disease. *Nutrients* **2023**, *15*, 664. [CrossRef]
15. Cory, H.; Passarelli, S.; Szeto, J.; Tamez, M.; Mattei, J. The Role of Polyphenols in Human Health and Food Systems: A Mini-Review. *Front. Nutr.* **2018**, *5*, 87. [CrossRef] [PubMed]
16. Sharma, E.; Attri, D.C.; Sati, P.; Dhyani, P.; Szopa, A.; Sharifi-Rad, J.; Hano, C.; Calina, D.; Cho, W.C. Recent updates on anticancer mechanisms of polyphenols. *Front. Cell Dev. Biol.* **2022**, *10*, 1005910. [CrossRef] [PubMed]
17. Liu, W.; Cui, X.; Zhong, Y.; Ma, R.; Liu, B.; Xia, Y. Phenolic metabolites as therapeutic in inflammation and neoplasms: Molecular pathways explaining their efficacy. *Pharmacol. Res.* **2023**, *193*, 106812. [CrossRef]
18. Munir, M.T.; Kay, M.K.; Kang, M.H.; Rahman, M.M.; Al-Harrasi, A.; Choudhury, M.; Moustaid-Moussa, N.; Hussain, F.; Rahman, S.M. Tumor-Associated Macrophages as Multifaceted Regulators of Breast Tumor Growth. *Int. J. Mol. Sci.* **2021**, *22*, 6526. [CrossRef]
19. Shah, R.; Ibis, B.; Kashyap, M.; Boussiotis, V.A. The role of ROS in tumor infiltrating immune cells and cancer immunotherapy. *Metabolism* **2024**, *151*, 155747. [CrossRef]
20. Khan, Y.H.; Uttra, A.M.; Qasim, S.; Mallhi, T.H.; Alotaibi, N.H.; Rasheed, M.; Alzarea, A.I.; Iqbal, M.S.; Alruwaili, N.K.; Khan, S.U.D.; et al. Potential role of phytochemicals against matrix metalloproteinase induced breast cancer; An explanatory review. *Front. Chem.* **2021**, *8*, 592152. [CrossRef]
21. Radisky, E.S.; Raeeszadeh-Sarmazdeh, M.; Radisky, D.C. Therapeutic potential of matrix metalloproteinase inhibition in breast cancer. *J. Cell. Biochem.* **2017**, *118*, 3531–3548. [CrossRef]
22. Kumar, G.B.; Nair, B.G.; Perry, J.J.P.; Martin, D.B.C. Recent insights into natural product inhibitors of matrix metalloproteinases. *MedChemComm* **2019**, *10*, 2024–2037. [CrossRef] [PubMed]
23. Rashid, Z.A.; Bardaweel, S.K. Novel Matrix Metalloproteinase-9 (MMP-9) Inhibitors in Cancer Treatment. *Int. J. Mol. Sci.* **2023**, *24*, 12133. [CrossRef]
24. Vandooren, J.; Knoops, S.; Aldinucci Buzzo, J.L.; Boon, L.; Martens, E.; Opdenakker, G.; Kolaczkowska, E. Differential inhibition of activity, activation and gene expression of MMP-9 in THP-1 cells by azithromycin and minocycline versus bortezomib: A comparative study. *PLoS ONE* **2017**, *12*, e0174853. [CrossRef] [PubMed]
25. Mori, K.; Uchida, T.; Yoshie, T.; Mizote, Y.; Ishikawa, F.; Katsuyama, M.; Shibanuma, M. A mitochondrial ROS pathway controls matrix metalloproteinase 9 levels and invasive properties in RAS-activated cancer cells. *FEBS J.* **2019**, *286*, 459–478. [CrossRef]
26. Bae, M.J.; Karadeniz, F.; Oh, J.H.; Yu, G.H.; Jang, M.S.; Nam, K.H.; Seo, Y.; Kong, C.S. MMP-Inhibitory Effects of Flavonoid Glycosides from Edible Medicinal Halophyte *Limonium tetragonum*. *J. Altern. Complement. Med.* **2017**, *2017*, 6750274. [CrossRef]
27. Lee, S.J.; Kim, M.M. Resveratrol with antioxidant activity inhibits matrix metalloproteinase via modulation of SIRT1 in human fibrosarcoma cells. *J. Life Sci.* **2011**, *88*, 465–472. [CrossRef] [PubMed]
28. Demeule, M.; Brossard, M.; Page, M.; Gingras, D.; Beliveau, R. Matrix metalloproteinase inhibition by green tea catechins. *BBA* **2000**, *1478*, 51–60. [CrossRef]
29. Park, W.H.; Kim, S.H.; Kim, C.H. A new matrix metalloproteinase-9 inhibitor 3,4-dihydroxycinnamic acid (caffeic acid) from methanol extract of Euonymus alatus: Isolation and structure determination. *Toxicology* **2005**, *207*, 383–390. [CrossRef]
30. Liuzzi, M.G.; Latronico, T.; Branà, M.T.; Gramegna, P.; Coniglio, M.G.; Rossano, R.; Larocca, M.; Riccio, P. Structure-dependent inhibition of gelatinases by dietary antioxidants in rat astrocytes and sera of multiple sclerosis patients. *Neurchem. Res.* **2011**, *36*, 518–527. [CrossRef]
31. Larocca, M.; Di Marsico, M.; Riccio, P.; Rossano, R. The in vitro antioxidant properties of *Muscari comosum* bulbs and their inhibitory activity on enzymes involved in inflammation, post-prandial hyperglycemia, and cognitive/neuromuscular functions. *J. Food Biochem.* **2018**, *42*, e12580. [CrossRef]
32. Scoditti, E.; Nestola, A.; Massaro, M.; Calabriso, N.; Storelli, C.; De Caterina, R.; Carluccio, M.A. Hydroxytyrosol suppresses MMP-9 and COX-2 activity and expression in activated human monocytes via PKCa and PKCb1 inhibition. *Aterosclerosis* **2014**, *232*, 17–24. [CrossRef] [PubMed]
33. Unachukwu, U.J.; Ahmed, S.; Kavalier, A.; Lyles, J.T.; Kennelly, E.J. White and green teas (*Camellia sinensis* var. sinensis): Variation in phenolic, methylxanthine, and antioxidant profiles. *J. Food Sci.* **2010**, *75*, C541–C548. [CrossRef] [PubMed]
34. Račková, L.; Košťálová, D.; Bezáková, L.; Fialová, S.; Bauerová, K.; Tóth, J.; Štefek, M.; Vanko, M.; Holková, I.; Obložinský, M. Comparative study of two natural antioxidants, curcumin and Curcuma longa extract. *J. Food Nutr. Res.* **2009**, *48*, 148–152.
35. Oliveira Calil, N.; Senra Gonçalves de Carvalho, G.; Zimmermann Franco, D.C.; da Silva, A.D.; Rezende Barbosa Raposo, N. Antioxidant activity of resveratrol analogs. *Lett. Drug Des. Discov.* **2012**, *9*, 8–11. [CrossRef]

36. Aldini, G.; Piccoli, A.; Beretta, G.; Morazzoni, P.; Riva, A.; Marinello, C.; Maffei Facino, R. Antioxidant activity of polyphenols from solid olive residues of c.v. Coratina. *Fitoterapia* **2006**, *77*, 121–128. [CrossRef] [PubMed]
37. Dell'Agli, M.; Fagnani, R.; Galli, G.V.; Maschi, O.; Gilardi, F.; Bellosta, S.; Crestani, M.; Bosisio, E.; De Fabiani, E.; Caruso, D. Olive oil phenols modulate the expression of metalloproteinase 9 in THP-1 cells by acting on nuclear factor-kappaB signaling. *J. Agric. Food Chem.* **2010**, *58*, 2246–2252. [CrossRef] [PubMed]
38. Nelson, K.K.; Melendez, J.A. Mitochondrial redox control of matrix metalloproteinases. *Free Radic Biol Med.* **2004**, *37*, 768–784. [CrossRef] [PubMed]
39. Iqbal, M.J.; Kabeer, A.; Abbas, Z.; Siddiqui, H.A.; Calina, D.; Sharifi-Rad, J.; Cho, W.C. Interplay of oxidative stress, cellular communication and signaling pathways in cancer. *Cell Commun. Signal.* **2024**, *22*, 7. [CrossRef]
40. Suzuki, T.; Ohishi, T.; Tanabe, H.; Miyoshi, N.; Nakamura, Y. Anti-inflammatory edffects of dietary polyphenols through inhibitory activity against metalloproteinases. *Molecules* **2023**, *28*, 5426. [CrossRef]
41. Tanabe, H.; Suzuki, T.; Ohishi, T.; Isemura, M.; Nakamura, Y.; Unno, K. Effects of Epigallocatechin-3-Gallate on Matrix Metalloproteinases in Terms of Its Anticancer Activity. *Molecules* **2023**, *28*, 525. [CrossRef]
42. Hider, R.C.; Liu, Z.D.; Khodr, H.H. Metal chelation of polyphenols. *Methods Enzymol.* **2001**, *335*, 190–203. [PubMed]
43. Le Nest, G.; Caille, O.; Woudstra, M.; Roche, S.; Guerlesquin, F.; Lexa, D. Zn–polyphenol chelation: Complexes with quercetin, (+)-catechin, and derivatives: I optical and NMR studies. *Inorganica Chim. Acta* **2004**, *357*, 775–784. [CrossRef]
44. Wei, Y.; Guo, M. Zinc-binding sites on selected flavonoids. *Biol. Trace Elem. Res.* **2014**, *161*, 223–230. [CrossRef] [PubMed]
45. Perron, N.R.; Brumaghim, J.L. A Review of the antioxidant mechanisms of polyphenol compounds related to iron binding. *Cell Biochem. Biophys.* **2009**, *53*, 75–100. [CrossRef] [PubMed]
46. Lakey-Beitia, J.; Burillo, A.M.; La Penna, G.; Hegde, M.L.; Rao, K.S. Polyphenols as potential metal chelation compounds against Alzheimer's disease. *JAD* **2021**, *82*, S335–S357. [CrossRef] [PubMed]
47. Zhong, H.; Wees, M.A.; Faure, T.D.; Carrillo, C.; Arbiser, J.; Bowen, J.P. The Impact of Ionization States of Matrix Metalloproteina-seInhibitors on Docking-Based Inhibitor Design. *ACS Med. Chem. Lett.* **2011**, *2*, 455–460. [CrossRef] [PubMed]
48. Ahmad, A.; Sayed, A.; Ginnebaugh, K.R.; Sharma, V.; Suri, A.; Saraph, A.; Padhye, S.; Sarkar, F.H. Molecular Docking and Inhibition of Matrix Metalloproteinase-2 by Novel Difluorinatedbenzylidene Curcumin Analog. *Am. J. Transl. Res.* **2015**, *7*, 298–308. [PubMed]
49. Chowdhury, A.; Nandy, S.K.; Sarkar, J.; Chakraborti, T.; Chakraborti, S. Inhibition of Pro-/Active MMP-2 by Green Tea Catechins and Prediction of Their Interaction by Molecular Docking Studies. *Mol. Cell. Biochem.* **2017**, *427*, 111–122. [CrossRef]
50. Sarkar, J.; Nandy, S.K.; Chowdhury, A.; Chakraborti, T.; Chakraborti, S. Inhibition of MMP-9 by Green Tea Catechins and Prediction of Their Interaction by Molecular Docking Analysis. *Biomed. Pharmacother.* **2016**, *84*, 340–347. [CrossRef]
51. Petraglia, T.; Latronico, T.; Fanigliulo, A.; Crescenzi, A.; Liuzzi, G.M.; Rossano, R. Antioxidant Activity of Polysaccharides from the Edible Mushroom Pleurotus eryngii. *Molecules* **2023**, *28*, 2176. [CrossRef] [PubMed]
52. Latronico, T.; Pati, I.; Ciavarella, R.; Fasano, A.; Mengoni, F.; Lichtner, M.; Vullo, V.; Mastroianni, C.M.; Liuzzi, G.M. In vitro effect of antiretroviral drugs on cultured primary astrocytes: Analysis of neurotoxicity and matrix metalloproteinase inhibition. *J. Neurochem.* **2018**, *144*, 271–284. [CrossRef] [PubMed]
53. Latronico, T.; Larocca, M.; Milella, S.; Fasano, A.; Rossano, R.; Liuzzi, G.M. Neuroprotective potential of isothiocyanates in an in vitro model of neuroinflammation. *Inflammopharmacology* **2021**, *29*, 561–571. [CrossRef]
54. Rossano, R.; Larocca, M.; Macellaro, M.; Bilancia, D.; Riccio, P. Unveiling a hidden biomarker of inflammation and tumor progression: The 65 kDa isoform of MMP-9 new horizons for therapy. *Curr. Issues Mol. Biol.* **2022**, *44*, 105–116. [CrossRef]

molecules

MDPI

Article

Genistein Prevents Apoptosis and Oxidative Stress Induced by Methylglyoxal in Endothelial Cells

Maria Liccardo, Luigi Sapio, Shana Perrella, Ivana Sirangelo * and Clara Iannuzzi

Department of Precision Medicine, Università degli Studi della Campania "Luigi Vanvitelli", Via L. De Crecchio 7, 80138 Naples, Italy; maria.liccardo@unicampania.it (M.L.); luigi.sapio@unicampania.it (L.S.); shana.perrella@unicampania.it (S.P.); clara.iannuzzi@unicampania.it (C.I.)
* Correspondence: ivana.sirangelo@unicampania.it

Abstract: Glycolytic overload promotes accumulation of the highly reactive dicarbonyl compounds, resulting in harmful conditions called dicarbonyl stress. Methylglyoxal (MG) is a highly reactive dicarbonyl species and its accumulation plays a crucial pathophysiological role in diabetes and its vascular complications. MG cytotoxicity is mediated by reactive oxygen species (ROS) generation, a key event underlying the intracellular signaling pathways leading to inflammation and apoptosis. The identification of compounds able to inhibit ROS signaling pathways and counteract the MG-induced toxicity is a crucial step for developing new therapeutic strategies in the treatment of diabetic vascular complications. In this study, the effect of genistein, a natural soybean isoflavone, has been evaluated on MG-induced cytotoxicity in human endothelial cells. Our results show that genistein is able to counteract the MG-induced apoptosis by restraining ROS production, thus inhibiting the MAPK signaling pathways and caspase-3 activation. These findings identify a beneficial role for genistein, providing new insights for its potential clinical applications in preserving endothelial function in diabetic vascular complications.

Keywords: methylglyoxal; oxidative stress; antioxidant activity; genistein; caspase-3; p38

1. Introduction

Methylglyoxal (MG) is a highly reactive dicarbonyl species spontaneously produced by the fragmentation of glycolytic intermediates, mainly glyceraldehyde 3-phosphate and dihydroxyacetone phosphate, formed during glucose metabolism and to a lesser degree from intermediates of lipid and protein metabolism [1,2]. Moreover, MG is a major intermediate of the non-enzymatic glycation reaction leading to the formation of the advanced glycation end-products (AGEs), which are associated with several aging-related diseases including diabetes, cancer, and neurodegenerative diseases [2–4]. For this reason, MG formation and its accumulation play a crucial pathophysiological role in these pathologies, especially in diabetes and its vascular complication, both microvascular (retinopathy, neuropathy, and nephropathy) and macrovascular (ischemic heart disease, cerebrovascular disease, and peripheral vascular diseases) [5–9]. Endothelial cell (EC) injury and its dysfunctions have been related to the pathogenesis of vascular complications in diabetes mellitus [10]. Oxidative stress in EC is considered the starting event that triggers the intracellular signaling transduction pathways that lead to inflammation and apoptosis [11,12].

In diabetic patients, due to their hyperglycemic state, plasma levels of MG and AGEs are strongly increased and accumulate abnormally in multiple tissues and organs, playing a prominent role in the pathogenesis of diabetes and correlated vascular complications [13,14]. The molecular mechanism by which physiopathological levels of MG induces endothelial dysfunction is not fully understood. However, several studies have evidenced that the MG cytotoxicity is mainly induced through mitochondrial membrane potential impairment and reactive oxygen species (ROS) generation, the key event underlying the intracellular signaling pathways leading to inflammation and apoptosis [15,16]. Specifically, the MG-induced oxidative stress causes endothelial cell apoptosis through the activation of the ROS-mediated MAPK (JNK, p38, and ERK1/2) signaling pathways and the activation of mitochondrial caspase-3 [17,18].

Significant efforts have been made to identify compounds able to inhibit ROS signaling pathways in order to counteract the MG-induced toxicity in EC with the aim of developing new potential therapeutic strategies in the treatment of diabetic vascular complications [18–21]. In this respect, plant polyphenols have attracted considerable interest as food supplements to ameliorate and prevent diabetic-related complications due to their multiple biological activities, including antioxidant, anti-inflammatory, and antidiabetic power [22–25].

Among these natural polyphenols, genistein is a soybean isoflavone and naturally occurring tyrosine kinase inhibitor, known for several biological and therapeutic properties due to its antioxidant power and ant-inflammatory capacity [26–29] (Figure 1). Genistein exhibits estrogen effects, neuroprotective action against ischemic injury, antilipogenic and hypolipidemic properties, as well as chemoprevention in cancer treatment [28,30,31]. Moreover, genistein is suitable in playing a protective role in diabetes and its vascular complication as result of its antioxidant, anti-inflammatory and hypoglycemic activity [32–37]. Genistein also plays a protective role in MG-induced immune dysfunction in diabetic patients, reducing the oxidative stress and DNA damage as well as ROS generation and apoptosis in mononuclear cells [38]. Plant extracts containing genistein also showed a protective effect against MG-induced glucotoxicity in EC by reducing ROS formation and apoptosis [39]. Moreover, it has been reported that genistein shows anti-glycation activity by trapping MG both in vitro and in vivo, thus inhibiting the formation of AGEs [40,41].

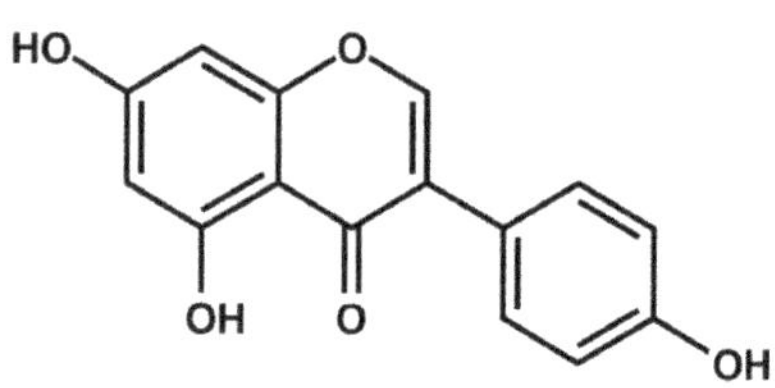

Figure 1. Chemical structure of genistein.

In light of these considerations, in this study, we have analyzed the effect of genistein on MG-induced apoptosis in a human endothelial cell model. Our results show that genistein is able to counteract the MG-induced cytotoxicity by restraining ROS production, thus inhibiting the ROS-mediated MAPK signaling pathways and the activation of caspase-3. These findings identify a beneficial role for genistein, providing a promising basis for further in vivo and pre-clinical studies with the aim of evaluating its possible use in diabetic vascular complications.

2. Results

2.1. Genistein Markedly Reduces the MG-Induced Toxicity in EA.HY926 Cells

Before testing the effect of the combination treatment of genistein plus MG in endothelial cells, the toxicity of both genistein and MG has been evaluated in EA.HY926 cells, a representative endothelial cell line, in order to optimize the experimental conditions. Specifically, EA.HY926 cells have been exposed to different concentrations of genistein (0–100 μM) and MG (0–2 mM) for 24 h and the cell viability has been assessed by the MTT assay (Figure 2). For genistein, after 24 h of treatment, significant cytotoxicity was observed only in the presence of 20, 50, and 100 μM (Figure 2A), IC50 = 100 μM, whereas, for MG, the toxicity was observed in the range of 250–2000 μM (IC50 = 400 μM). In particular, 250 μM MG promotes a 36% reduction in cell viability, while higher concentrations (500–2000 μM) reduce the cell viability by over 50% (Figure 2B).

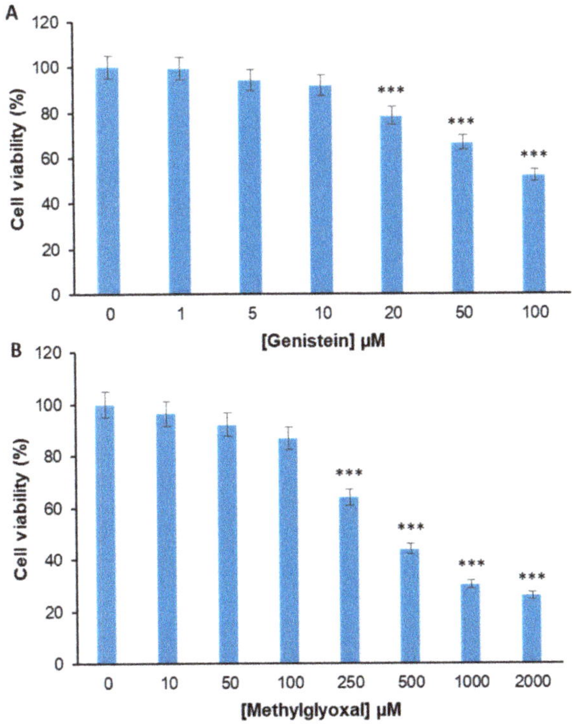

Figure 2. Evaluation of genistein and MG cytotoxicity in endothelial cells. Cell viability has been evaluated by MTT assay in EA.HY926 cells exposed for 24 h to increasing concentrations of genistein (0–100 µM) (**A**) and MG (0–2000 µM) (**B**). Data are expressed as average percentage of cell viability reduction ± SD relative to untreated cells from triplicate wells from 5 separate experiments. Other experimental details are described in the Materials and Methods section. *** $p < 0.05$ versus untreated cells.

Based on the above results, the working concentrations for our study were set at 0–20 µM for genistein and 250 µM for MG in order to minimize the cell death after 24 h of treatment.

To investigate the effect of genistein on MG-induced cytotoxicity, EA.HY926 cells pretreated for 2 h with genistein at different concentrations (0, 1, 5, 10, and 20 µM) were incubated with 250 µM MG and the cell viability was monitored by MTT assay after 24 h of treatment (Figure 3A).

Our results show that, while cells exposed to MG showed a strong reduction (about 40%) in cell viability after 24 h of treatment, the absence of toxicity was observed for cells treated in the presence of genistein in the range of 5–20 µM (about 80% cell viability). In particular, while the lower concentration of genistein (1 µM) only slightly affects the MG cytotoxicity, 5, 10, and 20 µM genistein are able to protect endothelial cells by MG toxicity. In order to evaluate modifications in cell morphology and cell number upon different treatments, cell samples were also analyzed by phase-contrast microscopy (Figure 3B). Similarly to the MTT assay, phase-contrast microscopy shows that cells exposed to MG for 24 h display both modifications in cell morphology and reductions in the cell number, whereas those pretreated with genistein exhibit no qualitative and quantitative alterations.

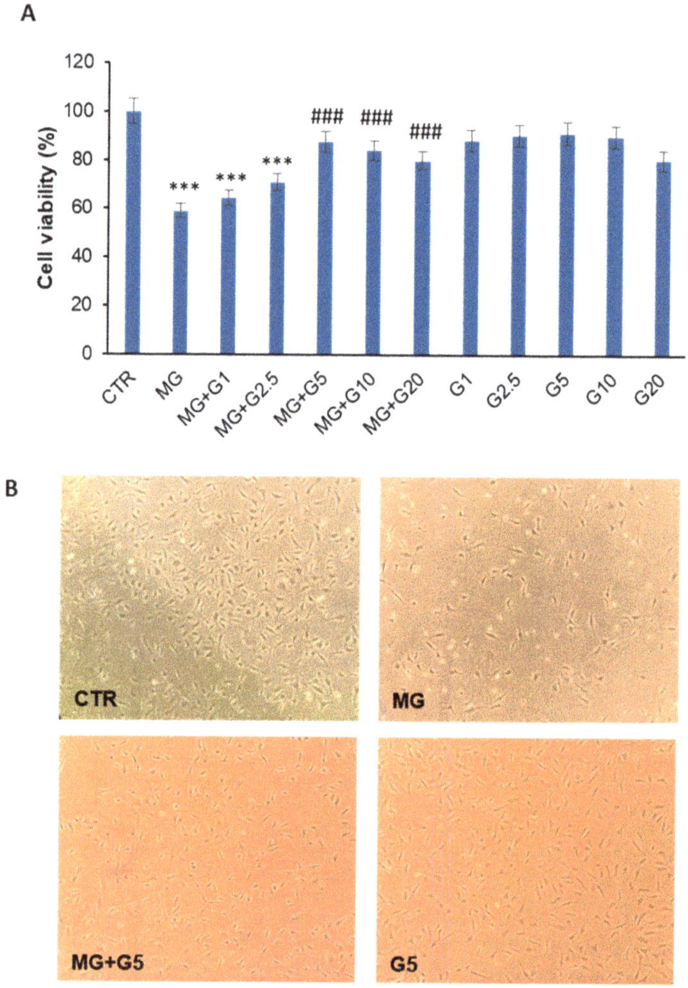

Figure 3. Effect of genistein on MG-induced cytotoxicity. (**A**) Cell viability evaluated by MTT assay in EA.HY926 cells exposed for 24 h to 250 µM MG (MG), pretreated for 2 h with 1 (MG + G1), 2.5 (MG + G2.5), 5 (MG + G5), 10 (MG + G10), and 20 (MG + G20) µM genistein. Data are expressed as average percentage of cell viability reduction ± SD relative to untreated cells (CTR) from triplicate wells from 5 separate experiments. *** $p < 0.05$ versus CTR, ### $p < 0.05$ versus MG. (**B**) Phase contrast microscopy images of EA.HY926 cells after 24 h of incubation. CTR: untreated cells; MG: cells exposed to 250 µM MG; MG + G5 cells pretreated with 5 µM genistein; G5: cells treated with 5 µM genistein. Other experimental details are described in the Materials and Methods section.

The protective effect of genistein on the MG-induced cytotoxicity has been further analyzed through the evaluation of the cell cycle distribution (Figure 4). Flow cytometric analysis (FACS) indicated that EA.HY926 cells treated with MG for 24 h showed a significant G0/G1 reduction (−14.6%) and a concomitant subG1 appearance (+13%) compared to untreated cells. Interestingly, pretreatment with genistein strongly mitigated the cell cycle alterations observed with MG. In particular, cells pretreated with genistein exhibited a higher percentage of G0/G1 (57.3% vs. 65.8%) and no subG1 occurrence compared to cells exposed to MG only. These data further confirm the cell protection observed for genistein by MTT assay and clearly suggest that genistein 5 μM strongly reduces the MG-induced toxicity in EA.HY926 endothelial cells.

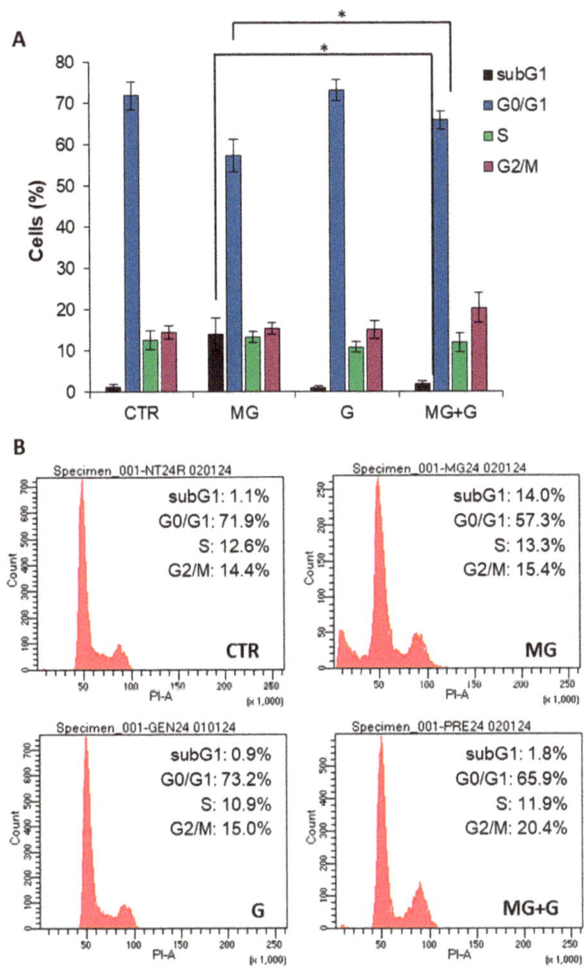

Figure 4. Evaluation of effects of genistein and MG, single and in combination, on cell cycle distribution. EA.HY926 cells were exposed for 24 h to 250 μM MG (MG) and underwent 2 h of pretreatment with 5 μM genistein (G), either alone or in combination (MG + G), for 24 h. Thereafter, the relative cell-cycle distribution was assessed using FACSCelestaTM employing PI as a DNA dye. (**A**) Quantitative analysis of multiple experiments. (**B**) Representative experiment. * $p < 0.05$.

2.2. Genistein Prevents MG-Induced ROS Production and Nrf2 Activation in EA.HY926 Cells

Oxidative stress has been identified as the main mechanism by which MG promotes cytotoxicity in endothelial cells, due to the increase in ROS production and subsequent mitochondrial membrane depolarization, a key event underlying the molecular signaling pathways leading to inflammation and apoptosis [42–44]. Genistein is known to possess a strong antioxidant activity mainly acting as a free radical scavenger [45–47]. In this respect, with the aim of identifying the molecular basis of the genistein cellular protection in MG toxicity, we have evaluated the effect of genistein in MG-associated oxidative damage. At first, the ability of genistein to reduce the ROS production associated with MG treatment has been tested (Figure 5). In particular, as MG is known to induce ROS production in endothelial cells [13,17,18], the intracellular ROS levels have been measured in EA.HY926 cells pretreated with 5 μM genistein and then treated with MG for 2 and 5 h, by the DCFH-DA fluorescence assay (Figure 5A). Our results show that treatment with MG 250 μM promotes an increase in the DCF fluorescence after both 2 and 5 h of incubation, indicative of ROS production. By contrast, in the sample pre-incubated for 2 h with genistein, the ROS levels were similar to those of untreated cells, thus suggesting that genistein is able to counteract the MG-induced ROS production in EA.HY926 cells.

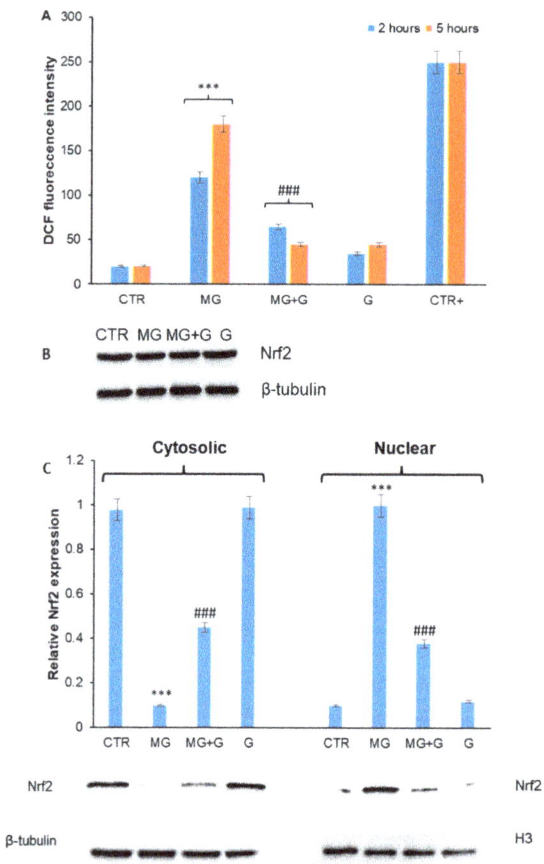

Figure 5. Role of genistein in MG-induced oxidative stress. EA.HY926 cells were exposed to MG (MG) and pretreated with genistein (MG + G), and it has been evaluated ROS production by DCFH-DA assay (**A**), Nrf2 expression (**B**), and Nrf2 nucleus translocation (**C**) by western blot analysis. Panel B shows Nrf2 expression in the total cell extract, while panel C shows the Nrf2 level for cytosolic and nuclear fractions. CTR: untreated cells; G: cells treated with genistein; CTR+: cells treated with 1.0 mM H_2O_2. Data are expressed as average ± SD from five independent experiments carried out in triplicate. MG and genistein concentrations were 250 and 5 μM, respectively. Other experimental details are described in the Materials and Methods section. *** $p < 0.05$ versus CTR, ### $p < 0.05$ versus MG.

Considering the protective effect of genistein in the MG-induced ROS production, we have evaluated the expression of Nrf2 in EA.HY926 cells exposed to MG in the absence and in the presence of genistein (Figure 5B,C). Nrf2 is a transcriptional regulator that modulates oxidative stress response and is known to be sequestered in the cytosol bound to the inhibitory protein Keap1. In oxidative stress conditions, Nrf2 dissociates from Keap1 and moves into the nucleus, thus inducing the expression of downstream antioxidant enzymes [48]. For this reason, Nrf2 expression has been evaluated in total cellular extracts as well as cytosolic and nuclear fractions. Analysis performed on the total extract shows that no variation in Nrf2 expression is observed in all experimental groups (Figure 5B). By contrast, the exposure of cells to MG resulted in Nrf2 translocation from cytosol to the nucleus, thus suggesting its activation. In cells pretreated with genistein, a significant reduction in Nrf2 translocation was observed (Figure 5C). The overall data suggest that the presence of genistein is able to impair the MG-induced ROS production in EA.HY926 cells.

2.3. Genistein Protects Endothelial Cells by the MG-Induced Apoptosis through the MAPK-Mediated Signaling Pathways

MG promotes apoptosis in endothelial cells mainly through the production of ROS able to promote oxidative stress-related pathways that underly the apoptotic cascade [28,48]. At the molecular level, MG is known to induce apoptosis in endothelial cells via mitochondrial dysfunction, ROS/MAPKs/Nf-kB signaling pathways, and ER stress [49]. In endothelial cells, the redox-sensitive family members of MAPKs triggered by MG are ERK, JNK, and p38 MAPKs [17,50]. Genistein is known to exert an antioxidant effect through the inhibition of JNK, ERK, p38 MAPKs, and Nf-kB activation [29,51,52]. Thus, to further investigate the genistein-mediated cell protection in response to MG treatment in endothelial cells, Western blotting analysis has been performed to identify putative effects of genistein on the MG-induced apoptosis. At first, the activation of caspase 3 was evaluated in EA.HY926 cells treated in the presence and absence of 5 μM genistein upon MG treatment for 24 h by measuring the cleaved active caspase 3 (C–C3) (Figure 6). As expected, while MG promotes caspase 3 cleavage, no activation was observed in cells pre-incubated with genistein, thus suggesting protection on MG-induced apoptosis in EA.HY926 cells.

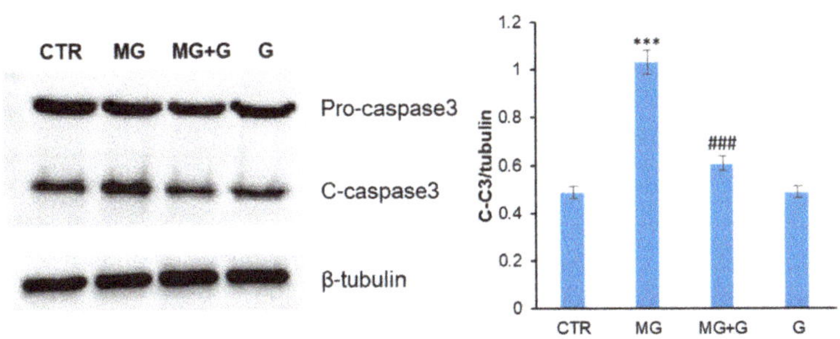

Figure 6. Effect of genistein on caspase 3 activation. EA.HY926 cells were exposed to MG (MG) and pretreated with genistein (MG + G) and caspase 3 activation has been evaluated by Western blot analysis. CTR: untreated cells; Gen: cells treated with genistein. Data are expressed as average ± SD from five independent experiments carried out in triplicate. MG and genistein concentrations were 250 and 5 μM, respectively. Other experimental details are described in the Materials and Methods section. *** $p < 0.05$ versus CTR, ### $p < 0.05$ versus MG.

To monitor whether the protection observed by genistein on MG-induced apoptosis occurred through the MAPK signaling pathways, the activation of ERK and p38 MAPKs was also evaluated in the same experimental groups (Figure 7). Western blot analysis suggests that MG promotes the activation of both ERK and p38 after 24 h of incubation in EA.HY926 cells, which is consistent with the observed caspase-3 activation as directly activated by p38 phosphorylation. By contrast, in cells pre-incubated with genistein, no activation of MAPKs is observed, thus suggesting that genistein is able to protect EA.HY926 cells by interfering with MAPK pro-apoptotic pathways.

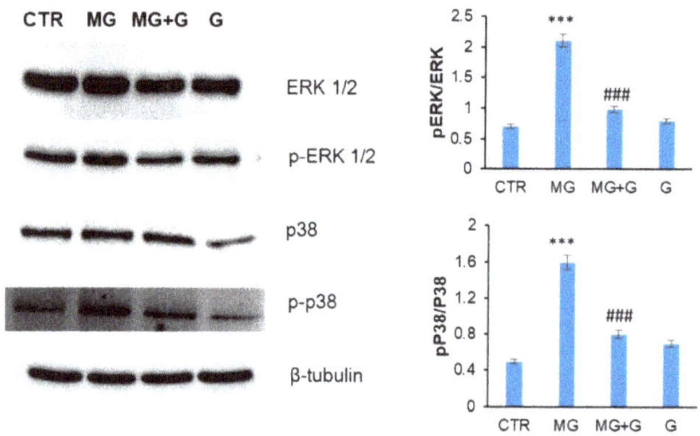

Figure 7. Effect of genistein MAPK activation. EA.HY926 cells were exposed to MG (MG) and pretreated with genistein (MG + G), and ERK1/2 activation and p38 MAPKs have been evaluated by Western blot analysis. CTR: untreated cells; G: cells treated with genistein. Data are expressed as average ± SD from five independent experiments carried out in triplicate. MG and genistein concentrations were 250 and 5 μM, respectively. Other experimental details are described in the Materials and Methods section. *** $p < 0.05$ versus CTR, ### $p < 0.05$ versus MG.

3. Discussion

High intracellular glucose concentration promotes the formation of highly reactive dicarbonyl compounds, which accumulate and lead to harmful conditions called dicarbonyl stress [53]. Methylglyoxal is the most reactive dicarbonyl metabolite of glucose and, in diabetic conditions, is known to trigger non-enzymatic glycation, which eventually lead to the irreversible overproduction of AGEs [54]. The molecular mechanism by which MG induces ROS production and oxidative stress in endothelial cells is mediated by the activation of NADPH oxidase, which induce superoxide generation and mitochondrial dysfunction, which eventually lead to endothelial dysfunction, inflammatory responses, and cell death [15,16,55]. In the present study, we show that genistein, a natural isoflavone, significantly counteracts the MG-induced toxicity in EA.HY926 endothelial cells. Indeed, genistein is able to restore the cell viability and damage following MG exposure, as indicated by the reduction in cell toxicity, cell cycle alterations, and morphological modifications. In particular, genistein protects endothelial cells by MG-induced apoptosis by suppressing ROS generation and MAPK signaling pathways.

The ROS generation is an early event in the MG-related toxicity, leading to inflammation and cell death [15,16,56]. In diabetes and other age-related diseases, the increase in MG and ROS levels also promotes the glycation of several biomolecules like DNA and proteins, leading to the formation of AGEs, highly toxic species also able to induce RAGE expression. As both circulating MG and its derivative hydroimidazolone-1 can bind RAGE, thus promoting ROS production in human endothelial cells [57], the suppression of ROS formation by phytoantioxidants can be considered a crucial approach useful in preventing MG-induced endothelial apoptosis. Indeed, Pang and coworkers have reported that polydatin, a glucoside of resveratrol, protects human umbilical endothelial cells by MG-induced apoptosis through the inhibition of ROS formation and the maintenance of mitochondrial membrane potential [20]. Similarly, the flavone apigenin and unripe Carica papaya have been shown to inhibit AGE-induced oxidative stress and inflammatory signaling through ERK1/2 and Nf-kB [58,59]. In the present study, we show that genistein markedly reduces intracellular ROS formation, thus inhibiting the ROS-mediated MAPK signaling pathways and the activation of mitochondrial caspase-3 in EA.HY926 cells.

The process of ROS-induced cellular production can be considered the link between oxidative stress and apoptosis. High concentrations of MG promote apoptosis in endothelial cells through mitochondrial dysfunction, ROS/MAPK/Nf-kB signaling pathways, and

ER stress [17,18,49]. The redox-sensitive family members of MAPKs triggered by MG in endothelial cells are mainly ERK, JNK, and p38 MAPKs [50]. We show that, while MG triggers cell apoptosis in endothelial cells by promoting the phosphorylation of ERK1/2 and p38, pretreatment with genistein decreases the MAPK phosphorylation and suppress apoptosis.

The ability of genistein to protect by MG-induced apoptosis can be ascribed to its strong antioxidant power. Genistein possesses natural antioxidant activity with different biological and pharmacological properties. Recently, many studies have been focused on the remedial roles of genistein for diabetes mellitus, renal, reproductive disorders, neurodegenerative dysfunctions, and malignancy [26–29]. As a strong antioxidative agent, genistein has been shown to inhibit oxidative stress and postpone the progression of diabetes-associated complication [32–37]. Several mechanisms have been proposed for MG in stimulating ROS production and promoting oxidative stress. In particular, NADPH oxidase has been suggested to be the major source of ROS generation [16,55,60]. Further mechanisms involve the production of hydrogen peroxide and superoxide anions during the reactions of protein glycation and the depletion of the glutathione content by the glyoxalase metabolic system [57,61,62]. Our results show that genistein inhibits ROS formation as indicated by the DCFH-DA assay, thus impairing defense mechanisms mediated by Nrf2. The ability of genistein to contrast ROS production could be ascribed both to its free radical scavenger activity and, also, to the increase in antioxidant enzyme expression. Further studies will be needed to better clarify the molecular mechanisms underlying the protective effect of genistein-MG-induced ROS production and oxidative stress.

The overall data suggest that the beneficial effect of genistein in the MG-related apoptosis is mainly associated with its antioxidant activity. Indeed, the ability of genistein in maintaining redox homeostasis seems to be the key mechanism in preventing inflammation and apoptosis from dicarbonyl stress in endothelial cells. Indeed, intracellular ROS generation can induce ER stress and NADPH oxidase activation that eventually drive endothelial cells to apoptosis by Akt/MAPKs/Nf-kB pathways. In this study, genistein is shown to protect endothelial cells by disrupting the signaling cascades that activate MAPKs (Figure 8). In this respect, our study identifies a promising beneficial activity for genistein in counteracting the dicarbonyl-related endothelial toxicity.

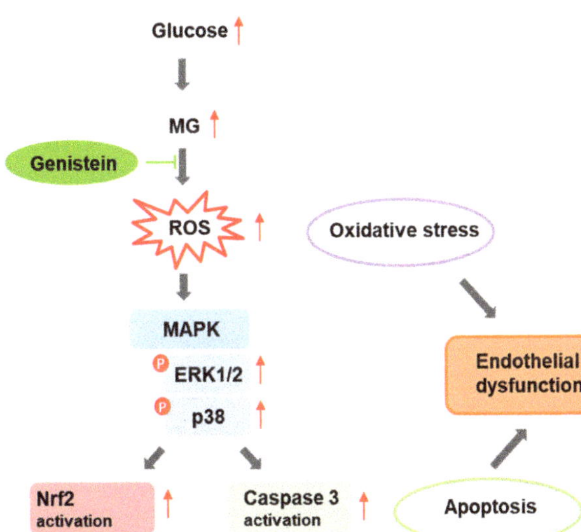

Figure 8. Schematic diagram for genistein protection in MG-induced endothelial cell apoptosis. Proposed mechanism for genistein in the MG-induced endothelial dysfunction. As depicted, genistein inhibits the production of ROS, thus preventing MAPKs and Nrf2 signaling pathways and protecting EA.HY926 endothelial cells by oxidative stress and apoptosis.

4. Materials and Methods

4.1. Materials

Genistein (G6649), methylglyoxal (M0252), 3-(4,5-dimethylthiazol-2-yl)-2,5-diphenyltetrazolium bromide (MTT) (Sigma-Aldrich Co., St. Louis, MO, USA). Antibodies: anti-caspase-3 (#9662); anti-p44/42 MAPK (ERK1/2) (#9107), anti-phospho-p44/42 MAPK (ERK1/2) (Thr202/Tyr204) (#4377), anti-NRF2 (#12721), anti-p38 MAPK# (#9212), anti-β-tubulin (#2146) (Cell Signaling Technology, Boston, MA, USA). Anti-Histone H3 (#06-755) (Merk Millipore, Burlington, MA, USA). Anti-phospho-p38 MAPK (Thr180/Tyr182) (Sigma-Aldrich Co., St. Louis, MO, USA). Secondary antibodies: anti-mouse (#7076) and anti-rabbit (#7074) (Cell Signaling Technology, Boston, MA, USA). All other chemicals were of analytical grade. Methylglyoxal was further purified by distillation under low pressure and its concentration was determined spectrophotometrically using the molar extinction coefficient at 284 nm: 12.3 M^{-1} cm^{-1}.

4.2. Cell Cultures and Treatments

EA.HY926 human endothelial cells (CRL-2922, ATCC Virginia, Manassas, VA, USA) were cultured using Dulbecco's Minimum Essential Medium (DMEM) containing 10% (*v*/*v*) fetal bovine serum, 100 U/mL penicillin, 100 mg/mL streptomycin, and 2.0 mM glutamine. Cell cultures were maintained at 37 °C in a humidified atmosphere containing 5.0% CO_2. Treatments with genistein were performed starting from a stock solution of 20 mM in DMSO diluted in cell media at 100 μM. The solution of genistein 100 μM in cell media was further diluted for different cell treatments. Treatments with MG were performed starting from a stock solution of 5 M in water that was diluted in cell media at 50 mM. The solution of MG 50 mM was further diluted for different cell treatments. For genistein–methylglyoxal experiments, cells were pretreated for 2 h with genistein before incubation with methylglyoxal.

4.3. MTT Assay

MTT assay was used to determine cellular metabolic activity as an indicator of cell viability through the ability of cells to reduce the tetrazolium salt (3-[4,5-dimethylthiazol-2-yl]-2,5-diphenyltetrazolium bromide, MTT) to blue formazan crystals. A stock MTT solution (5 mg/mL in phosphate-buffered solution, PBS) was prepared and filtered through a 0.22 μ Millipore® filter. After treatments, the culture medium of each cell culture was removed, and the culture was washed with PBS. The MTT stock solution was diluted ten times in cell medium (without red phenol) and incubated for 3 h at 37 °C. Then, the medium was removed, and the cells were treated with isopropyl alcohol and 0.1 M HCl for 20 min to dissolve the formazan crystals. Levels of reduced MTT were assayed by measuring the difference in absorbance between 570 and 690 nm. Data are expressed as the percentage reduction in MTT with respect to the control ± SD from five different experiments carried out in triplicate. The MTT reduction values were used for IC50 estimation by Microsoft Office Excel 2013 software.

4.4. Cell Cycle Analysis

Using Propidium Iodide (PI) as a well-known nuclear stain, EA.HY926 cells were analyzed according to their DNA content with the purpose of recognizing potential changes in cell cycle progression. Upon completing the treatment, cells were first detached by trypsin and then collected by centrifugation (1.300 RPM—5 min). Subsequently, pellets were washed once in PBS before being permeabilized with 70% ice-cold ethanol/PBS. Samples were finally stored at −20 °C until analysis. An appropriate volume of staining solution containing 15 μg/mL of PI plus 20 μg RNaseA in PBS was used to resuspend cells after the spin cycle (1.300 RPM—5 min). Thereafter, cells were analyzed using FACSCelestaTM (BD Bioscience, San Jose, CA, USA) acquiring at least 10 K events for each single sample. The percentage of cells with a DNA content equal to 2n (G0/G1 phase), 2n–4n (S phase), and 4n (G2/M phase) was defined by gating the respective subpopulations on a histogram plot. A similar procedure was also applied to assess the subG1 phase, namely cells exhibiting a DNA fragmentation (<2n). All biological replicates were carried out in triplicate.

4.5. Detection of Intracellular ROS

Intracellular ROS detection was performed using 2′,7′-dichlorofluorescin diacetate (DCFH-DA) assay. Cells cultured in 24-well plates were incubated with DCFH-DA 10 μM for 30 min and lysed with Tris-HCl 0.5 M, pH 7.6, 1% SDS. Controls were performed using untreated cells and cells exposed to 20 mM H_2O_2. In oxidative conditions, DCFH-DA was converted to the fluorescent molecule 2′,7′-dichlorofluorescein (DCF). DCF fluorescence intensity was recorded at a 530 nm emission wavelength upon an excitation wavelength of 488 nm using the Perkin Elmer Life Sciences LS 55 spectrofluorometer. Data are expressed as average ± SD from five different experiments carried out in triplicate.

4.6. Cellular Nuclear Extraction

Control and treated cells (1×10^6 cells) were harvested by centrifugation, washed with ice-cold PBS, resuspended in lysis buffer (10 mM HEPES pH 7.5, 10 mM KCl, 0.1 mM EDTA, 1 mM dithiothreitol, 0.5% Nonidet-40, and 0.5 mM PMSF, along with the protease and phosphatase inhibitor cocktail) and incubated on ice for 20 min, allowing them to swell. Then, tubes were vortexed, and cells were centrifuged at 12,000× *g* at 4 °C for 10 min. The supernatant, containing the cytosolic fraction, was transferred into a new tube. The pellet nuclei were washed threefold with the cell lysis buffer, resuspended in the nuclear extraction buffer (20 mM HEPES pH 7.5, 400 mM NaCl, 1 mM EDTA, 1 mM DTT, 1 mM PMSF with protease and phosphatase inhibitor cocktail), and incubated for 30 min on ice. After 15 min of centrifugation at 12,000× *g* at 4 °C, the supernatant (the nuclear extract) was recovered. The BioRad assay reagent (Bio-Rad, Hercules, CA, USA) was used to quantify the protein concentration. The nuclear and cytoplasmic extracts obtained were tested for cross-contamination through immunoblotting for HDAC1 and tubulin, respectively.

4.7. Immunoblotting

SDS-PAGE (10%) under reducing conditions was used to separate protein cellular extracts (25 μg) that were then transferred to a polyvinylidene difluoride membrane in transfer buffer (25 mM Tris, 192 mM glycine, 20% methanol, 0.1% SDS). The blots were incubated in 5% no-fat dry milk (A0530; AppliChem) blocker for 1 h at room temperature and then incubated overnight at 4 °C with specific primary antibodies. After 1 h of incubation with corresponding horseradish peroxidase-conjugated secondary antibodies, immunocomplexes were revealed using an enhanced chemiluminescence detection kit (Elabscience Biotechnology, Houston, TX, USA) and acquired using Chemi Doc XR (Biorad, Hercules, CA, USA). The relative intensity of protein bands was quantified using a Gel Doc XR System (Biorad, Hercules, CA, USA). Data analysis was performed by comparing each sample with the control, and the normalization was achieved using the housekeeping gene.

4.8. Statistical Analysis

Stata software (Version 13.0; StataCorp LP., College Station, TX, USA) was used for statistical analyses. For treatments that were significant from variance analysis (ANOVA), Tukey's test was performed. Results are represented as the mean ± SD Statistical significance was set at $p < 0.05$.

5. Conclusions

Genistein is shown to protect endothelial cells by MG-induced cytotoxicity by the inhibition of intracellular ROS production. Protection from oxidative stress leads to the inhibition of cellular apoptosis via the deactivation of ERK/p38 MAPK pathways. In this respect, our findings identify promising beneficial properties for the possible use of genistein as a therapeutic supplement aimed at preventing the risks of endothelial dysfunction and vascular complications in diabetes.

Author Contributions: Conceptualization, I.S., C.I. and M.L.; methodology, M.L., S.P. and L.S.; investigation, I.S., M.L., S.P., L.S. and C.I.; data analysis and interpretation, I.S., M.L., L.S. and C.I.; writing—original draft preparation, I.S. and C.I.; writing—review and editing, I.S., M.L. and C.I.; funding acquisition, C.I. All authors have read and agreed to the published version of the manuscript.

Funding: This work was supported by the "VALERE MOLTO" Program from the University of Campania, "L. Vanvitelli".

Institutional Review Board Statement: Not applicable.

Informed Consent Statement: Not applicable.

Data Availability Statement: The original contributions presented in the study are included in the article; further inquiries can be directed to the corresponding author.

Conflicts of Interest: The authors declare no conflicts of interest.

References

1. Allaman, I.; Bélanger, M.; Magistretti, P.J. Methylglyoxal, the dark side of glycolysis. *Front. Neurosci.* **2015**, *9*, 23. [CrossRef] [PubMed]
2. Lai, S.W.T.; Lopez Gonzalez, E.J.; Zoukari, T.; Ki, P.; Shuck, S.C. Methylglyoxal and Its Adducts: Induction, Repair, and Association with Disease. *Chem. Res. Toxicol.* **2022**, *35*, 1720–1746. [CrossRef] [PubMed]
3. Sirangelo, I.; Iannuzzi, C. Understanding the Role of Protein Glycation in the Amyloid Aggregation Process. *Int. J. Mol. Sci.* **2021**, *22*, 6609. [CrossRef] [PubMed]
4. Bellier, J.; Nokin, M.J.; Lardé, E.; Karoyan, P.; Peulen, O.; Castronovo, V.; Bellahcène, A. Methylglyoxal, a potent inducer of AGEs, connects between diabetes and cancer. *Diabetes Res. Clin. Pract.* **2019**, *148*, 200–211. [CrossRef] [PubMed]
5. Rahman, S.; Rahman, T.; Ismail, A.A.; Rashid, A.R. Diabetes-associated macrovasculopathy: Pathophysiology and pathogenesis. *Diabetes Obes. Metab.* **2007**, *9*, 767–780. [CrossRef] [PubMed]
6. Barrett, E.J.; Liu, Z.; Khamaisi, M.; King, G.L.; Klein, R.; Klein, B.E.K.; Hughes, T.M.; Craft, S.; Freedman, B.I.; Bowden, D.W.; et al. Diabetic Microvascular Disease: An Endocrine Society Scientific Statement. *J. Clin. Endocrinol. Metab.* **2017**, *102*, 4343–4410. [CrossRef] [PubMed]
7. Hanssen, N.M.J.; Westerink, J.; Scheijen, J.L.J.M.; van der Graaf, Y.; Stehouwer, C.D.A.; Schalkwijk, C.G.; SMART Study Group. Higher Plasma Methylglyoxal Levels Are Associated with Incident Cardiovascular Disease and Mortality in Individuals with Type 2 Diabetes. *Diabetes Care* **2018**, *41*, 1689–1695. [CrossRef] [PubMed]
8. Groener, J.; Oikonomou, D.; Cheko, R.; Kender, Z.; Zemva, J.; Kihm, L.; Muckenthaler, M.; Peters, V.; Fleming, T.; Kopf, S.; et al. Methylglyoxal and Advanced Glycation End Products in Patients with Diabetes—What We Know so Far and the Missing Links. *Exp. Clin. Endocrinol. Diabetes* **2019**, *127*, 497–504. [CrossRef] [PubMed]
9. Schalkwijk, C.G.; Stehouwer, C.D.A. Methylglyoxal, a Highly Reactive Dicarbonyl Compound, in Diabetes, Its Vascular Complications, and Other Age-Related Diseases. *Physiol. Rev.* **2020**, *100*, 407–461. [CrossRef]
10. Peng, Z.; Shu, B.; Zhang, Y.; Wang, M. Endothelial Response to Pathophysiological Stress. *Arterioscler. Thromb. Vasc. Biol.* **2019**, *39*, e233–e243. [CrossRef] [PubMed]
11. Darenskaya, M.A.; Kolesnikova, L.I.; Kolesnikov, S.I. Oxidative Stress: Pathogenetic Role in Diabetes Mellitus and Its Complications and Therapeutic Approaches to Correction. *Bull. Exp. Biol. Med.* **2021**, *171*, 179–189. [CrossRef] [PubMed]
12. Shaito, A.; Aramouni, K.; Assaf, R.; Parenti, A.; Orekhov, A.; Yazbi, A.E.; Pintus, G.; Eid, A.H. Oxidative Stress-Induced Endothelial Dysfunction in Cardiovascular Diseases. *Front. Biosci.* **2022**, *27*, 105. [CrossRef]
13. Sena, C.M.; Matafome, P.; Crisóstomo, J.; Rodrigues, L.; Fernandes, R.; Pereira, P.; Seiça, R.M. Methylglyoxal promotes oxidative stress and endothelial dysfunction. *Pharmacol. Res.* **2012**, *65*, 497–506. [CrossRef] [PubMed]
14. Papachristoforou, E.; Lambadiari, V.; Maratou, E.; Makrilakis, K. Association of Glycemic Indices (Hyperglycemia, Glucose Variability, and Hypoglycemia) with Oxidative Stress and Diabetic Complications. *J. Diabetes. Res.* **2020**, *2020*, 7489795. [CrossRef] [PubMed]
15. Yuan, T.; Yang, T.; Chen, H.; Fu, D.; Hu, Y.; Wang, J.; Yuan, Q.; Yu, H.; Xu, W.; Xie, X. New insights into oxidative stress and inflammation during diabetes mellitus-accelerated atherosclerosis. *Redox Biol.* **2019**, *20*, 247–260. [CrossRef]
16. Martens, R.J.H.; Broers, N.J.H.; Canaud, B.; Christiaans, M.H.L.; Cornelis, T.; Gauly, A.; Hermans, M.M.H.; Konings, C.J.A.M.; van der Sande, F.M.; Scheijen, J.L.J.M.; et al. Relations of advanced glycation endproducts and dicarbonyls with endothelial dysfunction and low-grade inflammation in individuals with end-stage renal disease in the transition to renal replacement therapy: A cross-sectional observational study. *PLoS ONE* **2019**, *14*, e0221058. [CrossRef] [PubMed]
17. Lee, J.H.; Parveen, A.; Do, M.H.; Kang, M.C.; Yumnam, S.; Kim, S.Y. Molecular mechanisms of methylglyoxal-induced aortic endothelial dysfunction in human vascular endothelial cells. *Cell Death Dis.* **2020**, *11*, 403. [CrossRef] [PubMed]
18. Wang, G.; Wang, Y.; Yang, Q.; Xu, C.; Zheng, Y.; Wang, L.; Wu, J.; Zeng, M.; Luo, M. Metformin prevents methylglyoxal-induced apoptosis by suppressing oxidative stress in vitro and in vivo. *Cell Death Dis.* **2022**, *13*, 29. [CrossRef] [PubMed]

19. Figarola, J.L.; Singhal, J.; Rahbar, S.; Awasthi, S.; Singhal, S.S. LR-90 prevents methylglyoxal-induced oxidative stress and apoptosis in human endothelial cells. *Apoptosis* **2014**, *19*, 776–788. [CrossRef]
20. Pang, N.; Chen, T.; Deng, X.; Chen, N.; Li, R.; Ren, M.; Li, Y.; Luo, M.; Hao, H.; Wu, J.; et al. Polydatin Prevents Methylglyoxal-Induced Apoptosis through Reducing Oxidative Stress and Improving Mitochondrial Function in Human Umbilical Vein Endothelial Cells. *Oxid. Med. Cell. Longev.* **2017**, *2017*, 7180943. [CrossRef]
21. Chu, P.; Han, G.; Ahsan, A.; Sun, Z.; Liu, S.; Zhang, Z.; Sun, B.; Song, Y.; Lin, Y.; Peng, J.; et al. Phosphocreatine protects endothelial cells from Methylglyoxal induced oxidative stress and apoptosis via the regulation of PI3K/Akt/eNOS and NF-κB pathway. *Vascul. Pharmacol.* **2017**, *91*, 26–35. [CrossRef]
22. Laddha, A.P.; Kulkarni, Y.A. Tannins and vascular complications of Diabetes: An update. *Phytomedicine* **2019**, *56*, 229–245. [CrossRef] [PubMed]
23. González, I.; Morales, M.A.; Rojas, A. Polyphenols and AGEs/RAGE axis. Trends and challenges. *Food Res. Int.* **2020**, *129*, 108843. [CrossRef]
24. Blahova, J.; Martiniakova, M.; Babikova, M.; Kovacova, V.; Mondockova, V.; Omelka, R. Pharmaceutical Drugs and Natural Therapeutic Products for the Treatment of Type 2 Diabetes Mellitus. *Pharmaceuticals* **2021**, *14*, 806. [CrossRef]
25. Huang, H.; Luo, Y.; Wang, Q.; Zhang, Y.; Li, Z.; He, R.; Chen, X.; Dong, Z. *Vaccinium* as Potential Therapy for Diabetes and Microvascular Complications. *Nutrients* **2023**, *15*, 2031. [CrossRef] [PubMed]
26. Rimbach, G.; De Pascual-Teresa, S.; Ewins, B.A.; Matsugo, S.; Uchida, Y.; Minihane, A.M.; Turner, R.; VafeiAdou, K.; Weinberg, P.D. Antioxidant and free radical scavenging activity of isoflavone metabolites. *Xenobiotica* **2003**, *33*, 913–925. [CrossRef]
27. Rüfer, C.E.; Kulling, S.E. Antioxidant activity of isoflavones and their major metabolites using different in vitro assays. *J. Agric. Food Chem.* **2006**, *54*, 2926–2931. [CrossRef]
28. Weng, L.; Zhang, F.; Wang, R.; Ma, W.; Song, Y. A review on protective role of genistein against oxidative stress in diabetes and related complications. *Chem. Biol. Interact.* **2019**, *310*, 108665. [CrossRef] [PubMed]
29. Goh, Y.X.; Jalil, J.; Lam, K.W.; Husain, K.; Premakumar, C.M. Genistein: A Review on its Anti-Inflammatory Properties. *Front. Pharmacol.* **2022**, *13*, 820969. [CrossRef]
30. Polkowski, K.; Mazurek, A.P. Biological properties of genistein. A review of in vitro and in vivo data. *Acta. Pol. Pharm.* **2000**, *57*, 135–155.
31. Rasheed, S.; Rehman, K.; Shahid, M.; Suhail, S.; Akash, M.S.H. Therapeutic potentials of genistein: New insights and perspectives. *J. Food Biochem.* **2022**, *46*, e14228. [CrossRef] [PubMed]
32. Wei, H.; Bowen, R.; Cai, Q.; Barnes, S.; Wang, Y. Antioxidant and antipromotional effects of the soybean isoflavone genistein. *Proc. Soc. Exp. Biol. Med.* **1995**, *208*, 124–130. [CrossRef] [PubMed]
33. Anderson, J.W.; Smith, B.M.; Washnock, C.S. Cardiovascular and renal benefits of dry bean and soybean intake. *Am. J. Clin. Nutr.* **1999**, *70*, 464S–474S. [CrossRef]
34. Si, H.; Liu, D. Phytochemical genistein in the regulation of vascular function: New insights. *Curr. Med. Chem.* **2007**, *14*, 2581–2589. [CrossRef] [PubMed]
35. Rizzo, G. The Antioxidant Role of Soy and Soy Foods in Human Health. *Antioxidants* **2020**, *9*, 635. [CrossRef] [PubMed]
36. Siriviriyakul, P.; Sriko, J.; Somanawat, K.; Chayanupatkul, M.; Klaikeaw, N.; Werawatganon, D. Genistein attenuated oxidative stress, inflammation, and apoptosis in L-arginine induced acute pancreatitis in mice. *BMC Complement. Med. Ther.* **2022**, *22*, 208. [CrossRef] [PubMed]
37. Jiang, T.; Dong, Y.; Zhu, W.; Wu, T.; Chen, L.; Cao, Y.; Yu, X.; Peng, Y.; Wang, L.; Xiao, Y.; et al. Underlying mechanisms and molecular targets of genistein in the management of type 2 diabetes mellitus and related complications. *Crit. Rev. Food Sci. Nutr.* **2023**, *27*, 1–13. [CrossRef]
38. Wu, H.J.; Chan, W.H. Genistein protects methylglyoxal-induced oxidative DNA damage and cell injury in human mononuclear cells. *Toxicol. In Vitro* **2007**, *21*, 335–342. [CrossRef]
39. Do, M.; Lee, J.H.; Wahedi, H.M.; Pak, C.; Lee, C.H.; Yeo, E.J.; Lim, Y.; Ha, S.K.; Choi, I.; Kim, S.Y. Lespedeza bicolor ameliorates endothelial dysfunction induced by methylglyoxal glucotoxicity. *Phytomedicine* **2017**, *36*, 26–36. [CrossRef] [PubMed]
40. Lv, L.; Shao, X.; Chen, H.; Ho, C.T.; Sang, S. Genistein inhibits advanced glycation end product formation by trapping methylglyoxal. *Chem. Res. Toxicol.* **2011**, *24*, 579–586. [CrossRef]
41. Wang, P.; Chen, H.; Sang, S. Trapping Methylglyoxal by Genistein and Its Metabolites in Mice. *Chem. Res. Toxicol.* **2016**, *29*, 406–414. [CrossRef] [PubMed]
42. Chang, T.; Wu, L. Methylglyoxal, oxidative stress, and hypertension. *Can. J. Physiol. Pharmacol.* **2006**, *84*, 1229–1238. [CrossRef] [PubMed]
43. Desai, K.M.; Chang, T.; Wang, H.; Banigesh, A.; Dhar, A.; Liu, J.; Untereiner, A.; Wu, L. Oxidative stress and aging: Is methylglyoxal the hidden enemy? *Can. J. Physiol. Pharmacol.* **2010**, *88*, 273–284. [CrossRef] [PubMed]
44. Seo, K.; Ki, S.H.; Shin, S.M. Methylglyoxal induces mitochondrial dysfunction and cell death in liver. *Toxicol. Res.* **2014**, *30*, 193–198. [CrossRef] [PubMed]
45. Savoia, P.; Raina, G.; Camillo, L.; Farruggio, S.; Mary, D.; Veronese, F.; Graziola, F.; Zavattaro, E.; Tiberio, R.; Grossini, E. Anti-oxidative effects of 17 β-estradiol and genistein in human skin fibroblasts and keratinocytes. *J. Dermatol. Sci.* **2018**, *92*, 62–77. [CrossRef]

46. Ruiz-Larrea, B.; Leal, A.; Martín, C.; Martínez, R.; Lacort, M. Effects of estrogens on the redox chemistry of iron: A possible mechanism of the antioxidant action of estrogens. *Steroids* **1995**, *60*, 780–783. [CrossRef] [PubMed]
47. Arora, A.; Nair, M.G.; Strasburg, G.M. Antioxidant activities of isoflavones and their biological metabolites in a liposomal system. *Arch. Biochem. Biophys.* **1998**, *356*, 133–141. [CrossRef] [PubMed]
48. Tang, D.; Xiao, W.; Gu, W.T.; Zhang, Z.T.; Xu, S.H.; Chen, Z.Q.; Xu, Y.H.; Zhang, L.Y.; Wang, S.M.; Nie, H. Pterostilbene prevents methylglyoxal-induced cytotoxicity in endothelial cells by regulating glyoxalase, oxidative stress and apoptosis. *Food Chem. Toxicol.* **2021**, *153*, 112244. [CrossRef] [PubMed]
49. Chan, C.M.; Huang, D.Y.; Huang, Y.P.; Hsu, S.H.; Kang, L.Y.; Shen, C.M.; Lin, W.W. Methylglyoxal induces cell death through endoplasmic reticulum stress-associated ROS production and mitochondrial dysfunction. *J. Cell. Mol. Med.* **2016**, *20*, 1749–1760. [CrossRef] [PubMed]
50. Akhand, A.A.; Hossain, K.; Mitsui, H.; Kato, M.; Miyata, T.; Inagi, R.; Du, J.; Takeda, K.; Kawamoto, Y.; Suzuki, H.; et al. Glyoxal and methylglyoxal trigger distinct signals for map family kinases and caspase activation in human endothelial cells. *Free Radic. Biol. Med.* **2001**, *31*, 20–30. [CrossRef] [PubMed]
51. Smolińska, E.; Moskot, M.; Jakóbkiewicz-Banecka, J.; Węgrzyn, G.; Banecki, B.; Szczerkowska-Dobosz, A.; Purzycka-Bohdan, D.; Gabig-Cimińska, M. Molecular action of isoflavone genistein in the human epithelial cell line HaCaT. *PLoS ONE* **2018**, *13*, e0192297. [CrossRef]
52. Xie, X.; Cong, L.; Liu, S.; Xiang, L.; Fu, X. Genistein alleviates chronic vascular inflammatory response via the miR-21/NF-κB p65 axis in lipopolysaccharide-treated mice. *Mol. Med. Rep.* **2021**, *23*, 192. [CrossRef] [PubMed]
53. Rabbani, N.; Thornalley, P.J. Hexokinase-2 Glycolytic Overload in Diabetes and Ischemia-Reperfusion Injury. *Trends Endocrinol. Metab.* **2019**, *30*, 419–431. [CrossRef] [PubMed]
54. Nigro, C.; Leone, A.; Fiory, F.; Prevenzano, I.; Nicolò, A.; Mirra, P.; Beguinot, F.; Miele, C. Dicarbonyl Stress at the Crossroads of Healthy and Unhealthy Aging. *Cells* **2019**, *8*, 749. [CrossRef] [PubMed]
55. Cepas, V.; Collino, M.; Mayo, J.C.; Sainz, R.M. Redox Signaling and Advanced Glycation Endproducts (AGEs) in Diet-Related Diseases. *Antioxidants* **2020**, *9*, 142. [CrossRef] [PubMed]
56. Irshad, Z.; Xue, M.; Ashour, A.; Larkin, J.R.; Thornalley, P.J.; Rabbani, N. Activation of the unfolded protein response in high glucose treated endothelial cells is mediated by methylglyoxal. *Sci. Rep.* **2019**, *9*, 7889. [CrossRef] [PubMed]
57. Ishibashi, Y.; Matsui, T.; Nakamura, N.; Sotokawauchi, A.; Higashimoto, Y.; Yamagishi, S.I. Methylglyoxal-derived hydroimidazolone-1 evokes inflammatory reactions in endothelial cells via an interaction with receptor for advanced glycation end products. *Diab. Vasc. Dis. Res.* **2017**, *14*, 450–453. [CrossRef] [PubMed]
58. Zhou, Q.; Cheng, K.W.; Gong, J.; Li, E.T.S.; Wang, M. Apigenin and its methylglyoxal-adduct inhibit advanced glycation end products-induced oxidative stress and inflammation in endothelial cells. *Biochem. Pharmacol.* **2019**, *166*, 231–241. [CrossRef] [PubMed]
59. Jarisarapurin, W.; Kunchana, K.; Chularojmontri, L.; Wattanapitayakul, S.K. Unripe Carica papaya Protects Methylglyoxal-Invoked Endothelial Cell Inflammation and Apoptosis via the Suppression of Oxidative Stress and Akt/MAPK/NF-κB Signals. *Antioxidants* **2021**, *10*, 1158. [CrossRef] [PubMed]
60. Kim, D.; Cheon, J.; Yoon, H.; Jun, H.S. Cudrania tricuspidata Root Extract Prevents Methylglyoxal-Induced Inflammation and Oxidative Stress via Regulation of the PKC-NOX4 Pathway in Human Kidney Cells. *Oxid. Med. Cell. Longev.* **2021**, *2021*, 5511881. [CrossRef]
61. Choudhary, D.; Chandra, D.; Kale, R.K. Influence of methylglyoxal on antioxidant enzymes and oxidative damage. *Toxicol. Lett.* **1997**, *93*, 141–152. [CrossRef] [PubMed]
62. Xia, F.; Wang, C.; Jin, Y.; Liu, Q.; Meng, Q.; Liu, K.; Sun, H. Luteolin protects HUVECs from TNF-α-induced oxidative stress and inflammation via its effects on the Nox4/ROS-NF-κB and MAPK pathways. *J. Atheroscler. Thromb.* **2014**, *21*, 768–783. [CrossRef] [PubMed]

molecules

MDPI

Article

Identification of Corn Peptides with Alcohol Dehydrogenase Activating Activity Absorbed by Caco-2 Cell Monolayers

Zhe Wang [1,2], Guanlong Li [2] and Xiaolan Liu [1,2,*]

1 College of Food Engineering, Harbin University of Commerce, Harbin 150076, China; wangzhe@qqhru.edu.cn
2 Key Laboratory of Corn Deep Processing Theory and Technology of Heilongjiang Province, College of Food and Bioengineering, Qiqihar University, Qiqihar 161006, China; 03580@qqhru.edu.cn
* Correspondence: 01275@qqhru.edu.cn; Tel./Fax: +86-04522738341

Abstract: Alcohol dehydrogenase (ADH) plays a pivotal role in constraining alcohol metabolism. Assessing the ADH-activating activity in vitro can provide insight into the capacity to accelerate ethanol metabolism in vivo. In this study, ADH-activating peptides were prepared from corn protein meal (CGM) using enzymatic hydrolysis, and these peptides were subsequently identified following simulated gastrointestinal digestion and their absorption through the Caco-2 cell monolayer membrane. The current investigation revealed that corn protein hydrolysate hydrolyzed using alcalase exhibited the highest ADH activation capability, maintaining an ADH activation rate of 52.93 ± 2.07% following simulated gastrointestinal digestion in vitro. After absorption through the Caco-2 cell monolayer membrane, ADH-activating peptides were identified. Among them, SSNCQPF, TGCPVLQ, and QPQQPW were validated to possess strong ADH activation activity, with EC_{50} values of 1.35 ± 0.22 mM, 2.26 ± 0.16 mM, and 2.73 ± 0.13 mM, respectively. Molecular Docking revealed that the activation of ADH occurred via the formation of a stable complex between the peptide and the active center of ADH by hydrogen bonds and hydrophobic interactions. The results of this study also suggest that corn protein hydrolysate could be a novel functional dietary element that helps protects the liver from damage caused by alcohol and aids in alcohol metabolism.

Keywords: ADH activating peptides; gastrointestinal digestion; Caco-2 cell monolayer; corn; identification

1. Introduction

China's corn production reached 288 million tons in 2023 [1], solidifying its position as one of the world's major corn-producing regions. In addition to being staple meal and an essential part of animal feed, corn can also be used as a raw material to make alcohol, corn oil, and corn starch [2–5]. Corn protein meal (CGM), a byproduct of wet milling of corn starch, contains approximately 60% protein [6]. The CGM faces challenges due to poor solubility and strong hydrophobicity, limiting its effective utilization in the food industry. By modifying CGM to create biologically active products, not only can the structure of corn industry be optimized, but it can also expand the corn industrial chain and enhance product value. This endeavor holds significant promise for the deep processing of CGM, offering a vital solution to agricultural challenges and contributing to the resolution of the "three dimensional rural issues".

The liver plays the key role in metabolizing alcohol within the human body, responsible for metabolizing 80–90% of alcohol through the alcohol dehydrogenase (ADH) pathway [7]. ADH (EC 1.1.1.1) is a zinc-containing metalloprotein enzyme that utilizes nicotinamide adenine dinucleotide (NAD^+) as a co-factor to convert alcohol into the toxic acetaldehyde, which is subsequently oxidized into non-toxic acetic acid with aldehyde dehydrogenase (ALDH) [8,9]. The activity of ADH in the liver is a major determinant of alcohol metabolism [10]. Short-term excessive alcohol consumption diminishes ADH activity in the liver, leading to the accumulation of excessive alcohol and contributing to the development of alcoholic liver disease (ALD), a significant global public health concern. Recent years have seen a rise in ethanol consumption, correlating with an increase in alcohol-related health issues, accounting for approximately 1/20 of global mortality [11]. In China, an estimated 62 million individuals suffer from alcohol-induced liver damage, with ALD being the leading cause of liver-related mortality [12]. While moderating alcohol consumption is an effective approach to prevent ethanol-induced liver damage, the identification of new bioactive agents for the prevention and treatment of ALD remains crucial.

Bioactive peptides, deemed natural agents promoting human health, demonstrate diverse biological activities. Besides nutritional value, they also exhibit antioxidative, hypotensive, anti-inflammatory, antimicrobial, and hepatoprotective properties, etc. [13–16]. Studies have reported that peptides derived from diverse sources, including corn [17,18], black soybeans [19], clam [20], chickpeas [21], and chicken [22], may enhance alcohol metabolism. Due to the need for bioactive peptides to exert their biological activity in the body, they must overcome two important physiological barriers: the complex enzymatic degradation in the gastrointestinal tract, and the low permeability of the intestinal epithelium. This process enables them to be absorbed in their intact form and to reach their target sites via the bloodstream [23]. Therefore, this study aims to isolate ADH-activating peptides from CGM and identify them after simulated gastrointestinal digestion and absorption through the Caco-2 cell monolayer membrane.

2. Results and Discussion

2.1. *Screening for Optimal Enzymatic Conditions*

CGM is a major byproduct generated during the wet milling process in the corn starch industry. It contains 60% protein, of which approximately 65% is zein and 30% is glutelin. Additionally, it is comprised of about 30% carbohydrates, with around 15% being starch and the remaining consisting of fats and fiber [24]. After the removal of starch by α-amylase, the protein content of corn protein was 87%. Then, the corn protein powder was prepared into a 5% suspension and subjected to enzymatic hydrolysis using six distinct proteases (alcalase, neutral, flavourzyme, protamex, papain, and trypsin). The effect of hydrolysates was assessed using molecular weight (MW) arrangement, degree of hydrolysis (DH), and ADH activation activity. As shown in Figure 1a, among these six enzymes, the corn protein hydrolysates produced by alcalase hydrolysis exhibited the highest ADH activation activity (55.2 $\pm$ 2.15%) at the concentration of 1 mg/mL, followed by neutral protease and trypsin. Peptides with varying degrees of activity may be produced as a result of unique enzyme-specific cleavage sites [25]. Xiao et al. [26] investigated the impact of chicken hydrolysates on ADH stability, and alcalase was also used to hydrolyze together with further bioactivity-oriented isolation and identification. The results suggested that ADH activation rate of corn protein hydrolysates (CPHs) obtained by alcalase was significantly higher than that of other five proteases.

Further investigation was conducted on the influence of alcalase addition on the ADH activation activity of CPHs. As shown in Figure 1b, at an alcalase addition of 400 U/g, the ADH activation rate of CPHs reached its maximum (50.3 $\pm$ 3.12%), and with an increase in enzyme addition, the activation rate of CPHs did not show a significant increase ($p < 0.05$). Therefore, the enzyme addition of 400 U/g was selected as the enzyme digestion condition of alcalase.

Subsequently, the impact of hydrolysis time of alcalase at an enzyme addition of 400 U/g on the ADH activation activity of CPHs was optimized. According to Figure 1c, after hydrolysis for three hours, the ADH activation activity reached its peak (55.4 $\pm$ 2.03%), and further extension of the hydrolysis time did not result in a significant increase ($p < 0.05$) in ADH activation activity. This observation suggests that within the initial 3 h period, the proteins were effectively hydrolyzed into peptides exhibiting maximal ADH activation activity.

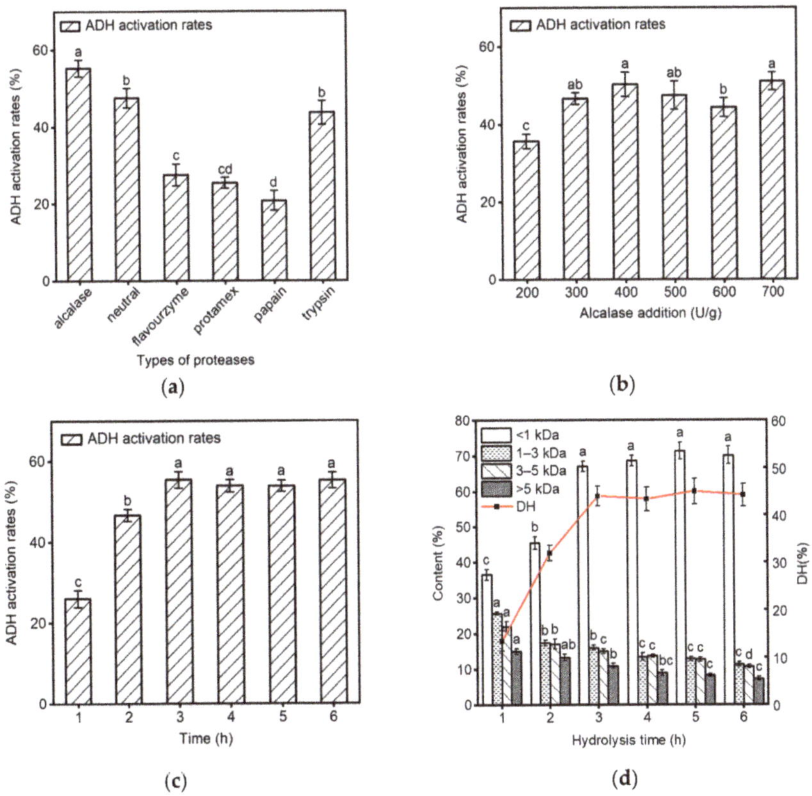

Figure 1. Screening for optimal enzymatic conditions. (**a**) ADH activation activity of CPHs prepared using different proteases; (**b**) ADH activation activity of hydrolysates with different alcalase addition; (**c**) ADH activation activity of hydrolysates with different hydrolysis time using alcalase; (**d**) DH and MW distribution of hydrolysates with different hydrolysis time using alcalase. Data are expressed as the mean ± SD, and differences were analyzed using Tukey's test, $n = 3$. The results without common superscript letters (a–d) were statistically different ($p < 0.05$).

To investigate the hydrolysis efficiency of alcalase on corn protein and to confirm the inference aforementioned, the DH and MW distribution of the hydrolysis products were determined at different hydrolysis times. As depicted in Figure 1d, the DH continuously increased with prolonged hydrolysis time. When the hydrolysis time reached 3 h, the degree of hydrolysis stabilized. Zheng et al. [27] reported a similar DH trend of corn glutelin. Likewise, with the extension of the hydrolysis time, the content of small peptides (<1 kDa) gradually increased, while the content of 1–3 kDa, 3–5 kDa, and >5 kDa components decreased gradually, stabilizing after 3 h of hydrolysis. This indicates that peptide components of <1 kDa may have higher ADH activation activity. Sonklin et al. [28] similarly found that peptide components of <1 kDa exhibit higher antioxidant activity. Considering the above conclusions, it is determined that using 400 U/g alcalase for 3 h of hydrolysis is the optimal condition for preparing ADH activating peptides.

2.2. Separation of ADH Activating Peptides by Ultrafiltration and RP-HPLC

The ADH-activating peptides were separated using ultrafiltration and RP-HPLC. Figure 2a shows ADH activation rates of the ultrafiltrated fractions. The MW of those four fractions were MW > 5 kDa, 3 kDa < MW < 5 kDa, 1 kDa < MW < 3 kDa, and MW < 1 kDa, respectively. The result demonstrated that the maximum ADH activity was seen in peptide fraction smaller than 1 kDa. With an increase in the molecular weight range, the associated ADH activation activity decreased. Due to their relative small size and flexibility in fitting into the three-dimensional structure of enzymes, small peptides may have better access to the active site of an enzyme [29]. Shi et al. [30] found that the ability of low molecular weight mushroom peptides (0–3 kDa) to activate ADH was higher than that of high molecular weight components (3–10 kDa). Zhao et al. [31] also reported the same result from mushroom foot peptides. Our findings were in agreement with these previous reports. Therefore, the MW < 1 kDa fraction was selected for further study due to its potent ADH activation activity.

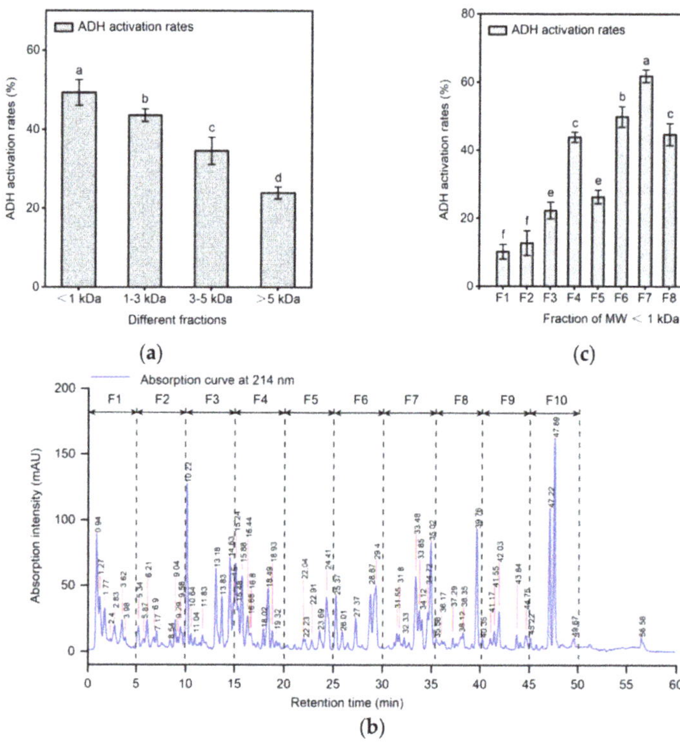

Figure 2. Separation of ADH-activating peptides by ultrafiltration and RP-HPLC. (**a**) The activation of ADH in various ultrafiltration fractions; (**b**) RP-HPLC chromatogram of pep-tides with MW < 1 kDa; (**c**) ADH activation activity of further isolated fractions with MW < 1 kDa (F1 to F10). Different letters above the error bar indicate significant differences ($p < 0.05$).

RP-HPLC is a frequently employed technique for peptide isolation and separation, offering insights into variations in hydrophobicity between fractions [32]. The fraction with MW less than 1 kDa was subsequently divided into ten fractions (F1–F10) using RP-HPLC (Figure 2b), and the results are illustrated in Figure 2c. The ADH activation rates of each fraction at the weight concentration of 1 mg/mL is exhibited in Figure 2b. It can be seen that F7 showed the highest ADH activation activity among all the fractions. The ADH activation activity of F7 was measured at 61.82 ± 1.83%, approximately four times higher than the lowest value (F1) and 20% higher than the second-highest level (F6). In reverse-phase chromatography, components with later elution times exhibit stronger hydrophobicity. Xiao et al. [26] utilized RP-HPLC to fractionate the ADH activating peptides, observing that fractions with a later elution time, exhibited greater ADH activation. Hence, F7 derived from MW < 1 kDa was selected for further analysis.

2.3. Stability of F7 in Gastrointestinal Digestion

Bioactive peptides need to be digested through the gastrointestinal tract and absorbed by the body in the form of active molecules before they can reach the target organs and exert physiological activity [33]. The rates of ADH activation ability of F7 after in vitro digestive stimulation was depicted in Figure 3. The results indicated no significant alteration in ADH activation activity following 90 min of pepsin digestion (F7-1). Subsequently, after an additional 4 h digestion with trypsin, the activity exhibited a significant decrease of 7.80% (F7-2). This result may be attributed to the insensitivity of smaller molecular weight peptides to pepsin, thereby allowing the preservation of active peptide activity [34]. Conversely, trypsin's action in reducing the overall hydrophobicity of the protein hydrolysis product led to a change in biological activity, consequently resulting in decreased ADH activation activity [35].

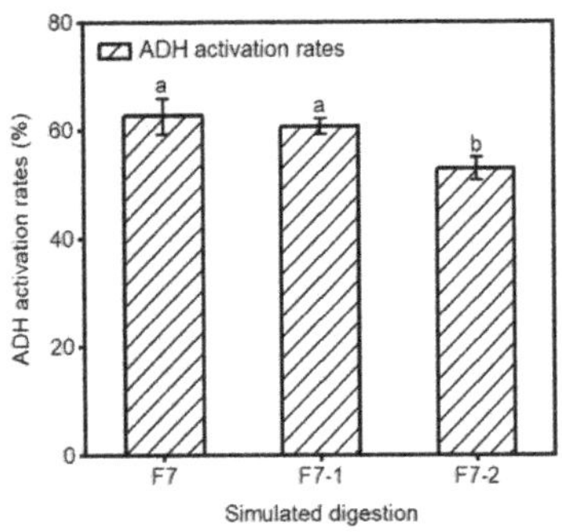

Figure 3. Effect of gastrointestinal digestion on the activity of ADH activating peptides. F7-1 represents the product after pepsin digestion; F7-2 represents the product further digested by trypsin. Different letters above the error bar indicate significant differences ($p < 0.05$).

2.4. Identification of Peptides Absorbed by Caco-2 Cells and Molecular Docking

The human small intestinal epithelium consists of numerous well-differentiated and polarized epithelial cells with tight junctions. This epithelium serves as a crucial physiological barrier against the external environment, and is primarily responsible for nutrient absorption [36]. The structure and function of Caco-2 cells are similar to differentiated intestinal epithelial cells, and Caco-2 cell monolayers are commonly used as an in vitro absorption model [37]. Following simulated gastrointestinal digestion, a portion of F7-2 was subjected to absorption experiments using Caco-2 monolayer membrane. Cytotoxicity of the F7-2 corn peptides was shown in Figure 4. At concentrations of 5–20 mg, F7-2 exhibited no cytotoxic effects on Caco-2 cells; instead, it demonstrated a significant proliferative impact at 15 and 20 mg/mL. For this reason, and to ensure the efficiency of doing Caco-2 absorption experiments, A concentration of 10 mg/mL was chosen for the absorption tests. Corn peptides (F7-2) were added to the AP side of Caco-2 cell monolayers and incubated for 2 h. The peptides absorbed in the BL side of Caco-2 cell monolayers were collected and identified using a Q Exactive MS/MS with F7 as the control. There were 13 peptide sequences (shown in Table 1) identified in both the F7 and the BL side, indicating that these peptides could not only withstand gastrointestinal digestion, but also endure cleavage by various brush border membrane peptidases, and finally can be transported intact across Caco-2 cell monolayers.

Molecular docking is a crucial tool for comprehending the structural molecular biology of ligand–biomolecule interactions and for simultaneously predicting the prominent binding modes [38]. To explore the binding mechanism of ADH-activating peptides to ADH receptors, molecular docking was performed with Autodock Vina. As shown in Table 1, 10 out of the 13 peptides that were fully absorbed by Caco-2 monolayers exhibited negative binding energy with ADH through molecular docking, of which TGCPVLQ,

SSNCQPF, and QPQQPW had relatively lower binding energy compared to other peptides. A lower binding energy indicates that the peptide is more stable within the active pocket of ADH [39].

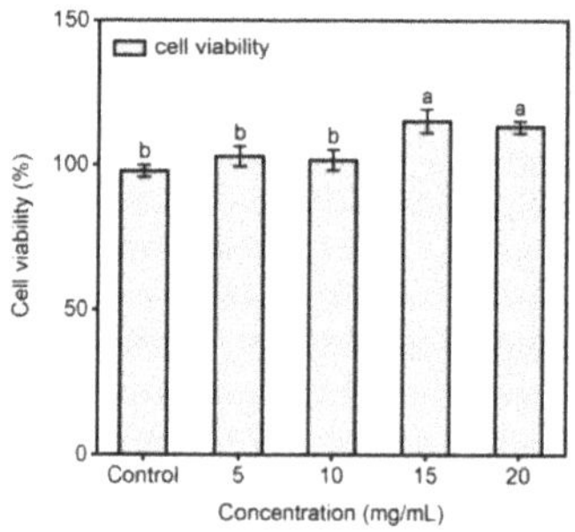

Figure 4. Viability of Caco-2 cells at different concentrations of corn peptide F7-2. Different letters above the error bar indicate significant differences. ($p < 0.05$).

Table 1. Peptide sequence from corn hydrolysates in both the control and BP side and binding energy with ADH.

No.	Peptide Sequence	Measured *m/z* (Da)	Charge	Calculated MW (Da)	Binding Energy (kcal/moL)
1	CENPILQ	408.6982	2	815.3847	−5.34
2	TGCPVLQ	717.3574	1	716.3527	−8.20
3	EVFEPF	767.3594	1	766.3538	0.95
4	SPFLGQ	648.3331	1	647.3279	−3.67
5	SSNCQPF	782.3066	1	781.3065	−8.69
6	DTPYSEF	429.6775	2	857.3443	−2.21
7	EVGDGVFE	426.1901	2	850.3709	1.26
8	TPYSEF	372.1648	2	742.3174	−3.42
9	FTPVLQ	704.3978	1	703.3905	−3.52
10	SVCENPAL	416.6947	2	831.3797	0.85
11	IFPQC	304.1461	2	606.2836	−5.65
12	TIFPQ	303.1659	2	604.3221	−3.49
13	QPQQPW	783.3767	1	782.3711	−7.31

The results of molecular docking were visualized by selecting three peptides with the lowest binding energy, as depicted in Figure 5. The detailed information of the three peptides combined with ADH were shown in Table 2. Three peptides successfully docked onto the hydrophobic cavity near the ADH active site, primarily binding to ADH through hydrogen bonds and hydrophobic interactions. Twelve hydrogen bonds were formed between SSNCQPF and ADH residues (His 44, Gly 181, Leu 182, Gly 183, Val 245, Ser 246, Val 247, Met 270, Glu 333, Gly 339, Arg 340). Ten hydrogen bonds were formed between QPQQPW and ADH residues (His 44, Asp 53, Gly 181, Asp 201, Lys 206, Glu 333, Gly 335, Arg 340). Eleven hydrogen bonds were formed between TGCPVLQ and ADH residues (His 44, Thr 45, His 66, Thr 157, Gly 181, Asp 201, Lys 206, Val 245, Gly 335, Arg 340). Especially, The aromatic ring in the Phe of SSNCQPF formed a π-stack with Phe 221. QPQQPW formed salt bridges and π-cation interactions with His 44 in ADH. TGCPVLQ established salt bridge with Lys 206 in ADH. The demonstration of all these binding sites suggests that these three peptides bind well to ADH. Xiao et al. [40] reported the binding energy between peptide KPC and ADH enzyme was −6.6 kcal/mol. In this regard, the three peptides identified in this study showed relatively strong interactions with ADH.

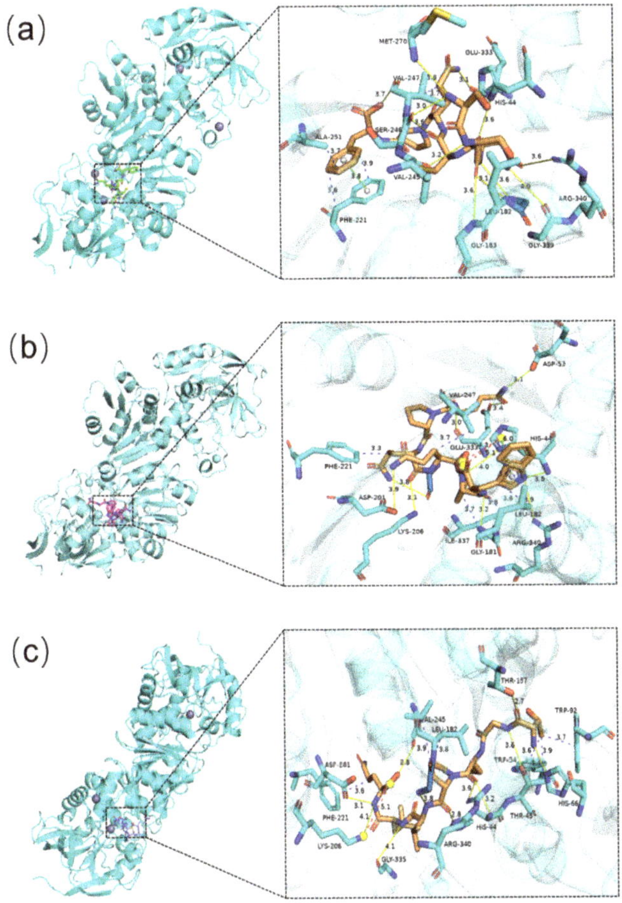

Figure 5. Molecular docking result of three peptides. ADH was selected as receptor and ligands were (**a**) SSNCQPF, (**b**) QPQQPW, and (**c**) TGCPVLQ. Hydrogen bonds, hydrophobic interactions, π-cation interactions, and salt bridges are indicated in yellow, blue, green, and red, respectively.

Table 2. The detailed information of the three peptides combined with ADH.

Sequence	Binding Energy (kcal/moL)	Hydrogen Bonds	Hydrophobic Interaction	Salt Bridge	π-Stacking	π-Cation Interaction
SSNCQPF	−8.69	His 44 (3.55 Å), Gly 181 (3.56 Å), Leu 182 (3.11 Å), Gly 183 (3.63 Å), Val 245 (3.20 Å), Ser 246 (2.96 Å), Val 247 (3.56 Å, 3.72 Å), Met 270 (3.82 Å), Glu 333 (3.12 Å), Gly 339 (4.04 Å), Arg 340 (3.64 Å)	Phe 221 (3.62 Å, 3.90 Å), Val 247 (3.52 Å), Ala 251 (3.67 Å)	-	Phe 221 (3.78 Å)	-

Table 2. *Cont.*

Sequence	Binding Energy (kcal/moL)	Hydrogen Bonds	Hydrophobic Interaction	Salt Bridge	π-Stacking	π-Cation Interaction
QPQQPW	−7.31	His 44 (3.98 Å, 3.47 Å), Asp 53 (3.13 Å), Gly 181 (3.15 Å), Asp 201 (3.89 Å), Lys 206 (3.06 Å), Glu 333 (2.99 Å, 3.42 Å), Gly 335 (3.79 Å), Arg 340 (3.87 Å)	Leu 182 (3.84 Å, 3.58 Å, 3.58 Å), Phe 221 (3.27 Å), Val 247 (3.67 Å), Ile 337 (3.69 Å)	His 44 (4.31 Å, 5.14 Å)	-	His 44 (5.97 Å)
TGCPVLQ	−8.20	His 44 (2.80 Å, 3.15 Å), Thr 45 (3.64 Å), His 66 (3.45 Å), Thr 157 (2.72 Å), Gly 181 (3.85 Å), Asp 201 (3.09 Å), Lys 206 (4.06 Å), Val 245 (2.79 Å), Gly 335 (4.06 Å), Arg 340 (3.92 Å)	Thr 45 (3.89 Å), Trp 54 (3.61 Å), Trp 92 (3.69 Å), Leu 182 (3.78 Å), Phe 221 (3.57 Å), Val 245 (3.93 Å)	Lys 206 (5.09 Å)	-	-

Note: individual '-' indicates no relevant receptor residues interacting with the peptide.

2.5. *In Vitro Activity Verification of Peptides*

The three aforementioned peptides, which possess potential to activate ADH, were synthesized and subsequently evaluated for their in vitro activating activity, as detailed in Table 3. Three peptides with low binding energy, as confirmed using molecular docking, all exhibit ADH activating activity. The EC_{50} values were 1.35 ± 0.22 mM (SSNCQPF), 2.26 ± 0.16 mM (TGCPVLQ), and 2.73 ± 0.13 mM (QPQQPW), which were slightly lower than that reported in the previous literature [21], indicating a slightly more potent effect on ADH. This is consistent with earlier results of molecular docking binding energy, showing that the lower the molecular docking binding energy, the lower the EC_{50} value. Therefore, SSNCQPF, TGCPVLQ, and QPQQPW formed stable complexes with the active center of ADH though combined multiple chemical bonds, which consequently activated the enzymatic function of ADH.

Table 3. ADH activation concentration for EC_{50} of synthetic peptides in vitro.

Sequence	ADH Activation EC_{50} (mM)
SSNCQPF	1.35 ± 0.22 [b]
TGCPVLQ	2.26 ± 0.16 [a]
QPQQPW	2.73 ± 0.13 [a]

Note: different letters ([a], [b]) represent significant differences in values.

3. Materials and Methods

3.1. *Materials and Chemicals*

Corn gluten meal (CGM) was manufactured by FuFeng Development Co., Ltd. with a recorded total protein content of 66.56% (*w*/*w*) (Qiqihar, Heilongjiang, China). Alcalase (23,000 U/g), Neutral (21,000 U/mL), flavourzyme (20,000 U/g), Protamex (38,000 U/g), Papain (11,000 U/g) and Trypsin (21,000 U/g) were obtained from Novo Nordisk (Bagsvaerd, Denmark). Caco-2 cells were purchased from JianCheng Co., Ltd. (Nanjing, Jiangsu, China). Fetal bovine serum (FBS) were bought from GE Healthcare (Pittsburgh, PA, USA). Dulbecco's modified eagle medium (DMEM), Penicillin-streptomycin, and phosphate-buffered saline (PBS) were bought from Gibco U.S. Biotechnology Co. (Grand Island, NY, USA). Ultrafiltion membranes were bought from Pall (Port Washington, NY, USA).

3.2. Removal of Starch from CGM

Starch was extracted from the CGM using predetermined protocols [24]. The CGM was immersed in distilled water (10%, *w*/*v*), and 0.1 M HCl was used to bring the pH down to 6.5. Then, α-amylase (30 U/g of protein) was added, and the mixture was incubated at 65 °C for 120 min. Subsequently, the mixture was centrifuged at 4000 r/min for 15 min after being heated in boiling water for 15 min to deactivate the enzyme. Three rounds of distilled water washings were performed on the resultant precipitate. After drying, the pretreatment CGM was obtained and ready for the preparation of corn protein hydrolysates.

3.3. Preparation of Maize Protein Hydrolysates with ADH-Activating Activity

The corn protein, from which starch has been removed, was suspended in water to prepare a 5% (*w*/*v*, protein/water) suspension. It was adjusted to the optimal pH for each enzyme using 0.1 M HCl. The hydrolysis was carried out at the optimal temperature and pH for each protease (Table 4), with an enzyme/substrate ratio of 400 U/g and a reaction time of 240 min. After the hydrolysis reaction, the hydrolysate mixture was heated in a boiling water bath for 15 min to terminate the reaction, followed by centrifugation at 4000 r/min for 15 min. The hydrolysis products were collected and freeze-dried for use in subsequent experiments and tests.

Table 4. Optimal enzymatic conditions.

Enzyme	pH	Temperature (°C)
Alcalase	8.5	60
Neutral	7.0	45
Flavourzyme	7.5	50
Protamex	7.0	55
Papain	8.0	50
Trypsin	8.0	37

Based on the results of the optimal enzyme experiments, a suitable enzyme was selected to study the effect of enzyme addition and hydrolysis time on the ADH activation activity of maize protein hydrolysates. The experimental range of enzyme additions were 200, 300, 400, 500, 600, and 700 U/g (by protein base), the substrate concentration was 5%, and the hydrolysis time was 4 h. Other hydrolysis conditions were selected according to the results of the optimal enzyme experiments. After optimizing the optimal enzyme addition, experiments on enzyme digestion time were performed. The hydrolysis times were 1, 2, 3, 4, 5, and 6 h, and other hydrolysis conditions were as above. Following hydrolysis, the enzyme was deactivated in a boiling water bath for 15 min. The supernatant obtained after centrifugation at 4000 r/min for 15 min was freeze-dried and utilized to assess the ADH activation activity, aiming to determine the enzymatic digestion time.

3.4. Separation and Purification of ADH-Activating Peptides

3.4.1. Ultrafiltration

Peptides collected from the optimal enzymatic hydrolysis conditions were ultrafiltered sequentially using ultrafiltration centrifuge tube with 5 kDa, 3 kDa, and 1 kDa. All recovered fractions (MW > 5 kDa; 3 kDa < MW < 5 kDa; 1 kDa < MW < 3 kDa; and MW < 1 kDa) were freeze-dried and stored at −80 °C for use.

3.4.2. High-Performance Liquid Chromatography in Reverse Phase (RP-HPLC)

The fraction with the highest ADH-activating activity were separated using semi-preparative HPLC instrument equipped with a reversed-phase C18 column (Hypersil GOLD PREP C18 column; 5 μm,150 mm × 10 mm). The gradient elution was carried out at a flow rate of 1 mL/min with 20 mmol/L ammonium formate in deionized water (pH = 8, corrected with ammonium hydroxide) as elution A and 20 mmol/L ammonium formate in 75% aqueous acetonitrile solution as elution B. Elution conditions were as follows: 0–10 min, 5% B; 10–50 min, 5–100% B; and 50–60 min return to initial conditions. The column temperature was maintained at 30 °C. The elution was collected every 5 min measured at 214 nm. Ultimately, 10 fractions (F1–F10) were detected, and their ADH activating activity were detected as described later.

3.5. Determination of ADH Activation Activity In Vitro

The ADH activation activity in vitro was detected using an improved assay kit method. The specific steps are as follows: 50 μL of the sample solution was mixed with 150 μL of the assay reagent (including buffer, NAD^+, and ethanol); after equilibrating at 37 °C for 5 min, 50 μL of ADH (0.2 U/mL) was added to initiate the reaction. The absorbance at 340 nm was measured using a Varioskan Flash (Thermo Fisher Scientific, Waltham, MA, USA) full-wavelength scanning multimode reader, with readings taken every 10 s for 10 min. Distilled water was used as a negative control in place of the sample. The reaction kinetics curve was fitted, and the first derivative of the curve at 0 min was calculated as the initial reaction rate. The initial reaction rate of the sample was recorded as V_S, while that of the negative control was recorded as V_0. The ADH activation rate of the sample can be calculated using the following equation:

$$\text{ADH activation activity (\%)} = \frac{V_S - V_0}{V_0} \times 100 \quad (1)$$

3.6. Simulated Gastrointestinal Digestion In Vitro

The corn active peptide was dissolved in distilled water (3%, *w*/*v*), adjusted to pH = 2.0 with 1 mol/L HCl, and then pepsin (enzyme activity: 3000 units/mg) was added at an enzyme-to-substrate ratio of 1:50 (*w*/*w*). The mixture was incubated in a constant temperature oscillator at 37 °C for 90 min to simulate gastric digestion. Subsequently, the pH was adjusted to 7.0 with 1 mol/L NaOH, pancreatin (enzyme activity: 300 units/mg) was added at an enzyme-to-substrate ratio of 1:50 (*w*/*w*), and the mixture was incubated at 37 °C for 4 h to simulate intestinal digestion. After digestion, the sample was kept in a boiling water bath for 10 min to deactivate the enzymes and then cooled to room temperature. The hydrolysate was centrifuged at 4000 r/min for 10 min, the supernatant was collected, and the solution was freeze-dried under vacuum at −20 °C. The sample was resuspended to 1 mg/mL, and its ADH activation activity was determined.

3.7. Transport Experiment

3.7.1. Cell Culture

Caco-2 cells were cultured in DMEM supplemented with 10% fetal bovine serum, 1% nonessential amino acid solution, and 1% penicillin-streptomycin solution (10,000 units/mL penicillin and 10,000 μg/mL streptomycin) at 37 °C in an atmosphere of 5% CO^2 and 90% relative humidity. Stock cultures were cultured in 75 cm^2 tissue culture flasks and sub-cultured at 80% to 90% confluency using 0.25% trypsin and 0.02% EDTA solution. Cells from passages 20–30 were suspended at a density of 2×10^5 cells/mL and added to the apical (AP) side of a 12-well Transwell plate (12 mm diameter, pore size 0.4 μm, growth surface area 1.12 cm^2, Corning Inc., Corning, NY, USA). The cell culture medium on both the apical and basolateral sides was initially replaced every 2 days for the first week, and subsequently replaced daily. The integrity of the cell monolayer was assessed by measuring transepithelial electrical resistance (TEER) using a Millicell-ERS-2 system (Millipore, Billerica, MA, USA). Only Caco-2 cell monolayers with TEER values exceeding 300 $\Omega \cdot cm^2$ were eligible for transport studies.

3.7.2. Transport Experiments

The Caco-2 cell membrane was washed twice with Hank's buffer. Subsequently, 0.5 mL and 1.5 mL of fresh Hank's buffer were added to the apical (AP) and basolateral (BL) sides, respectively. The cells were then cultured at 37 °C with 5% CO^2 for 30 min. The Hank's buffer was then removed from both sides. A 0.5 mL solution of corn peptides (concentration 10 mg/mL, dissolved in Hank's buffer) was added to the AP side, while 1.5 mL of fresh Hank's buffer was added to the BL side. After 2 h of culture at 37 °C with 5% CO_2, samples from the BL sides were collected, vacuum freeze-dried, and stored at −20 °C. The BL side sample contained corn peptides that were absorbed by the Caco-2 monolayer membrane.

3.8. Peptide Identification

Following desalting and freeze-drying, the corn peptide sample was dissolved in mobile phase A, filtered through a 0.22 μm filter membrane, and separated using the EASY-nLC 1200 ultra-high-performance liquid chromatography system. Mobile phase A comprised a 0.1% formic acid aqueous solution, while mobile phase B was a 0.1% formic acid and 80% acetonitrile aqueous solution. The liquid phase gradient was programmed as follows: from 0 to 80 min, the composition changed from 8% to 30% B; from 70 to 85 min, it transitioned from 30% to 40% B; and from 85 to 90 min, it reached 100% B, while maintaining a flow rate of 300 nL/min. Following separation by the ultra-high-performance liquid chromatography system, the peptide segments were ionized in the NSI ion source and subsequently analyzed using an orbitrap mass spectrometer. The ion source voltage was set to 2.8 kV, and both the precursor ions and their secondary fragments were detected and analyzed using the high-resolution Orbitrap. The primary mass spectrometry scan range was 200–2000 m/z, with a scan resolution of 70,000, and the secondary scan resolution was 17,500. The data acquisition mode employed a data-dependent scan (Full MS/DD-MS^2) program. This program selected the top 20 peptides precursor ions with the highest signal intensity from the primary scan, sequentially subjected them to the HCD collision pool, fragmented using 27% collision energy, and then underwent secondary mass spectrometry analysis. To optimize mass spectrometry utilization, the automatic gain control (AGC) was set to 1E5, the signal threshold to 2E4 ions/s, the maximum injection time to 45 ms, and the dynamic exclusion time for tandem mass spectrometry scans to 15 s, reducing repeated scanning of precursor ions.

3.9. Molecular Docking

The crystal structure of yeast ADH (PDB ID: 5ENV) was obtained from the PDB database (http://www.rcsb.org/pdb/ accessed on 17 December 2023). The receptor protein pre-processing (deletion of water molecules and excess ligands, addition of hydrogen atoms) was accomplished using PyMOL 2.4 (Schrödinger Inc., New York, NY, USA). The two-dimensional and three-dimensional structures of the peptide were constructed using ChemDraw 19.0 and Chem 3D 19.0. The peptide and ADH were preprocessed using AutoDock Vina 1.1.2 for molecular docking. Docking boxes for receptors were identified using POCASA software and the relevant literature [21]. The dimensions of the protein ADH docking box were 58 Å × 52 Å × 40 Å with a grid spacing of 0.375 Å. The coordinates of the docking box were: x:y:z: −50.23:44.167:−22.274. The analysis of the molecular docking results from the molecular simulations provided the docking energy of the enzyme's active site. The lowest docking energy corresponds to the most stable molecular docking structure. The visualization of the molecular docking results was conducted using PyMOL 2.4.

3.10. Synthesis of Peptides

The three purified peptides (purity > 98%) were synthesized by Qiangyao Biotechnology Co., Ltd. (Wuhan, China) using Fmoc solid-phase synthesis.

3.11. Statistical Analysis

All data were obtained from three parallel experiments, and the results were expressed as the mean ± standard deviation. The data were analyzed using SPSS 27.0.1 for one-way analysis of variance (ANOVA), and the Duncan multiple comparison test was employed for significance analysis. A *p*-value of less than 0.05 was considered to indicate a significant difference in the data.

4. Conclusions

In this study, the ADH activation rate was employed as the indicator to hydrolyze corn protein under optimal conditions using alkaline protease. The digestion and transport of corn ADH-activating peptides were investigated. By means of peptide sequence identification and molecular docking confirmation, three new peptides (SSNCQPF, TGCPVLQ, and QPQQPW) were identified that have the potential to activate ADH. All of these peptides could closely bind to the active center of ADH through multiple hydrogen bonds and hydrophobic interactions. The results of in vitro activity demonstrated that all three peptides exhibited varying degrees of the ADH activation. Among them, SSNCQPF exhibited the most notable ADH activation (EC_{50}: 1.35 ± 0.22 mM). Therefore, it was demonstrated that corn gluten meal is an excellent source for the preparation of peptides with high activation ability to ADH, and the peptides released from corn protein are promising to intervene in ALD by activating ADH. Further research should focus on elucidating the in vivo effects of these peptides, exploring their pharmacokinetics, and assessing their therapeutic potential in ALD. Understanding the mechanisms underlying the activation of ADH by these peptides will be crucial for developing targeted interventions. Moreover, investigating the broader implications of these findings in the context of liver metabolism and related diseases could open new avenues for therapeutic strategies in ALD.

Author Contributions: Writing—original draft preparation, Z.W.; writing—reviewing and editing, X.L.; visualization, G.L. All authors have read and agreed to the published version of the manuscript.

Funding: This study was funded by National Key Research and Development Project of China (No.2021YFD2100904), China Central Government Guided Heilongjiang Provincial Local Science and Technology Development Program (ZY23QY06), the Innovation Scientific Research Fund for Graduate Students Harbin University of Commerce (YJSCX2020-637HSD), the Fundamental Research Funds of Department of Education of Heilongjiang Province (145209309) and Open project of Heilongjiang Key Laboratory of Corn Deep Processing Theory and Technology (No. SPKF202025).

Institutional Review Board Statement: Not applicable.

Informed Consent Statement: Not applicable.

Data Availability Statement: Data are contained within the article.

Conflicts of Interest: The authors declare no conflict of interest.

References

1. National Bureau of Statistics of China. Grain Yield Statistics Reports for the China. 2023 [EB/OL]. Available online: https://www.stats.gov.cn/sj/zxfb/202312/t20231211_1945417.html (accessed on 25 December 2023).
2. Ranum, P.; Pena-Rosas, J.P.; Garcia-Casal, M.N. Global maize production, utilization, and consumption. *Ann. N. Y. Acad. Sci.* **2014**, *1312*, 105–112. [CrossRef] [PubMed]
3. Li, X.L.; Wu, K.N.; Hao, S.H.; Yue, Z.; Ran, Z.; Ma, J.L. Mapping cropland suitability in China using optimized MaxEnt model. *Field Crop. Res.* **2023**, *302*, 109064. [CrossRef]
4. Silva, W.R.; Carvalho, F.R.; Silva, R.B.; Pereira, R.A.N.; Avila, C.L.S.; Devries, T.J.; Pereira, M.N. Fibrous coproducts of corn and citrus as forage and concentrate sources for dairy cows. *J. Dairy Sci.* **2022**, *105*, 8099–8114. [CrossRef]
5. Lobos, N.E.; Wattiaux, M.A.; Broderick, G.A. Effect of rumen-protected lysine supplementation of diets based on corn protein fed to lactating dairy cows. *J. Dairy Sci.* **2021**, *104*, 6620–6632. [CrossRef]
6. Landry, J.; Delhaye, S.; Di Gioia, L. Protein distribution in gluten products isolated during and after wet-milling of maize grains. *Cereal Chem.* **1999**, *76*, 503–505. [CrossRef]
7. Cederbaum, A.I. Alcohol Metabolism. *Clin. Liver Dis.* **2012**, *16*, 667–685. [CrossRef] [PubMed]
8. Jelski, W.; Szmitkowski, M. Alcohol dehydrogenase (ADH) and aldehyde dehydrogenase (ALDH) in the cancer diseases. *Clin. Chim. Acta* **2008**, *395*, 1–5. [CrossRef] [PubMed]
9. Hyun, J.; Han, J.; Lee, C.; Yoon, M.; Jung, Y. Pathophysiological Aspects of Alcohol Metabolism in the Liver. *Int. J. Mol. Sci.* **2021**, *22*, 5717. [CrossRef] [PubMed]

10. Plapp, B.V.; Charlier, H.A.; Ramaswamy, S. Mechanistic implications from structures of yeast alcohol dehydrogenase complexed with coenzyme and an alcohol. *Arch. Biochem. Biophys.* **2016**, *591*, 35–42. [CrossRef]
11. WHO. Global Status Report on Alcohol and Health. 2018. Available online: https://www.who.int/substance_abuse/publications/global_alcohol_report/en/ (accessed on 23 December 2023).
12. Xiao, J.; Wang, F.; Wong, N.-K.; He, J.; Zhang, R.; Sun, R.; Xu, Y.; Liu, Y.; Li, W.; Koike, K.; et al. Global liver disease burdens and research trends: Analysis from a Chinese perspective. *J. Hepatol.* **2019**, *71*, 212–221. [CrossRef] [PubMed]
13. Zhang, Q.; Yu, Z.; Zhao, W. Identification and action mechanism of novel antioxidative peptides from copra meal protein. *LWT—Food Sci. Technol.* **2023**, *188*, 115425. [CrossRef]
14. Iram, D.; Sansi, M.S.; Zanab, S.; Vij, S.; Ashutosh; Meena, S. In silico identification of antidiabetic and hypotensive potential bioactive peptides from the sheep milk proteins—A molecular docking study. *J. Food Biochem.* **2022**, *46*, e14137. [CrossRef]
15. Li, G.; Liu, X.; Miao, Z.; Hu, N.; Zheng, X. Preparation of corn peptides with anti-adhesive activity and its functionality to alleviate gastric injury induced by helicobacter pylori infection in vivo. *Nutrients* **2023**, *15*, 3467. [CrossRef]
16. Wang, Z.; Xing, L.; Cai, J.; Toldrá, F.; Hao, Y.; Zhang, W. Identification of hepatoprotective peptides from porcine liver and its interaction with ethanol metabolizing enzymes in vitro. *Food Biosci.* **2023**, *55*, 115425. [CrossRef]
17. Wang, X.J.; Liu, X.L.; Zheng, X.Q.; Qu, Y.; Shi, Y.G. Preparation of corn glycopeptides and evaluation of their antagonistic effects on alcohol-induced liver injury in rats. *J. Funct. Foods* **2020**, *66*, 103776. [CrossRef]
18. Ma, Z.L.; Hou, T.; Shi, W.; Liu, W.W.; Ibrahim, S.A.; He, H. Purification and identification of corn peptides that facilitate alcohol metabolism by semi-preparative high-performance liquid chromatography and nano liquid chromatography with electrospray ionization tandem mass spectrometry. *J. Sep. Sci.* **2016**, *39*, 4234–4242. [CrossRef] [PubMed]
19. Ren, J.; Li, S.; Song, C.; Sun, X.; Liu, X. Black soybean-derived peptides exerted protective effect against alcohol-induced liver injury in mice. *J. Funct. Foods* **2021**, *87*, 104828. [CrossRef]
20. Gao, J.; Zhang, C.; Qin, X.; Cao, W.; Chen, J.; Li, Y.; Zheng, H.; Lin, H.; Chen, Z. Hepatoprotective effect of clam (*Corbicula fluminea*) protein hydrolysate on alcohol-induced liver injury in mice and partial identification of a hepatoprotective peptide from the hydrolysate. *Food Sci. Technol.* **2022**, *42*, e61522. [CrossRef]
21. Zan, R.; Zhu, L.; Wu, G.; Zhang, H. Identification of Novel Peptides with Alcohol Dehydrogenase (ADH) Activating Ability in Chickpea Protein Hydrolysates. *Foods* **2023**, *12*, 1574. [CrossRef] [PubMed]
22. Xiao, C.Q.; Zhou, F.B.; Zheng, L.; Cai, Y.J.; Su, G.W.; Luo, D.H.; Zhao, M.M. Chicken breast-derived alcohol dehydrogenase-activating peptides in response to physicochemical changes and digestion simulation: The vital role of hydrophobicity. *Food Res. Int.* **2020**, *136*, 109592. [CrossRef] [PubMed]
23. Bröer, S. Intestinal Amino Acid Transport and Metabolic Health. *Annu. Rev. Nutr.* **2023**, *43*, 73–99. [CrossRef]
24. Zheng, X.Q.; Liu, X.L.; Yu, S.F.; Wang, X.J.; Ma, Y.Q.; Yang, S.; Jing, S.S. Effects of Extrusion and Starch Removal Pretreatment on Zein Proteins Extracted from Corn Gluten Meal. *Cereal. Chem.* **2014**, *91*, 496–501. [CrossRef]
25. Giansanti, P.; Tsiatsiani, L.; Low, T.Y.; Heck, A.J.R. Six alternative proteases for mass spectrometry-based proteomics beyond trypsin. *Nat. Protoc.* **2016**, *11*, 993–1006. [CrossRef] [PubMed]
26. Xiao, C.; Zhao, M.; Zhou, F.; Gallego, M.; Gao, J.; Toldrá, F.; Mora, L. Isolation and identification of alcohol dehydrogenase stabilizing peptides from Alcalase digested chicken breast hydrolysates. *J. Funct. Foods* **2020**, *64*, 103617. [CrossRef]
27. Zheng, X.Q.; Wang, J.T.; Liu, X.L.; Sun, Y.; Zheng, Y.J.; Wang, X.J.; Liu, Y. Effect of hydrolysis time on the physicochemical and functional properties of corn glutelin by Protamex hydrolysis. *Food Chem.* **2015**, *172*, 407–415. [CrossRef] [PubMed]
28. Sonklin, C.; Laohakunjit, N.; Kerdchoechuen, O. Assessment of antioxidant properties of membrane ultrafiltration peptides from mungbean meal protein hydrolysates. *Peerj* **2018**, *6*, e5337. [CrossRef]
29. Gao, S.; Shi, J.; Wang, K.; Tan, Y.; Hong, H.; Luo, Y. Protective effects of oyster protein hydrolysates on alcohol-induced liver disease (ALD) in mice: Based on the mechanism of anti-oxidative metabolism. *Food Funct.* **2022**, *13*, 8411–8424. [CrossRef] [PubMed]
30. Shi, Y.; Yu, F.; Wu, Y.; Dai, L.; Feng, Y.; Chen, S.; Wang, G.; Ma, H.; Li, X.; Dai, C. Identification of a novel peptide that activates alcohol dehydrogenase from crucian carp swim bladder and how it protects against acute alcohol-induced liver injury in mice. *J. Pharmaceut. Biomed.* **2022**, *207*, 114426. [CrossRef] [PubMed]
31. Zhao, R.-J.; Huo, C.-Y.; Qian, Y.; Ren, D.-F.; Lu, J. Ultra-high-pressure processing improves proteolysis and release of bioactive peptides with activation activities on alcohol metabolic enzymes in vitro from mushroom foot protein. *Food Chem.* **2017**, *231*, 25–32. [CrossRef] [PubMed]
32. Acquah, C.; Chan, Y.W.; Pan, S.; Agyei, D.; Udenigwe, C.C. Structure-informed separation of bioactive peptides. *J. Food Biochem.* **2019**, *43*, e12765. [CrossRef]
33. Ahmed, T.; Sun, X.; Udenigwe, C.C. Role of structural properties of bioactive peptides in their stability during simulated gastrointestinal digestion: A systematic review. *Trends Food Sci. Technol.* **2022**, *120*, 265–273. [CrossRef]
34. Huang, P.; Zhao, W.; Cai, L.; Liu, Y.; Wu, J.; Cui, C. Enhancement of functional properties, digestive properties, and in vitro digestion product physiological activity of extruded corn gluten meal by enzymatic modification. *J. Sci. Food Agric.* **2024**, *104*, 3477–3486. [CrossRef] [PubMed]
35. Li, G.; Liu, X.; Miao, Z.; Zheng, X. Purification and structural characterization of three novel anti-adhesive peptides against Helicobacter pylori from corn gluten meal. *J. Funct. Foods* **2024**, *112*, 105992. [CrossRef]

36. Marchiando, A.M.; Graham, W.V.; Turner, J.R. Epithelial Barriers in Homeostasis and Disease. *Annu. Rev. Pathol.-Mech.* **2010**, *5*, 119–144. [CrossRef] [PubMed]
37. Iftikhar, M.; Iftikhar, A.; Zhang, H.; Gong, L.; Wang, J. Transport, metabolism and remedial potential of functional food extracts (FFEs) in Caco-2 cells monolayer: A review. *Food Res. Int.* **2020**, *136*, 109240. [CrossRef] [PubMed]
38. Stanzione, F.; Giangreco, I.; Cole, J.C. Use of molecular docking computational tools in drug discovery. *Prog. Med. Chem.* **2021**, *60*, 273–343.
39. Igbokwe, C.J.; Feng, Y.; Louis, H.; Benjamin, I.; Quaisie, J.; Duan, Y.; Tuly, J.A.; Cai, M.; Zhang, H. Novel antioxidant peptides identified from coix seed by molecular docking, quantum chemical calculations and invitro study in HepG2 cells. *Food Chem.* **2024**, *440*, 138234. [CrossRef] [PubMed]
40. Xiao, C.; Toldrá, F.; Zhou, F.; Mora, L.; Luo, L.; Zheng, L.; Luo, D.; Zhao, M. Chicken-derived tripeptide KPC (Lys-Pro-Cys) stabilizes alcohol dehydrogenase (ADH) through peptide-enzyme interaction. *LWT* **2022**, *161*, 113376. [CrossRef]

molecules

MDPI

Article

Concentrations, Sources and Health Risk Assessment of Polycyclic Aromatic Hydrocarbons in Chinese Herbal Medicines

Deyan Cao [1,2], Zhu Zhu [1,2], Siyuan Zhao [1,2], Xi Zhang [3], Jianzai Lin [1,2], Junji Wang [1,2], Qinghong Zeng [1,2] and Meilin Zhu [1,3,*]

1 School of Public Health, Ningxia Medical University, Yinchuan 750004, China; jolly2025@163.com (D.C.)
2 Key Laboratory of Environmental Factors and Chronic Disease Control, Ningxia Medical University, Yinchuan 750004, China
3 College of Basic Medical Sciences, Ningxia Medical University, Yinchuan 750004, China
* Correspondence: 20140127@nxmu.edu.cn

Abstract: The determination and evaluation of 16 polycyclic aromatic hydrocarbons (PAHs) in seven Chinese herbal medicines (CHMs) were conducted through a rapid and straightforward extraction and purification method, coupled with GC-MS. A sample-based solid-phase extraction (SPE) pretreatment technique, incorporating isotopic internal standards, was employed for detecting various medicinal parts of CHMs. The assay exhibited linearity within the range of 5 to 500 ng/mL, with linear coefficients (R^2) for PAHs exceeding 0.999. The recoveries of spiked standards ranged from 63.37% to 133.12%, with relative standard deviations (RSDs) ranging from 0.75% to 14.54%. The total PAH content varied from 176.906 to 1414.087 μg/kg. Among the 16 PAHs, phenanthrene (Phe) was consistently detected at the highest levels (47.045–168.640 μg/kg). Characteristic ratio analysis indicated that oil, coal, and biomass combustion were the primary sources of PAHs in CHMs. The health risk associated with CHMs was assessed using the lifetime carcinogenic risk approach, revealing potential health risks from the consumption of honeysuckle, while the health risks of consuming *Lycium chinense* berries were deemed negligible. For the other five CHMs (glycyrrhizae, *Coix lacryma*, ginseng, lotus seed, seed of *Sterculia lychnophora*), the health risk from consumption fell within acceptable ranges. Furthermore, sensitivity analyses utilizing Monte Carlo exposure assessment methods identified PAH levels in CHMs as health risk sensitizers. It is crucial to recognize that the consumption of herbal medicines is not a continuous process but entails potential health risks. Hence, the monitoring and risk assessment of PAH residues in CHMs demand careful attention.

Keywords: Chinese herbal medicine; polycyclic aromatic hydrocarbon; health risk assessment; Monte Carlo simulation

1. Introduction

In contrast to synthetic drugs, which effectively treat various diseases but often come with side effects, Chinese herbal medicines (CHMs) have gained increasing interest in recent years due to their fewer side effects [1,2]. They have become widely accepted worldwide and are utilized by numerous pharmaceutical companies as a top-tier resource for discovering natural bioactive compounds [3]. According to estimates from the World Health Organization (WHO), over 80% of the global population primarily relies on traditional medicine, a substantial portion of which involves the use of plant extracts or their active ingredients [4]. The commercialization of medicinal plants has experienced significant growth, with herbs finding applications in various fields such as phytochemicals, pharmaceuticals, nutraceuticals, herbal remedies, food supplements, perfumes, cosmetics, and food flavoring [5]. However, a key issue in the development and application of herbal medicines is ensuring their quality and safety. This concern extends not only to toxic ingredients but also to residual contaminants, including pesticides [6–8], heavy metals [9,10], polycyclic aromatic hydrocarbons (PAHs) [11,12], and other environmental pollutants.

PAHs constitute a sizable group of semi-volatile organic compounds, characterized by two or more fused aromatic carbocycles, and are designated as environmental pollutants due to their carcinogenic properties [13,14]. Notably, the U.S. Environmental Protection Agency has identified 16 PAHs as priority control pollutants, commonly known as the 16 EPA-PAHs. The primary sources of PAHs stem from incomplete combustion and thermal decomposition of organic matter, encompassing fossil fuels and biomass [15–18]. This occurs predominantly when proteins, lipids, and carbohydrates undergo incomplete combustion at elevated temperatures (usually between 300 and 600 °C) [19]. Exposure to PAHs in the population, especially among non-smokers, occurs through various routes, including the atmosphere, water, air, food, and the food chain [20–22]. Studies have indicated that plants can accumulate PAHs from the environment, leading to the contamination of crops and herbs [23]. There are two primary pathways for PAH accumulation in plants [24]: one involves the uptake of PAHs from the soil through the plant's root system, while the other involves the absorption of atmospheric PAHs through the above-ground parts of the plant [25]. Consequently, these processes result in the accumulation and presence of harmful substances in Chinese medicinal plants, significantly impacting the quality of CHMs. PAHs have been associated with a range of adverse health effects, potentially disrupting male and female hormones, causing endocrine disruption, infertility, and immunosuppression, as well as exhibiting carcinogenic and teratogenic properties [26–28]. It is crucial to safeguard herbal medicines and herbal products from such contaminants, ensuring they are either protected or controlled for use within safe levels [29]. Given the increasingly stringent testing standards set by the European Union, Japan, and other countries, international organizations have yet to formulate a fully unified standard for CHMs. Therefore, there is a pressing need to establish a method based on the detection and analysis of harmful substances like PAHs in CHMs, coupled with health risk assessments.

2. Results and Discussion

2.1. Optimization of Extraction Conditions

GC-MS offers high chromatographic resolution, with mass spectra providing elevated mass selectivity and a wealth of structural information. In this experiment, an Agilent DB-5 capillary column was employed to explore various temperature ramping programs. Based on the test results, we opted to use the earlier-described temperature program, allowing for baseline separation and sample detection within a 32 min timeframe. Furthermore, GC-MS, utilizing an electron ionization energy of 70 eV, was our instrument for the quantitative analysis of PAHs. Figure 1 illustrates the total ion chromatograms of the blank matrix with added target analytes and each internal standard. The data in Figure 1 and Table 1 indicate consistent retention times for standards and their isotopes, with no matrix interference. Isomers (Phe/Ant, Flt/Pyr, BaA/Chry, and BbF/BkF/BaP) could be baseline separated within the 32 min window.

The experiment investigated ultrasonic extraction and oscillatory extraction as pretreatment methods, utilizing acetonitrile and n-hexane/acetone as extraction solvents for comparison. The effectiveness of the three parallel methods was assessed through recovery comparisons. Our outcomes demonstrated that ultrasonic and acetonitrile extraction proved more effective compared to oscillatory and n-hexane/acetone extractions. While ultrasonic extraction outperformed oscillatory extraction, recoveries for analytes with small molecular weights (Naph, Acy, Ace, Flu) were not particularly high, necessitating repeated extraction or a change in the extraction solvent. Analytes with low recoveries, likely due to their volatility and low boiling points, were possibly lost during evaporation. None of the analytes exhibited a relative standard deviation (RSD) exceeding 15%. Table 2 presents the recoveries of the 16 PAHs (with 5 PAHs-D as a surrogate, $n = 6$) in the seven

CHMs, showcasing recovery ranges for the three spiked concentrations (n = 6) from 63.37% to 133.12%.

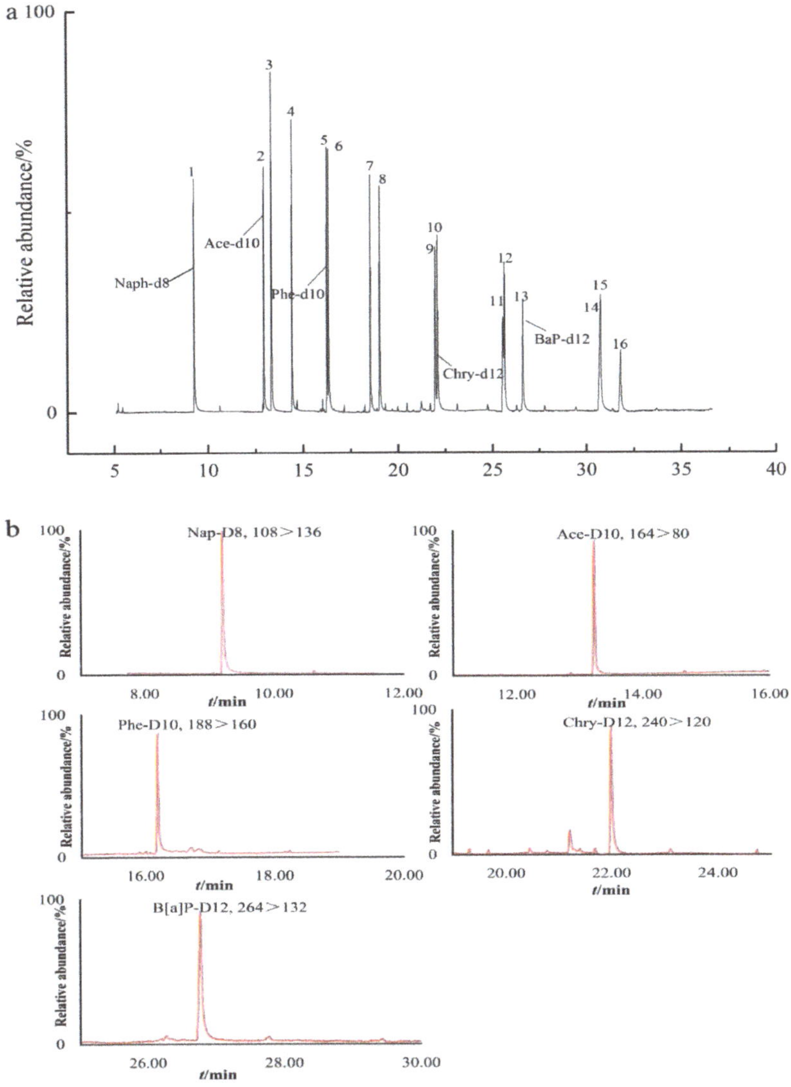

Figure 1. TIC of PAH standard (**a**) and their deuterated PAH standard (**b**). Naphthalene (1), Acenaphthylene (2), Acenaphthene (3), Fluorene (4), Phenanthrene (5), Anthracene (6), Fluoranthene (7), Pyrene (8), Benz[a]anthracene (9), Chrysene (10), Benzo[b]fluorathene (11), Benzo[k]fluorathene (12), Benzo[a]pyrene (13), Indeno[1,2,3-c,d]pyrene (14), Dibenz[a,h]anthracene (15), Benzo[g,h,i]perylene (16). Peak identification numbers correspond to compounds reported in Table 1.

Table 1. Optimized parameters for analysis of PAHs using GC-MS.

Number	Compound	Abbreviation	CAS Number	Molecular Weight	Retention Time (min)	Ion Pair for Quantitative Analysis (*m*/*z*)	Ion Pair for Qualitative Analysis (*m*/*z*)
1	Naphthalene	Naph	91-20-3	128.17	9.273	128 > 102	128 > 77
2	Acenaphthylene	Acy	208-96-8	152.20	12.925	152 > 151	152 > 126
3	Acenaphthene	Ace	83-32-9	154.20	13.326	153 > 152	153 > 127
4	Fluorene	Flu	86-73-7	166.22	14.417	166 > 165	166 > 163
5	Phenanthrene	Phe	85-01-8	178.23	16.345	178 > 176	178 > 152
6	Anthracene	Ant	120-12-7	178.22	16.358	178 > 176	178 > 152
7	Fluoranthene	Flt	206-44-0	202.25	18.549	202 > 200	202 > 150
8	Pyrene	Pyr	129-00-0	202.26	19.028	202 > 200	202 > 150
9	Benz[a]anthracene	BaA	56-55-3	228.30	22.102	228 > 226	228 > 202
10	Chrysene	Chry	218-01-9	228.29	21.971	228 > 226	228 > 202
11	Benzo[b]fluorathene	BbF	205-99-2	252.31	25.551	252 > 250	252 > 224
12	Benzo[k]fluorathene	bkF	207-08-9	252.31	25.640	252 > 250	252 > 224
13	Benzo[a]pyrene	BaP	50-32-8	252.31	26.618	252 > 250	252 > 224
14	Indeno[1,2,3-c,d]pyrene	Ind(cd)P	193-39-5	276.00	30.699	276 > 274	276 > 248
15	Dibenz[a,h]anthracene	DahA	53-70-3	278.35	30.744	278 > 276	278 > 274
16	Benzo[g,h,i]perylene	BghiP	191-24-2	276.33	31.798	276 > 274	276 > 272
17	Naphthalene-d8	Naph-d8	1146-65-2	136.22	9.232	136 > 108	136 > 134
18	Acenaphthene-d10	Ace-d10	15067-26-2	164.17	13.255	164 > 162	136 > 134
19	Phenanthrene-d10	Phe-d10	1517-22-2	188.29	16.213	188 > 160	164 > 160
20	Chrysene-d12	Chry-d12	1719-03-5	240.36	22.016	240 > 236	240 > 212
21	Benzo[a]pyrene-d12	BaP-d12	63466-71-7	266.399	26.788	264 > 260	264 > 232

Table 2. Spiked recoveries of 16 PAHs in 7 CHMs (5 PAHs-D as a proxy, n = 6).

Analyte	Spiked 20 µg/kg		Spiked 100 µg/kg		Spiked 400 µg/kg	
	Mean Recovery (%)	RSD (%)	Mean Recovery (%)	RSD (%)	Mean Recovery (%)	RSD (%)
Naph	66.97	3.06	79.21	4.38	67.2	0.75
Acy	84.67	4.26	89.37	5.94	84.1	1.54
Ace	63.62	4.91	82.66	3.94	85.4	1.13
Flu	63.37	10.96	76.89	2.28	74.0	2.35
Phe	133.12	2.76	116.05	1.95	108.1	5.21
Ant	116.72	14.54	119.15	9.47	98.9	2.66
Flt	104.45	2.10	104.91	2.74	107.1	3.26
Pyr	113.30	3.27	107.19	3.63	106.4	2.44
BaA	89.27	8.44	121.45	1.04	108.6	4.82
Chry	85.50	13.09	117.37	2.86	99.1	2.50
BbF	119.23	3.89	121.85	6.37	117.0	7.98
bkF	119.15	3.00	117.51	4.49	111.5	2.06
BaP	112.71	14.33	117.96	5.34	113.1	6.39
Ind(cd)P	100.96	4.38	109.01	1.08	114.1	6.18
DahA	106.04	3.31	111.17	5.20	114.2	3.40
BghiP	107.02	0.69	114.56	2.60	116.1	6.00

Note: Relative standard deviation (RSD).

2.2. Performance of the Method

To ensure experimental accuracy, we utilized an analytical calibration curve. Regression equations were derived using five-point concentrations ranging from 5 to 500 ng/mL. As shown in Table 3, the 16 PAHs exhibited favorable linearity, with a correlation coefficient (R2) of 0.999. The limit of detection (LOD) and limit of quantification (LOQ) were determined by adding different amounts of mixed standards to blank samples, and the signal-to-noise ratio (S/N) $\geq$ 3 was used to determine the LOD, and the signal-to-noise ratio (S/N) ≥ 10 was used to determine the LOQ. The LODs for the 16 analytes ranged from 4.96 to 8.14 μg/kg, and the LOQs ranged from 9.9 to 16.3 μg/kg (Table 3).

Table 3. Analytical performance of the proposed method.

Analyte	Linearity Range (ng/mL)	LOD (μg/kg)	LOQ (μg/kg)	Regression Equation	Coefficient of Determination (R^2)
Naph	5–500	6.81	13.6	$y = 0.453x + 0.053$	0.9990
Acy	5–500	5.20	10.4	$y = 0.626x - 0.013$	0.9990
Ace	5–500	5.68	11.4	$y = 0.759x + 0.029$	0.9996
Flu	5–500	5.14	10.3	$y = 0.517x + 0.037$	0.9997
Phe	5–500	6.35	12.7	$y = 0.527x + 0.003$	0.9997
Ant	5–500	4.96	9.9	$y = 0.571x - 0.020$	0.9998
Flt	5–500	5.13	10.3	$y = 0.474x + 0.009$	0.9999
Pyr	5–500	5.33	10.7	$y = 0.346x + 0.025$	0.9998
BaA	5–500	5.60	11.2	$y = 0.207x + 0.032$	0.9998
Chry	5–500	5.54	11.1	$y = 0.458x + 0.009$	0.9992
BbF	5–500	6.99	14.0	$y = 0.276x + 0.005$	0.9996
bkF	5–500	6.75	13.5	$y = 0.502x + 0.017$	0.9997
BaP	5–500	8.14	16.3	$y = 0.338x + 0.016$	0.9993
Ind(cd)P	5–500	5.72	11.4	$y = 0.440x - 0.018$	0.9998
DahA	5–500	6.25	12.5	$y = 0.245x + 0.224$	0.9997
BghiP	5–500	7.70	15.4	$y = 0.235x + 0.019$	1.0000

Note: Limit of detection (LOD); limit of quantification (LOQ).

2.3. Application to Real Samples

The aforementioned method was applied to monitor the presence of 16 PAHs in seven CHMs, yielding results with an RSD of less than 15%. Calibration curves for mixed working standard solutions were constructed by plotting the peak area ratios of the quantitative ion pairs of each standard substance to the five internal standard substances. Extracts were analyzed by GC-MS, and the outcomes are detailed in Table 4. The composition of herbs is extremely complex, with a wide range of amino acids, volatile oils, sugars, and vitamins. Currently, Florisil solid-phase extraction (SPE) columns, PSA SPE columns, and C18 SPE columns are used for cleanup in many studies. PSA can chelate with metal ions and can effectively remove fatty acids, organic acids, and some polar colors and sugar substances. The functional group of C18 contains 10% carbon, has hydrophobic effect, and has adsorption effect on neutral and non-polar components, such as aromatic oils and fat-soluble vitamins. PSA filler removes water, coloring, and other substances. The experimental purification was carried out using a combination of three commonly used adsorbents to investigate the effect of purification of herbal medicines, the purification effects of the seven herbs are depicted in Figure 2.

Table 4. PAH content in selected CHMs (µg/kg).

Analyte	Glycyrrhizae Radix et Rhizoma	RSD%	Honeysuckle	RSD%	*Coix lacryma*	RSD%	Ginseng Radix et Rhizoma	RSD%	Lotus Seed	RSD%	Seed of *Sterculia lychnophora*	RSD%	*Lycium chinense*	RSD%
Naph	42.313	0.39	36.019	0.25	21.666	0.36	60.153	0.59	37.674	0.40	40.844	0.26	48.876	0.31
Acy	13.600	0.73	13.609	0.32	22.913	0.37	nd	-	18.681	0.55	27.400	0.38	11.378	0.32
Ace	31.851	0.10	25.060	0.56	24.754	0.50	33.842	0.47	27.726	0.52	38.175	0.39	39.044	0.12
Flu	24.576	0.21	25.191	0.51	nd	-	nd	-	nd	-	12.241	0.37	18.880	0.40
Phe	153.753	0.05	160.906	0.21	82.388	0.58	103.442	0.22	80.887	0.30	168.640	0.12	47.045	0.18
Ant	154.849	0.07	nd	-	147.707	0.29	121.351	0.29	117.332	0.23	112.253	0.15	nd	-
Flt	97.790	0.26	52.555	0.17	27.020	0.83	18.462	0.76	21.164	0.64	41.773	0.25	nd	-
Pyr	99.612	0.23	42.752	0.22	nd	-	18.190	0.83	11.859	0.21	52.345	0.40	11.683	0.19
BaA	31.207	0.39	nd	-	nd	-	nd	-	nd	-	nd	-	nd	-
Chry	50.084	0.12	76.563	0.21	nd	-	nd	-	nd	-	nd	-	nd	-
BbF	148.469	0.08	941.504	0.08	nd	-	nd	-	nd	-	462.085	0.04	nd	-
bkF	nd	-	nd	-	nd	-	nd	-	22.594	0.48	nd	-	nd	-
BaP	nd	-	39.928	0.24	nd	-	nd	-	nd	-	nd	-	nd	-
Ind(cd)P	nd	-	nd	-	nd	-	nd	-	nd	-	nd	-	nd	-
DahA	nd	-	nd	-	nd	-	nd	-	nd	-	nd	-	nd	-
BghiP	nd	-	nd	-	nd	-	nd	-	nd	-	nd	-	nd	-
∑16PAHs	848.102	-	1414.087	-	326.448	-	355.440	-	337.917	-	955.758	-	176.906	-

Note: Mean content (µg/kg, n = 3, RSD ≤ 15%); nd, not detected; -, excluded from calculation.

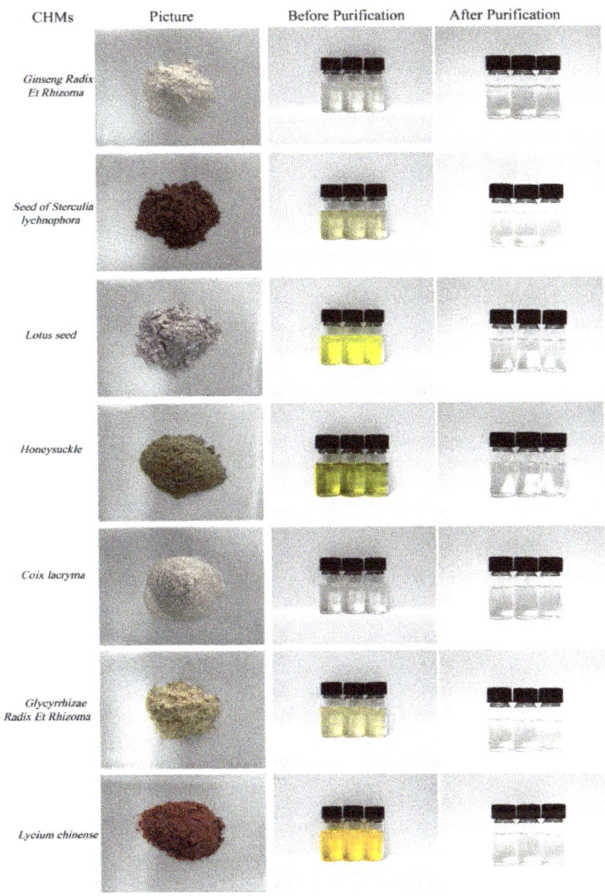

Figure 2. CHMs of seven types and the purification performances compared before and after cleanup.

The contamination levels of the 16 EPA-PAHs were assessed in seven selected CHMs at varying levels, revealing contamination rates of 62.50% for honeysuckle, 56.25% for the *seed of Sterculia lychnophora*, 68.75% for glycyrrhizae, 35.83% for *Coix lacryma*, 37.50% for ginseng, 50% for lotus seed, and 37.50% for *Lycium chinense* berries. Statistically different contamination rates of PAHs in seven Chinese herbal medicines (chi-squared test χ^2 $p < 0.05$). Glycyrrhizae exhibited the highest contamination, while *Coix lacryma* showed the least contamination. The contamination rates were ranked from high to low, as follows: glycyrrhizae > honeysuckle > seed of *Sterculia lychnophora* > lotus seed > ginseng > *Lycium chinense* berries > *Coix lacryma*. Upon analyzing the total detected concentrations of each herbal medicine type, root and stem samples had the highest levels, followed by flowers, while fruit samples exhibited the lowest levels, and similar trends were observed in the TEQ. This may be due to the fact that root herbs are exposed to soil for a long period of time relative to flowers and fruits and absorb PAHs through soil and water, resulting in a large amount of PAHs being enriched in the roots, where the highest mass concentrations of PAHs were found. Fruits and seeds, on the other hand, develop later, are exposed to PAHs in the environment for a shorter period of time, and are enriched with low amounts of PAHs. In addition, different types of herbs may come from different places of origin, so the concentration of PAHs in the environment of the herbs' place of origin is another important factor affecting the concentration of PAHs.

Furthermore, concentration levels of individual PAHs in different herbs were analyzed. Among the 16 PAHs, Phe emerged as the most prevalent and severe contaminant, displaying the highest contamination levels and was detected in all tested samples, with total mass concentrations of PAHs-Phe ranging from 47.045 to 797.061 μg/kg. Notably, PHAs-Ant and Naph also exhibited higher concentration ranges, with values of nd~653.492 μg/kg and 21.666~287.545 μg/kg, respectively. BbF was detected in only three herbs (glycyrrhizae, honeysuckle, and the seed of *Sterculia lychnophora*), with the highest level recorded at 941.504 μg/kg. BaP, a representative carcinogenic PAH, was found only in *honeysuckle* at 39.928 μg/kg. From the results of PAH assay monomers, it was found that lower molecules of PAHs had higher levels of contamination in herbal medicines, which was related to the higher solubility and faster transfer rate of low-ring PAHs from soil to plants.

In 2017, China's National Standard for Food Safety set limits for BaP in various food products, with a 5 μg/kg limit for cereals and their products, meat and meat products, aquatic animals and their products, and oils and fats and their products. *Honeysuckle* exceeded this limit, and European Union regulations also established limits for PAHs (PAH4) at 35 μg/kg. Glycyrrhizae, honeysuckle, and the seed of *Sterculia lychnophora* were also found to exceed these limits, as indicated by (EU) no. 1, Standard No. 835/2011 for different food products, such as processed cereals or dietary foods for special medical purposes intended for infants.

2.4. Distributional Characteristics of PAHs

Figure 3 illustrates the distribution of PAHs with different ring numbers based on concentration and TEQ, respectively. A comparison between the two reveals a discrepancy ($p < 0.05$). When considering total concentration for analysis, PAHs with 2–3 rings predominated, representing over 60% of 4-ring PAHs in most samples. PAHs with larger molecules (5–6 rings) accounted for less than 10% of all samples, aligning with findings reported by Ishizaki and Kataoka [1]. In other words, small-molecule PAHs were predominant among the detected PAHs in the CHMs. In contrast, when evaluated using toxic equivalent concentration (Figure 3b), no large-molecule PAHs were detected. The percentage of 5–6-ring PAHs in the samples was 0, and the percentage of total toxic equivalents of 2–3-ring PAHs exceeded 90%. This outcome can be attributed to the higher toxicity of large-molecule PAHs compared to small-molecule PAHs. However, the detection rate of large-molecule PAHs was low, resulting in a higher overall toxic equivalent concentration for small-molecule PAHs. The substantial proportion of low-molecular-weight (2–3-ring) PAHs suggests the potential presence of recent contamination in the affected area. Herbs not only actively absorb harmful pollutants in the soil during the growth process, but may also passively absorbing pollution caused by anthropogenic activities such as haze weather that emits PAHs.

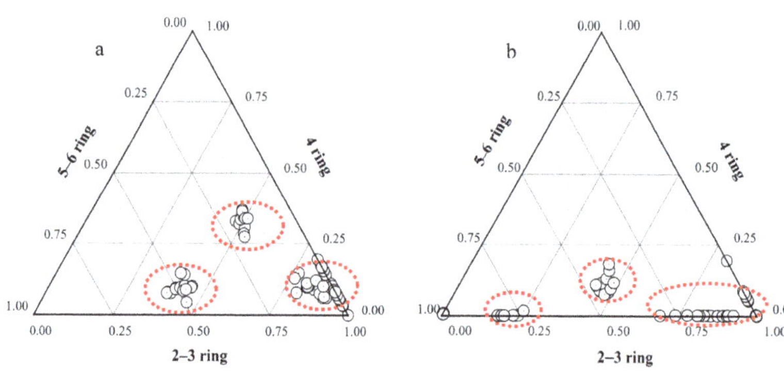

Figure 3. Distribution of PAHs with different ring numbers in herbal medicine samples. (**a**) Concentration percentage. (**b**) Toxic equivalent concentration percentage.

2.5. PAH Source Analysis

Natural sources of PAHs encompass volcanic eruptions and brush burning, which directly contributes to background PAH levels in the environment. However, the primary origins of PAHs are anthropogenic, stemming from incomplete combustion of petroleum fuels, biomass, and the natural volatilization or leakage of petroleum fuels [30]. The diverse range of PAH sources renders them ubiquitous environmental pollutants. Analyzing the sources of PAHs in CHMs is crucial, not only for understanding the pollution origins, but also for effective PAH control. Several methods, such as the characteristic marker method, characteristic compound ratio method, and multivariate statistical method, are employed for source analysis. Among these, the characteristic compound ratio method is frequently utilized.

This method relies on significantly different concentration ratios of PAHs from various pollution sources. Due to the volatility of PAHs in the gas phase, the ratio method primarily utilizes data from PAHs adsorbed on particulate matter to infer potential sources. Four ratios distinguish PAH sources from crude oil pollution, gasoline combustion, biomass combustion, and coal combustion. The judgment criteria involve the dominance of 2–3-ring PAHs, indicating petroleum source pollution, while the dominance of four or more rings PAHs suggests high-temperature combustion sources. In this study, the herbal samples were judged to be primarily influenced by petroleum sources of PAHs [31–36]. The monomer distribution of PAHs provides insights into their potential sources. Three characteristic ratios—Flt/(Flt + Pyr), BaA/(BaA + Chr), and Ant/(Ant + Phe)—were employed in this study to characterize the pollution sources of CHMs.

Empirical results for seven CHM samples indicated that Flt/(Flt + Pyr) ratios mainly ranged between 0.4 and 0.60, reflecting characteristics of biomass and fossil fuel combustion, as well as vegetation, grass, and coal combustion sources. BaA/(BaA + Chr) ratios exceeding 0.2 exhibited dual characteristics of oil-fired and biomass-fired coal, while Ant/(Ant + Phe) ratios greater than 0.1 displayed dual characteristics of biomass-fired coal and fossil fuels. Overall, these characteristic ratios suggested that the main sources of PAHs in CHMs are combustion sources, including biomass combustion and coal combustion. Plotting crossplots BaA/(BaA + Chr), Flt/(Flt + Pyr), and Ant/(Ant + Phe) (Figure 4) helped us to visualize the possible sources of PAH contamination in traditional Chinese medicine. Data points from the characteristic ratios indicated that oil sources, biomass, and coal combustion collectively contribute to herbal medicine pollution, posing significant environmental impacts and constituting the most substantial pollution source among the identified sources. However, the vast variety of herbal medicine types, coupled with the absorption, enrichment, transportation, and transformation of PAH compounds within the plant [37], as well as potential production and processing-related PAH production, necessitate further investigation to determine the precise sources of PAHs in CHMs.

2.6. Health Risk Assessment

PAHs are commonly detected in CHM samples, and their potential human health risks should not be ignored. The results of the ILCR evaluation of PAH contamination in seven CHMs are presented in Figure 5 and Table S3. Except for *Lycium barbarum*, the ILCR health risk values associated with the intake of herbs across different age groups primarily fell within the range of 1.0×10^{-6} to 1.0×10^{-4}, indicating a potential carcinogenic risk. However, the calculated risk levels remained within the acceptable range. Notably, in four age groups, the carcinogenic risk linked to the ingestion of *Lycium barbarum* was negligible (ILCR $< 1.0 \times 10^{-6}$). Although the health risk of PAHs to humans is within acceptable limits, the potential carcinogenic risk should be emphasized and the quality and safety control of traditional Chinese medicines need to be further strengthened.

Regarding the extent of PAH contamination in CHMs, statistically significant differences in cancer risk among seven Chinese herbs ($p < 0.05$), and the order of carcinogenicity risk is as follows: honeysuckle > *Coix lacryma* > lotus seed > glycyrrhizae > ginseng > seed of *Sterculia lychnophora* > *Lycium chinense*. This suggests a higher risk of carcinogenicity associated with the consumption of honeysuckle as a raw material for CHMs and its products. As a consequence, there is a need to enhance the detection and control of residual PAH contamination in honeysuckle raw materials. The highest cancer risk from honeysuckle consumption in adults may be attributed to prolonged dietary exposure to herbs among the analyzed parameters. Moreover, research indicates that children exhibit greater sensitivity to pollutants, posing health risks, underscoring the importance of prioritizing children's health concerns in this context [38].

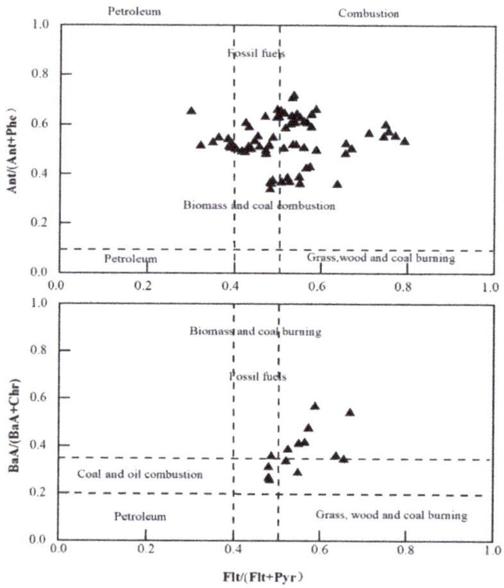

Figure 4. Crossplot of Flt/(Flt + Pyr), BaA/(BaA + Chr) and Ant/(Ant + Phe) characteristic ratios.

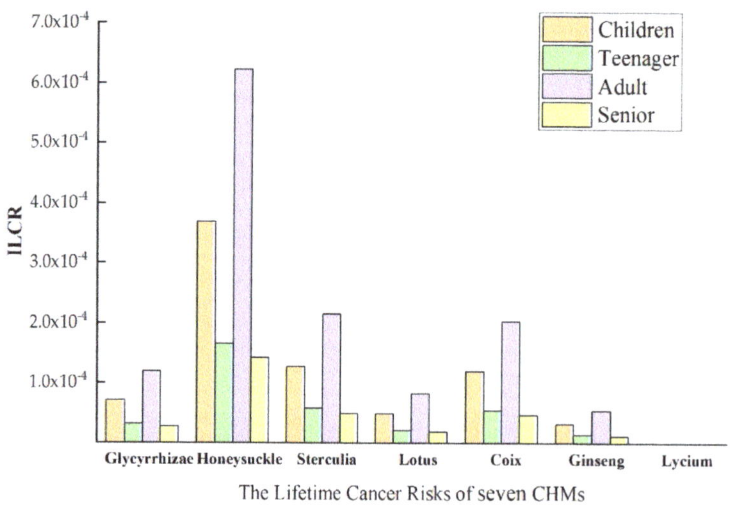

Figure 5. Health risk assessment of herbal medicines (ILCR).

Table 5 shows the results of the probabilistic assessment of the Monte Carlo carcinogenic risk of herbal medicines. To determine the carcinogenic risk, a lognormal distribution was fitted to the ILCR values of the herbs. The 90th percentile ILCR values for glycyrrhizae, *Coix lacryma*, ginseng, lotus seed, seed of *Sterculia lychnophora*, and *Lycium chinense* were 4.69×10^{-5}, 1.48×10^{-4}, 1.20×10^{-4}, 1.45×10^{-4}, 5.92×10^{-4}, and 2.25×10^{-6}, respectively. None of these values exceeded the maximum acceptable level of 1.0×10^{-4}. However, the 90th percentile ILCR value for *honeysuckle* was 1.37×10^{-3}, surpassing the maximum acceptable level. Consequently, it can be inferred that the intake of the herbal medicines analyzed in this study by residents leads to varying degrees of carcinogenic risk, with significant differences in risk levels ($p < 0.05$); however, no unacceptable carcinogenic risk occurred.

Table 5. Statistics of probabilistic estimation of lifetime carcinogenic risk values.

CHMs	Distribution	Parameters		10%	50%	90%
		Mean	SD			
Glycyrrhizae Radix et Rhizoma	Lognormal	2.00×10^{-5}	5.66×10^{-5}	1.29×10^{-6}	7.55×10^{-6}	4.59×10^{-5}
Honeysuckle	Lognormal	6.04×10^{-4}	1.48×10^{-4}	2.96×10^{-5}	2.00×10^{-4}	1.37×10^{-4}
Coix lacryma	Lognormal	7.04×10^{-5}	4.47×10^{-4}	1.84×10^{-4}	1.68×10^{-5}	1.48×10^{-4}
Ginseng Radix et Rhizoma	Lognormal	5.64×10^{-5}	1.81×10^{-4}	1.66×10^{-4}	1.43×10^{-5}	1.20×10^{-4}
Lotus seed	Lognormal	6.56×10^{-5}	1.98×10^{-4}	2.45×10^{-4}	1.93×10^{-5}	1.45×10^{-4}
seed of *Sterculia lychnophora*	Lognormal	2.70×10^{-4}	9.79×10^{-4}	1.08×10^{-5}	7.86×10^{-5}	5.92×10^{-4}
Lycium chinense	Lognormal	5.85×10^{-7}	1.41×10^{-4}	-1.03×10^{-4}	5.53×10^{-7}	2.25×10^{-4}

Note: Standard deviation (SD).

The sensitivity analysis results, presented in Figure 6, highlight the predominant factors influencing total carcinogenic risks associated with herbal medicines. The TEQ of PAHs emerged as the most influential factor, contributing significantly to health risks at 79.2%. Following closely was the exposure duration (ED) related to the dietary exposure time to herbal medicines, accounting for 11.9% of the total risk. The findings from the sensitivity analysis underscore TEQ as the most critical factor, emphasizing that the residual amount of PAHs in Chinese herbal medicines plays a pivotal role in shaping the overall carcinogenic risk. High enrichment of PAHs in herbal medicines increases the likelihood of human intake and increases the risk of human dietary exposure; therefore, the concentration of PAHs in herbal medicines is a key factor in the development of carcinogenic risk. This suggests that effective monitoring of PAH concentrations in herbal medicines could prove instrumental in mitigating the health risks posed by PAHs to the population.

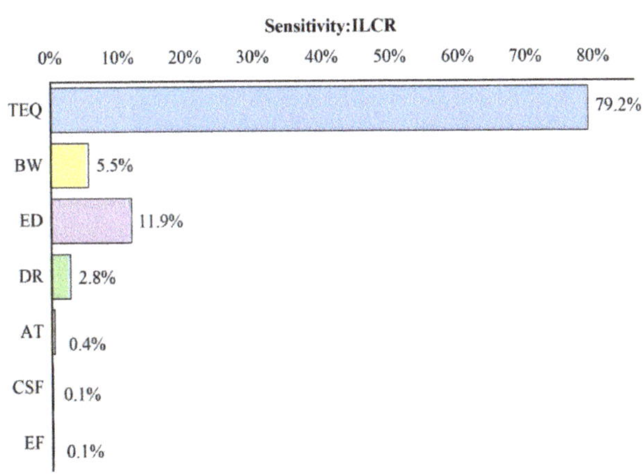

Figure 6. Sensitivity analysis of the carcinogenic risk of PAHs ingested from CHMs.

3. Experimental Methods

3.1. Instrumentation

PAHs were separated on an Agilent DB-5MS capillary column (60 m × 0.25 mm × 0.25 μm), using a 7890-5977 Gas Chromatography Mass Spectrometer (GC-MS) (Agilent, Santa Clara, CA, USA). The flow rate of the column was 1.5 mL/min; the temperature of the injection port was 300 °C; the temperature of the interface was 310 °C; the ion source was 250 °C; the four-stage rod was 150 °C; the injection volume was 2 μL; the split ratio was 5:1; and the solvent excision time was 5 min (determined according to the peak time of cyclohexane). The measurement was performed in SIM mode. Temperature increase procedure: 80 °C for 0.5 min, then 10 °C/min to 200 °C for 0 min, then 20 °C/min to 260 °C for 0 min, then 5 °C/min to 290 °C for 2 min, then 5 °C/min to 315 °C for 8 min. Other instruments used for sample preparation are as follows: 8000C Multi-function pulverizer (Yongkang Red Sun Electromechanical Co., Ltd., Zhejiang, China), AL204 Electronic balance (METTLER TOLEDO INSTRUMENTS, Shanghai, China), Sorvall ST 16R High-speed Freezing Centrifuge (Thermo Fisher, Waltham, MA, USA), VJL-Eortex Mixer (Shanghai Jinlan Instrument Manufacturing Co., Shanghai, China), HGC-96ANitrogen Blower (Tianjin Hengao Technology Development Co., Tianjin, China), and KQ5200DE Type ultrasonic instrument (Kunshan Ultrasonic Instrument Co., Kunshan, China). The optimized parameters for the analysis of 16 PAHs and 5 deuterated PAHs (PAHs-D) using GC-MS with selected ion monitoring mode (SIM) are listed in Table 1.

3.2. Chemicals and Solutions

The 16 PAHs' standard mixture (1000 mg/L, purity ≥ 96%, hexane/acetone 1:1) and 5 PAHs' isotope internal standard mixture (2000 mg/L, purity ≥ 98% methylene dichloride) were purchased from ANPEL Laboratory Technologies Inc. (Shanghai, China), including naphthalene, acenaphthylene, acenaphthene, fluorene, phenanthrene, anthracene, fluoranthene, pyrene, chrysene, benzo[a]anthracene, benzo[b]fluoranthene, benzo[k]fluoranthene, benzo[a]pyrene, indeno[1,2,3-cd]pyrene, dibenzo[a,h], anthracene, benzo[g,h,i]perylene, naphthalene-d8, acenaphthene-d10, phenanthrene-d10, chrysene-d12, and perylene-d12. PAH standard and isotope internal standard stock solutions were diluted in cyclohexane at a concentration of 10 μg/L. Three types of SPE columns, ProElut C18 (1 g/6 mL), ProElut Florisil (1 g/6 mL), and ProElut PSA (500 mg/6 mL) (Dikma Technologies, Radnor, PA, USA), were used for sample purification. All organic solvents, including acetonitrile, cyclohexane, and ethyl acetate were of HPLC grade (Dikma Technologies Inc., Radnor, PA, USA).

3.3. Sample Pretreatment

Thirty-five samples of CHMs (ginseng radix et rhizoma, seed of *Sterculia lychnophora*, lotus seed, honeysuckle, *Coix lacryma*, glycyrrhizae radix et rhizoma, *Lycium chinense*; 5 each) were purchased and analyzed, all of which were locally grown and harvested. The samples were stored in a cool, ventilated, and dry environment after purchase from the local market. Individual samples were weighed appropriately (>200 g) and crushed into powder using a high-speed pulverizer, which passed through a 60-mesh sieve, then collected into labeled sample bags and stored at −20 °C for further analysis.

3.3.1. Extraction

Weigh 0.500 g of herbal powder in a 15 mL centrifuge tube; add 100 μL of mixed isotope internal standard; mix thoroughly; add 5 mL of acetonitrile; seal with a lid; carry out ultrasonic extraction for 30 min; leave it to cool and then centrifuge it at 4000 r/min for 5 min. Transfer the supernatant to a clean 15 mL centrifuge tube. Precipitate, then add 5 mL of acetonitrile. Carry out ultrasonic extraction once again, followed by centrifugation, then combine the supernatant. Add 3 g of anhydrous sodium sulfate to absorb water by sufficient shaking. Perform centrifugation once again; take the extracted liquid nitrogen, blowing at 40 °C to a volume of about 0.2 mL, then add 2 mL of cyclohexane to dilute and dissolve, and then proceed to the next step of purification. Extraction of PAHs and the following purification methods reference the methods of Cui, Z. et al. [39].

3.3.2. Purify

For root and stem samples (ginseng radix et rhizoma, glycyrrhizae radix et rhizoma) the extraction solvent was cyclohexane, the purification column was PSA, and the activation and elution solution was cyclohexane. For the fruits and seeds (seed of *Sterculia lychnophora*, lotus seed, *Coix lacryma*, *Lycium chinense*) the purification column was a C18 column, activated with cyclohexane, with acetonitrile as eluent. For flowers (*honeysuckle*), the purification columns were C18, Florisil, PSA, acetonitrile, cyclohexane–ethyl acetate (49:1, v/v), and cyclohexane as eluents. Purification process: First, activate the column with 10 mL of activation solution. After all the liquid passes through the column, replace the bottom with a clean centrifuge tube and transfer the sample extract to the column. Then, add 2 mL of cyclohexane to wet-clean the centrifuge tube and load until all the liquid passes through the column. Add 5 mL of eluent solution to wash the column, collect the liquid, and then nitrogen-blow and concentrate the liquid at 40 °C to approx. 0.5 mL. Finally, take the sample solution for GC-MS analysis.

4. Health Risk Assessment

4.1. Toxic Equivalent Content of PAHs

Traditionally, carcinogenic toxicity assessments have been used to evaluate the health risk of an individual's exposure to a carcinogen. The toxic equivalency assessment of PAHs is based on the toxic equivalency factor (TEF) of benzo [a]pyrene, which is set to 1, and the toxic equivalency of other PAHs is calculated according to the formula of the conversion factor [40–42]. The formula for calculating the toxic equivalent quotient (TEQ) concentration of PAHs (TEQBaP) is as follows (1):

$$TEQ_{BaP} = \frac{C_i \times TEF_i}{1000} \quad (1)$$

where TEQ_{BaP} denotes the toxic equivalence of PAHs converted to benzo[a]pyrene, ng/kg; C_i denotes the concentration of PAHs in CHMs, μg/kg; TEF_i denotes the TEF value of PAHs in CHMs.

4.2. Carcinogenic Risk Assessment

According to the U.S. EPA-recommended Incremental Lifetime Cancer Risk (ILCR) model [43], an ILCR greater than 10^{-4} is considered an unacceptable cancer risk, and results between 10^{-6} and 10^{-4} are classified as potential cancer risks. On the other hand, for ILCR values that are less than 10^{-6}, the cancer risk associated with those are considered negligible [44]. The formula for calculating ILCR is shown in Equation (2):

$$ILCR = \frac{TEQ_{BaP} \times DR \times CSF \times EF \times ED}{BW \times AT} \quad (2)$$

where DR denotes the daily intake of herbal medicines, which was obtained from the study sample using a questionnaire, CSF denotes the dietary carcinogenicity slop e factor for BaP [45,46]. EF denotes the exposure frequency, taken as 365 d; ED denotes the exposure time; BW is body weight; AT is the average lifetime of the carcinogen [47,48] (see Supplementary Materials Table S1).

4.3. Probabilistic Assessment and Sensitivity Analysis

Uncertainty analysis in health risk assessment consists of two main parts: determining the probabilistic outcome and assessing the contribution of each variable to the outcome. Monte Carlo simulation was used to analyze the uncertainty of the results [49]. First, the type of best-fit probability distribution of the exposure factor was simulated by the Anderson–Darling test and chi-squared test, stable exposure distribution results were obtained through 10,000 iterations, and probabilistic risks were assessed by using values of different orders of magnitude (e.g., 10th, 50th, and 90th percentiles) of the exposure distribution results. The extent to which the exposure factors influenced the results was assessed by performing sensitivity analyses. Positive values indicate that the exposure factor is positively associated with health risk; conversely, it is negatively associated [50–52].

Determination of the best-fitting distribution for each parameter, Monte Carlo simulation, and sensitivity analysis were performed using Crystal Ball.

5. Conclusions

The GC-MS method was employed for the quantitative detection of EPA-PAHs in seven CHMs, and its successful application in analyzing 24 Chinese herbal medicine samples in 2016, with an RSD < 15%, affirmed the method's accuracy. Extraction and purification steps catered to the diverse components of CHMs were specifically designed, encompassing roots, stems, flowers, fruits, and leaves.

The method's pretreatment eliminated the need for extensive equipment such as GPS purifiers or bulky laboratory instruments like Soxhlet extraction, reducing solvent usage in the extraction and purification processes and shortening the extraction time. Consequently, this method emerges as a simple, accurate, sensitive, and efficient tool, holding promise for quality control and potential health risk assessment of PAHs in traditional Chinese medicines. Its applicability for online determination of PAHs offers valuable insights for drug safety monitoring and PAH risk management. Experimental results underscore widespread PAH contamination in most herbal medicines, with higher contamination levels observed in flowers and seeds compared to roots and fruits. PAH sources in herbal medicines reveal a combination of fuel oil, biomass, and coal combustion, with wood or coal combustion potentially contributing significantly to PAHs found in root herbs. While honeysuckle consumption may pose a health risk, the consumption of the other six herbs presents negligible health risks.

PAH content in herbs is a contributing factor to health risk sensitivity. Given the potential long-term nature of herbal medicine consumption, especially for patients with chronic diseases, stringent quality control measures are essential. This study not only provides an effective method for quantitative analysis and quality monitoring of PAHs in CHMs, but also offers insights into assessing the pollution levels and health risks associated with herbal medicines, serving as a reference for the safe export and quality development of ethnomedicine.

Supplementary Materials: The following supporting information can be downloaded at: https://www.mdpi.com/article/10.3390/molecules29050972/s1, Table S1. Calculation parameters for lifetime cancer risk assessment. Table S2. Commonly used PAHs characterization ratios and their sources. Table S3. Health Risk Assessment Results of 7 Chinese Herbal Medicines (ILCR).

Author Contributions: M.Z. conceived and designed the experiments. D.C. and S.Z. performed the experiments. X.Z., J.L. and J.W. analyzed the data. Z.Z. and Q.Z. provided the analysis tools. D.C. wrote the manuscript. M.Z. critically revised the manuscript. All authors have read and agreed to the published version of the manuscript.

Funding: The present study was financially supported by the National Natural Science Foundation of China (grant nos. 21966025 and 21667023), Ningxia Natural Science Foundation (2023AAC0314), the Ningxia Key Research and Development Program (grant no. 2019BFG02020).

Institutional Review Board Statement: This study does not involve experimental research on humans or animals.

Informed Consent Statement: The study was approved by the institutional research ethics committee of Ningxia Medical University, and written informed consent was obtained from each participant.

Data Availability Statement: The datasets used and/or analyzed during the current study are available from the corresponding author on reasonable request.

Acknowledgments: We are grateful for the National Natural Science Foundation of China, Ningxia Natural Science Foundation, the Ningxia Key Research and Development Program.

Conflicts of Interest: The authors declare that they have no conflicts of interest.

References

1. Ishizaki, A.; Sito, K.; Kataoka, H. Analysis of contaminant polycyclic aromatic hydrocarbons in tea products and crude drugs. *Anal. Methods Adv. Methods Appl.* **2011**, *3*, 299–305. [CrossRef] [PubMed]
2. Hwang, H.-J.; Lee, S.-H.; Kim, Y.-Y.; Shin, H.-S. Polycyclic aromatic hydrocarbon risk assessment and analytical methods using QuEchERS pretreatment for the evaluation of herbal medicine ingredients in Korea. *Foods* **2021**, *10*, 2200. [CrossRef] [PubMed]
3. Jiang, Y.; David, B.; Tu, P.; Barbin, Y. Recent analytical approaches in quality control of traditional Chinese medicines—A review. *Anal. Chim. Acta* **2010**, *657*, 9–18. [CrossRef] [PubMed]
4. Akerele, O. WHO Guidelines for the Assessment of Herbal Medicine. *Fitoterapia* **1992**, *63*, 99–104. [CrossRef]
5. Yu, L.; Cui, Z.; Cao, Y.; Zhang, J.; Sun, H. Investigation of 15 polycyclic aromatic hydrocarbons in selected medicinal herbs used as health food additives by ultra-performance liquid chromatography. *J. Liq. Chromatogr. Relat. Technol.* **2015**, *38*, 1783–1788. [CrossRef]
6. Harris, E.S.J.; Cao, S.; Littlefield, B.A.; Craycroft, J.A.; Scholten, R.; Kaptchuk, T.; Fu, Y.; Wang, W.; Liu, Y.; Chen, H. Heavy metal and pesticide content in commonly prescribed individual raw Chinese Herbal Medicines. *Sci. Total Environ.* **2011**, *409*, 4297–4305. [CrossRef]
7. Ling, Y.C.; Teng, H.C.; Cartwright, C. Supercritical fluid extraction and clean-up of organochlorine pesticides in Chinese herbal medicine. *J. Chromatogr. A* **1999**, *835*, 145–157. [CrossRef]
8. Tong, H.; Tong, Y.; Xue, J.; Liu, D.; Wu, X. Multi-residual Pesticide Monitoring in Commercial Chinese Herbal Medicines by Gas Chromatography–Triple Quadrupole Tandem Mass Spectrometry. *Food Anal. Methods* **2014**, *7*, 135–145. [CrossRef]
9. Liu, Y.; Wu, J.; Wei, W.; Xu, R. Simultaneous determination of heavy metal pollution in commercial traditional Chinese medicines in China. *J. Nat. Med.* **2013**, *67*, 887–893. [CrossRef]
10. Ting, A.; Chow, Y.; Tan, W. Microbial and heavy metal contamination in commonly consumed traditional Chinese herbal medicines. *J. Tradit. Chin. Med.* **2013**, *33*, 119–124. [CrossRef]
11. Yu, L.; Cao, Y.; Zhang, J.; Cui, Z.; Sun, H. Isotope dilution-GC-MS/MS analysis of 16 polycyclic aromatic hydrocarbons in selected medicinal herbs used as health food additives. *Food Addit. Contam. Part A Chem. Anal. Control Expo. Risk Assess.* **2012**, *29*, 1800–1809. [CrossRef] [PubMed]
12. Ravindra, K.; Sokhi, R.; Van Grieken, R. Atmospheric polycyclic aromatic hydrocarbons: Source attribution, emission factors and regulation. *Atmos. Environ.* **2008**, *42*, 2895–2921. [CrossRef]
13. Zhang, W.; Zhang, S.; Yue, D.; Wan, C.; Ye, Y.; Wang, X. Characterization and loading estimation of polycyclic aromatic hydrocarbons in road runoff from urban regions of Beijing, China. *Environ. Toxicol. Chem.* **2008**, *27*, 31–37. [CrossRef] [PubMed]
14. Keyte, I.J.; Harrison, R.M.; Lammel, G. Chemical reactivity and long-range transport potential of polycyclic aromatic hydrocarbons—A review. *Chem. Soc. Rev.* **2013**, *42*, 9333–9391. [CrossRef]
15. Zhang, Y.; Tao, S. Global atmospheric emission inventory of polycyclic aromatic hydrocarbons (PAHs) for 2004. *Atmos. Environ.* **2009**, *43*, 812–819. [CrossRef]
16. Chen, B.H.; Chen, Y.C. Formation of polycyclic aromatic hydrocarbons in the smoke from heated model lipids and food lipids. *J. Agric. Food Chem.* **2001**, *49*, 5238–5243. [CrossRef]
17. Crone, T.J.; Tolstoy, M. Magnitude of the 2010 Gulf of Mexico oil leak. *Science* **2010**, *330*, 634. [CrossRef]
18. Xia, Z.; Duan, X.; Tao, S.; Qiu, W.; Liu, D.; Wang, Y.; Wei, S.; Wang, B.; Jiang, Q.; Lu, B. Pollution level, inhalation exposure and lung cancer risk of ambient atmospheric polycyclic aromatic hydrocarbons (PAHs) in Taiyuan, China. *Environ. Pollut.* **2013**, *173*, 150–156. [CrossRef]
19. Kim, Y.-Y.; Patra, J.-K.; Shin, H.-S. Evaluation of analytical method and risk assessment of polycyclic aromatic hydrocarbons for fishery products in Korea. *Food Control* **2022**, *131*, 108421. [CrossRef]
20. Boll, E.S.; Christensen, J.H.; Holm, P.E. Quantification and source identification of polycyclic aromatic hydrocarbons in sediment, soil, and water spinach from Hanoi, Vietnam. *J. Environ. Monit.* **2008**, *10*, 261–269. [CrossRef] [PubMed]
21. Rey-Salgueiro, L.; Martínez-Carballo, E.; García-Falcón, M.S.; Simal-Gándara, J. Effects of a chemical company fire on the occurrence of polycyclic aromatic hydrocarbons in plant foods. *Food Chem.* **2008**, *108*, 347–353. [CrossRef]
22. Sampaio, G.R.; Guizellini, G.M.; da Silva, S.A.; de Almeida, A.P.; Pinaffi-Langley, A.C.C.; Rogero, M.M.; de Camargo, A.C.; Torres, E.A. Polycyclic aromatic hydrocarbons in foods: Biological effects, legislation, occurrence, analytical methods, and strategies to reduce their formation. *Int. J. Mol. Sci.* **2021**, *22*, 6010. [CrossRef]
23. Stadler, R.H.; Lineback, D.R. *Process-Induced Food Toxicants. Occurrance, Formation, Mitigation and Health Risks*; John Wiley & Sons, Inc.: Hoboken, NJ, USA, 2009. [CrossRef]
24. Zelinkova, Z.; Wenzl, T. The occurrence of 16 EPA PAHs in food—A review. *Polycycl. Aromat. Compd.* **2015**, *35*, 248–284. [CrossRef]
25. Wennrich, L.; Popp, P.; Zeibig, M. Polycyclic aromatic hydrocarbon burden in fruit and vegetable species cultivated in allotments in an industrial area. *Int. J. Environ. Anal. Chem.* **2002**, *82*, 667–690. [CrossRef]

26. Wei, M.-C.; Jen, J.-F. Determination of polycyclic aromatic hydrocarbons in aqueous samples by microwave assisted headspace solid-phase microextraction and gas chromatography/flame ionization detection. *Talanta* **2007**, *72*, 1269–1274. [CrossRef]
27. Edokpayi, J.N.; Odiyo, J.O.; Popoola, O.E.; Msagati, T.A. Determination and distribution of polycyclic aromatic hydrocarbons in rivers, sediments and wastewater effluents in Vhembe District, South Africa. *Int. J. Environ. Res. Public Health* **2016**, *13*, 387. [CrossRef]
28. Zhou, Q.; Lei, M.; Wu, Y.; Yuan, Y. Magnetic solid phase extraction of typical polycyclic aromatic hydrocarbons from environmental water samples with metal organic framework MIL-101 (Cr) modified zero valent iron nano-particles. *J. Chromatogr. A* **2017**, *1487*, 22–29. [CrossRef] [PubMed]
29. De Smet, P.G.; Keller, K.; Hänsel, R.; Chandler, R. *Adverse Effects of Herbal Drugs*; Springer: Berlin/Heidelberg, Germany, 1997. [CrossRef]
30. Yunker, M.B.; Backus, S.M.; Graf Pannatier, E.; Jeffries, D.S.; Macdonald, R.W. Sources and Significance of Alkane and PAH Hydrocarbons in Canadian Arctic Rivers. *Estuar. Coast. Shelf Sci.* **2002**, *55*, 1–31. [CrossRef]
31. Yunker, M.B.; Macdonald, R.W.; Vingarzan, R.; Mitchell, R.H.; Goyette, D.; Sylvestre, S. PAHs in the Fraser River basin: A critical appraisal of PAH ratios as indicators of PAH source and composition [Review]. *Org. Geochem.* **2002**, *33*, 489–515. [CrossRef]
32. Torre-Roche, R.J.D.L.; Lee, W.Y.; Campos-Diaz, S.I. Soil-borne polycyclic aromatic hydrocarbons in El Paso, Texas: Analysis of a potential problem in the United States/Mexico border region. *J. Hazard. Mater.* **2009**, *163*, 946–958. [CrossRef]
33. Akyüz, M.; Cabuk, H. Gas–particle partitioning and seasonal variation of polycyclic aromatic hydrocarbons in the atmosphere of Zonguldak, Turkey. *Sci. Total Environ.* **2010**, *408*, 5550–5558. [CrossRef]
34. Lv, M.; Luan, X.; Liao, C.; Wang, D.; Liu, D.; Zhang, G.; Jiang, G.; Chen, L. Human impacts on polycyclic aromatic hydrocarbon distribution in Chinese intertidal zones. *Nat. Sustain.* **2020**, *3*, 878–884. [CrossRef]
35. Zhou, D.-B.; Han, F.; Ding, L.; Song, W.; Lv, Y.-N.; Hu, Y.-Y.; Liu, Y.-X.; Sheng, X.; Zheng, P. Magnetic C60 nanospheres based solid-phase extraction coupled with isotope dilution gas chromatography–mass spectrometry method for the determination of sixteen polycyclic aromatic hydrocarbons in Chinese herbal medicines. *J. Chromatogr. B* **2020**, *1144*, 122076. [CrossRef] [PubMed]
36. Yu, B.; Zhang, D.; Tan, L.-H.; Zhao, S.-P.; Wang, J.-W.; Yao, L.; Cao, W.-G. Polycyclic aromatic hydrocarbons in traditional Chinese medicines: An analytical method based on different medicinal parts, levels, distribution, and sources. *RSC Adv.* **2017**, *7*, 4671–4680. [CrossRef]
37. Dorine, D.; Philippe, B.; Geneviève, C. Challenges in Tracing the Fate and Effects of Atmospheric Polycyclic Aromatic Hydrocarbon Deposition in Vascular Plants. *Environ. Sci. Technol.* **2013**, *47*, 3967–3981. [CrossRef]
38. Cai, Y.; Zhang, Y.; Ren, L.; Li, C.; Wang, G.; Huang, H. Characteristics and Ecological Risk Assessment of Heavy Metal Pollution in Vegetable Soils of Guangzhou Urban Districts. *Guangdong Agric. Sci.* **2019**, *46*, 73–78. [CrossRef]
39. Cui, Z.; Ge, N.; Zhang, A.; Liu, Y.; Zhang, J.; Cao, Y. Comprehensive determination of polycyclic aromatic hydrocarbons in Chinese herbal medicines by solid phase extraction and gas chromatography coupled to tandem mass spectrometry. *Anal. Bioanal. Chem.* **2015**, *407*, 1989–1997. [CrossRef]
40. Nisbet, I.C.; Lagoy, P.K. Toxic equivalency factors (TEFs) for polycyclic aromatic hydrocarbons (PAHs). *Regul. Toxicol. Pharmacol.* **1992**, *16*, 290–300. [CrossRef]
41. Madill, R.E.; Brownlee, B.G.; Josephy, P.D.; Bunce, N.J. Comparison of the Ames Salmonella assay and Mutatox genotoxicity assay for assessing the mutagenicity of polycyclic aromatic compounds in porewater from Athabasca oil sands mature fine tailings. *Environ. Sci. Technol.* **1999**, *33*, 2510–2516. [CrossRef]
42. Stroomberg, G.J.; Knecht, J.A.D.; Ariese, F.; Gestel, C.A.V.; Velthorst, N.H. Pyrene metabolites in the hepatopancreas and gut of the isopod Porcellio scaber, a new biomarker for polycyclic aromatic hydrocarbon exposure in terrestrial ecosystems. *Environ. Toxicol. Chem. Int. J.* **1999**, *18*, 2217–2224. [CrossRef]
43. U.S. Environmental Protection Agency. *Risk Assessment Guidance for Superfund. Volume I Human Health Evaluation Manual. (Part A)*; Interim Final; U.S. Environmental Protection Agency: Washington, DC, USA, 2023.
44. Jiang, D.; Xin, C.; Li, W.; Chen, J.; Li, F.; Chu, Z.; Xiao, P.; Shao, L. Quantitative analysis and health risk assessment of polycyclic aromatic hydrocarbons in edible vegetable oils marketed in Shandong of China. *Food Chem. Toxicol.* **2015**, *83*, 61–67. [CrossRef]
45. Zhao, Z.; Zhang, L.; Cai, Y.; Chen, Y. Distribution of polycyclic aromatic hydrocarbon (PAH) residues in several tissues of edible fishes from the largest freshwater lake in China, Poyang Lake, and associated human health risk assessment. *Ecotoxicol. Environ. Saf.* **2014**, *104*, 323–331. [CrossRef] [PubMed]
46. Maliszewska-Kordybach, B.; Smreczak, B.; Klimkowicz-Pawlas, A. Concentrations, sources, and spatial distribution of individual polycyclic aromatic hydrocarbons (PAHs) in agricultural soils in the Eastern part of the EU: Poland as a case study. *Sci. Total Environ.* **2009**, *407*, 3746–3753. [CrossRef]
47. Phillips, L.; Moya, J. The evolution of EPA's Exposure Factors Handbook and its future as an exposure assessment resource. *J. Expo. Sci. Environ. Epidemiol.* **2013**, *23*, 13–21. [CrossRef]
48. Phillips, L.J.; Moya, J. Exposure factors resources: Contrasting EPA's Exposure Factors Handbook with international sources. *J. Expo. Sci. Environ. Epidemiol.* **2014**, *24*, 233–243. [CrossRef]
49. Jiang, C.; Zhao, Q.; Zheng, L.; Chen, X.; Li, C.; Ren, M. Distribution, source and health risk assessment based on the Monte Carlo method of heavy metals in shallow groundwater in an area affected by mining activities, China. *Ecotoxicol. Environ. Saf.* **2021**, *224*, 112679. [CrossRef] [PubMed]

50. Guo, G.; Zhang, D.; Wang, Y. Probabilistic human health risk assessment of heavy metal intake via vegetable consumption around Pb/Zn smelters in Southwest China. *Int. J. Environ. Res. Public Health* **2019**, *16*, 3267. [CrossRef]
51. Yang, S.; Zhao, J.; Chang, S.X.; Collins, C.; Xu, J.; Liu, X. Status assessment and probabilistic health risk modeling of metals accumulation in agriculture soils across China: A synthesis. *Environ. Int.* **2019**, *128*, 165–174. [CrossRef] [PubMed]
52. Hu, W.; Wang, H.; Dong, L.; Huang, B.; Borggaard, O.K.; Hansen, H.C.B.; He, Y.; Holm, P.E. Source identification of heavy metals in peri-urban agricultural soils of southeast China: An integrated approach. *Environ. Pollut.* **2018**, *237*, 650–661. [CrossRef]

molecules

MDPI

Article

Analytical Assessment of the Antioxidant Properties of the Coneflower (*Echinacea purpurea* L. Moench) Grown with Various Mulch Materials

Celestina Adebimpe Ojo [1], Kinga Dziadek [1], Urszula Sadowska [2], Joanna Skoczylas [1] and Aneta Kopeć [1,*]

[1] Department of Human Nutrition and Dietetics, Faculty of Food Technology, University of Agriculture in Krakow, Balicka 122, 30-149 Krakow, Poland; celestina.adebimpe.ojo@student.urk.edu.pl (C.A.O.); kinga.dziadek@urk.edu.pl (K.D.); joannaskoczylas7@gmail.com (J.S.)

[2] Faculty of Mechanisation and Energy Technologies in Agriculture, University of Agriculture in Krakow, Majora Łupaszki 6, 30-198 Krakow, Poland; urszula.sadowska@urk.edu.pl

* Correspondence: aneta.kopec@urk.edu.pl

Abstract: Antioxidants are added to foods to decrease the adverse effect of reactive species that create undesirable compounds that destroy essential nutrients and, therefore, lower the nutritional, chemical and physical properties of foods. This study was carried out to determine the antioxidant properties of flowers and plant stems with leaves of *Echinacea purpurea* grown with mulches of different colours and thicknesses. Coneflowers were grown in the Experimental Station of the Agricultural University in Kraków, Poland. The mulching materials used were black, green and brown colours of 100 g/m^2 and 80 g/m^2 density. In plant material, e.g., flowers or plant stems plus leaves the proximate analysis, the total polyphenol content and the ability to scavenge free radicals (ABTS, DPPH and FRAP) were determined. The results show that flower samples had a higher content of compound proteins, ash and phenolic compounds. The mulching colour and density did not affect the proximate analysis of the *E. purpurea* plant. Based on the result of this study, *E. purpurea* is a potential source of natural antioxidants and can be used to improve the antioxidant activity of various food products as well as in cosmetics within the pharmaceutical industry.

Keywords: antioxidant properties; coneflower; chemical composition; *Echinacea purpurea*

1. Introduction

Echinacea purpurea L. (Moench) is a native and ornamental plant in the Atlantic geographical drainage area of the United States of America and Canada, but not including Mexico. The cultivation of this plant is also popular in Europe [1,2]. Based on some data in the literature, it is reported that *E. purpurea* is the last one in a list of the ten most frequently cultivated herbs in Poland [3,4].

E. purpurea is known as an "anti-infectious" agent because of an excellent potential to treat viral and bacterial infections and has been used to treat several infectious conditions ranging from simple acne and ulcers to mild septicemias [5–7]. *Echinacea purpurea* (*E. purpurea*) contains a number of chemical components including alkamides, polyalkenes, polyalkynes, chicoric and caftaric acids, as well as caffeic acid derivatives, glycoproteins and polysaccharides [4,8]. Some of the substances mentioned above exhibit an immunomodulatory effect [4,5]. *E. purpurea* volatile oil contains borneol, bornyl acetate, pentadeca-8-(Z)-en-2-one, germacrene D, caryophyllene and caryophyllene epoxide. Isobutyl amides of C11–C16 straight-chain fatty acids with olefinic or acetylenic bonds (or both) are found in the above parts of the *E.purpureae* herb, such as isomeric dodeca-(2E, 4E, 8Z, 10E/Z)-tetraenoic acid isobutyl amides, among others. The major active compound of the phenolic acid class found in the aerial parts is caffeic acid, with a concentration range of 1.2–3.1%. Chicoric acid methyl ester and other derivatives are also present [8]. Over the years, the immunomodulatory properties of *E. purpurea* have been reported, too. It was described that extract of *E. purpurea* considerably inhibited the growth of pathogenic yeasts such as *Candida albicans* [9–11] and bacteria (*Streptococcus pyogenes*, *Haemophilus influenzae* and *Legionella pneumophila*). It was also reported that the aqueous extract of *E. purpurea* has an antiviral activity for herpes simplex virus 1 (HSV-1) and herpes simplex virus 2 (HSV-2) [10–12].

According to the literature, *E. purpurea* requires high water capacity and well-oxygenated soil for root development [13]. Soil water deficit significantly inhibits plants' growth and development, consequently reducing yield. Negative temperatures during winter are also a threat. Moreover, windy and low-temperature springs also limit plant growth during the initial growing season [4]. *E. purpurea* is one of Poland's ten most frequently cultivated herbs [3,4] and, traditionally, Poland's climatic conditions enable its cultivation throughout the country. However, in recent years, and due to global climate change, Polish agriculture has been negatively affected by uneven rainfall and drought periods, leading to weaker production [13,14].

Synthetic (inorganic) mulching material is frequently used in agriculture to protect crops from adverse environmental conditions including severe weather (low or high temperature, low rainfall), birds and insect crop damage [15]. However, mulching affects not only plant growth and yield but also plant chemical composition [16,17].

This study, therefore, aims to evaluate how different colours and thicknesses of synthetic mulching affect the chemical composition and antioxidant properties of *E. purpurea* flowers and leaves plus stems.

2. Results

2.1. Crude Protein Fat and Ash Content in Flowers, Plant Stems plus Leaves of E.purpurea

The obtained results indicated that flowers were richer in protein then the plant stems plus leaves (Table 1), showing that the different mulching materials with respect to colour did not significantly affect the protein content in *E. purpurea*. Among the flower samples, a significantly higher content of protein was assessed in samples F1 and F2; however, the lowest protein level was observed in sample F0. The result of the protein content in plant stems plus leaves showed the same trend as in the flower. Among this group, samples SL1 and SL2 were characterised by the highest content of protein, while sample SL0 was characterised by the lowest. Furthermore, it was observed that application at 100 g/m^2 of mulch material in flower and plant stem plus leaves resulted in a higher protein content in *E. purpurea* irrespective of colour (Table 1) (the exceptions were the samples F3 and F6).

A comparison between sample F0 and SL0 showed that flowers were characterised by a higher crude fat level than plant stems plus leaves, with the exception of sample SL6 (Table 1). Among the flower samples, the highest content of fat was found in sample F0, compared to F1 and F5. Samples SL1, SL2, SL3 and SL5 showed the lowest fat content compared to the SL6 sample. Ash content in *E. purpurea* was richer in flowers than the plant stems plus leaves (Table 1). Among the flower samples, the highest content of ash was assessed in sample F1, while the lowest ash level was observed in sample F3. Between the plant stems plus leaves, samples SL2 and SL6 were characterised by a higher content of ash; therefore, the sample SL1 was characterised by the lowest. Different mulch applications with respect to colours and thicknesses did not significantly affect the ash content in *E. purpurea*.

Table 1. Protein, crude fat and ash content in the flower and the mixture of plant stems plus leaves of *E. purpurea* [g/100 g D.M.].

Sample Name	Protein	Crude Fat	Ash
F0	13.21 ± 0.11 ^{de}	1.99 ± 0.13 ^{c}	8.59 ± 0.06 ^{g}
F1	17.10 ± 0.89 ^{fg}	1.02 ± 0.25 ^{ab}	9.09 ± 0.65 ^{h}
F2	17.47 ± 0.08 ^{g}	1.23 ± 0.04 ^{bc}	8.32 ± 0.07 ^{efg}
F3	15.40 ± 0.48 ^{e}	1.45 ± 0.23 ^{bc}	7.74 ± 0.08 ^{d}
F4	15.58 ± 0.01 ^{e}	1.47 ± 0.38 ^{bc}	8.39 ± 0.05 ^{efg}
F5	15.98 ± 0.23 ^{ef}	0.64 ± 0.04 ^{a}	7.87 ± 0.15 ^{de}
F6	15.67 ± 0.38 ^{ef}	1.67 ± 0.00 ^{bc}	8.05 ± 0.11 ^{def}
SL0	7.20 ± 0.76 ^{a}	0.85 ± 0.11 ^{ab}	6.99 ± 0.37 ^{c}
SL1	12.00 ± 0.01 ^{de}	0.60 ± 0.02 ^{a}	5.45 ± 0.10 ^{a}
SL2	12.04 ± 1.68 ^{de}	0.62 ± 0.08 ^{a}	8.36 ± 0.00 ^{efg}
SL3	9.34 ± 0.86 ^{b}	0.60 ± 0.03 ^{a}	6.25 ± 0.02 ^{b}
SL4	11.38 ± 0.27 ^{cd}	1.25 ± 0.04 ^{abc}	7.91 ± 0.09 ^{de}
SL5	8.90 ± 0.42 ^{b}	0.73 ± 0.11 ^{a}	6.04 ± 0.28 ^{b}
SL6	10.12 ± 0.24 ^{bc}	2.06 ± 0.15 ^{c}	8.51 ± 0.02 ^{fg}
	Part of the plant		
Flower	15.77 ± 3.94 ^{B}	1.35 ± 0.56 ^{A}	8.29 ± 0.48 ^{B}
Stems plus leaves	10.14 ± 1.83 ^{A}	0.96 ± 0.52 ^{A}	7.07 ± 1.17 ^{A}
p-value	0.001	0.065	0.001
	Mulching material colour		
Control	10.20 ± 3.50 ^{A}	1.42 ± 0.67 ^{A}	7.79 ± 0.95 ^{A}
Black	14.02 ± 2.59 ^{A}	1.09 ± 0.39 ^{A}	7.71 ± 1.49 ^{A}
Green	13.60 ± 3.65 ^{A}	0.80 ± 0.27 ^{A}	7.65 ± 1.02 ^{A}
Brown	12.63 ± 3.15 ^{A}	1.45 ± 0.74 ^{A}	7.64 ± 0.91 ^{A}
p-value	0.264	0.097	0.996
	Thickness of mulching material g/100 m^2		
Control	10.20 ± 3.50 ^{A}	1.42 ± 0.67 ^{A}	7.79 ± 0.95 ^{A}
100	13.89 ± 3.18 ^{A}	0.92 ± 0.52 ^{A}	7.53 ± 1.35 ^{A}
80	12.94 ± 3.03 ^{A}	1.31 ± 0.54 ^{A}	7.795 ± 0.86 ^{A}
p-value	0.150	0.153	0.827

Mean values in the same column with different letters (a–h) are statistically different ($p < 0.05$); mean values with different capital letters (A,B) in the same column are statistically different; differences are between the experiment factors, e.g., the part of the plant, colour of the mulching material, and thickness of the mulching material; results are expressed as mean ± SD; D.M., dry matter; F0, flower control (cultivated without mulching material); F1, flower cultivated with 100 g/m^2 black mulching material; F2, flower cultivated with 100 g/m^2 green mulching material; F3, flower cultivated with 100 g/m^2 brown mulching material; F4, flower cultivated with 80 g/m^2 black mulching material; F5, flower cultivated with 80 g/m^2 green mulching material; F6, flower cultivated with 80 g/m^2 brown mulching material; SL0, plant stems plus leaves—control (cultivated with no mulching material); SL1, plant stems plus leaves cultivated with 100 g/m^2 black mulching material, SL2, plant stems plus leaves cultivated with 100 g/m^2 green mulching material, SL3, plant stems plus leaves cultivated with 100 g/m^2 brown mulching material; SL4, plant stems plus leaves cultivated with 80 g/m^2 black mulching material; SL5, plant stems plus leaves planted with 80 g/m^2 green mulching material; SL6, plant stems plus leaves planted with 80 g/m^2 brown mulching material.

2.2. Total Polyphenolic Content and Antioxidant Activity

The results obtained indicated that flowers were richer in total polyphenols and had higher antioxidant activity measured with all methods than plants stems plus leaves (Table 2). The different mulching materials with respect to thickness and colours did not significantly affect the total polyphenol content in *E. purpurea*. Among the flower samples, the significantly higher content of these compounds was assessed in samples F2, F4, F5 and F6, respectively. However, the lowest total polyphenol level was observed in sample F1. The result of total polyphenol content in plant stems plus leaves showed the same trend as the flower. Among this group, the highest content of total polyphenols was recorded in samples SL2 and SL4; therefore, sample SL1 showed the lowest value.

Table 2. Total polyphenol concentration and antioxidant activity of flowers and a mixture of plant stems plus leaves of coneflower.

Sample Name	Total Polyphenols (mg/100 g D.M.) *	DPPH (μmol Trolox/1 g D.M.)	ABTS (μmol Trolox/1 g D.M.)	FRAP (μmol Trolox/1 g D.M.)
F0	12,881.59 ± 547.89 [g]	690.27 ± 40.75 [f]	547.70 ± 17.09 [h]	1680.71 ± 73.22 [f]
F1	9728.05 ± 232.05 [e]	609.74 ± 27.52 [e]	356.36 ± 7.99 [f]	1501.03 ± 12.05 [e]
F2	14,222.89 ± 233.76 [h]	860.06 ± 10.60 [h]	331.66 ± 31.55 [ef]	2319.29 ± 23.52 [i]
F3	10,821.37 ± 182.08 [f]	660.21 ± 27.52 [ef]	314.67 ± 9.13 [e]	2141.01 ± 11.63 [g]
F4	13,941.18 ± 0.00 [h]	937.17 ± 30.80 [i]	512.66 ± 56.92 [g]	2220.66 ± 67.23 [h]
F5	13,663.35 ± 441.41 [h]	824.41 ± 18.58 [h]	511.51 ± 23.60 [gh]	2116.97 ± 26.34 [g]
F6	14,129.95 ± 156.91 [h]	752.51 ± 29.29 [g]	588.66 ± 12.06 [i]	2294.97 ± 33.79 [i]
SL0	5363.62 ± 70.56 [c]	351.93 ± 12.60 [bc]	189.46 ± 18.46 [d]	798.95 ± 6.99 [d]
SL1	3256.16 ± 26.52 [a]	269.36 ± 1.40 [a]	45.63 ± 4.47 [a]	438.78 ± 20.28 [a]
SL2	6395.19 ± 7.91 [d]	373.40 ± 37.35 [c]	109.76 ± 14.32 [b]	595.36 ± 4.46 [c]
SL3	4270.80 ± 54.41 [b]	268.23 ± 4.32 [a]	45.92 ± 6.56 [a]	457.23 ± 3.85 [a]
SL4	6673.99 ± 75.66 [d]	432.54 ± 28.51 [d]	201.02 ± 9.04 [d]	779.71 ± 8.88 [d]
SL5	4257.47 ± 425.34 [b]	316.69 ± 9.00 [ab]	80.46 ± 5.92 [ab]	515.39 ± 11.72 [b]
SL6	4843.00 ± 25.34 [c]	389.09 ± 8.04 [cd]	151.97 ± 5.72 [c]	602.70 ± 8.72 [c]
		Part of plant		
flower	12,770 ± 1706 [B]	762 ± 115 [B]	452 ± 110 [B]	2039 ± 305 [B]
Steams plus leaves	5009 ± 1168 [A]	343 ± 60 [A]	118 ± 62 [A]	598 ± 137 [A]
p-value	<0.001	<0.001	<0.001	<0.001
		Colour of mulching material		
Control	9123 ± 4125 [A]	521 ± 197 [A]	368 ± 197 [A]	1239 ± 473 [A]
Black	8399 ± 4109 [A]	562 ± 265 [A]	279 ± 177 [A]	1235 ± 747 [A]
Green	9634 ± 4578 [A]	593 ± 267 [A]	258 ± 184 [A]	1386 ± 872 [A]
brown	8516 ± 4317 [A]	517 ± 210 [A]	275 ± 214 [A]	1374 ± 885 [A]
p-value	0.580	0.554	0.041	0.761
		Thickness g/m^2		
control	9122 ± 4125 [A]	521 ± 199 [A]	368 ± 197 [B]	1239 ± 473 [A]
100	8116 ± 3955 [A]	507 ± 230 [A]	201 ± 140 [A]	1242 ± 810 [A]
80	9585 ± 4519 [A]	609 ± 249 [A]	341 ± 208 [AB]	1422 ± 818 [A]
p-value	0.580	0.554	0.041	0.761

Mean values in the same column with different letters (a–i) are statistically different ($p < 0.05$); mean values with different capital letters (A,B) in the same column are statistically different; differences are between the experiment factors, e.g., the part of the plant, the colour of the mulching material and the thickness of the mulching material; results are expressed as mean ± SD; D.M., dry matter; F0, flower control (cultivated without mulching material); F1, flower cultivated with 100g/m^2 black mulching material; F2, flower cultivated with 100 g/m^2 green mulching material; F3, flower cultivated with 100 g/m^2 brown mulching material; F4, flower cultivated with 80 g/m^2 black mulching material; F5, flower cultivated with 80 g/m^2 green mulching material; F6, flower cultivated with 80 g/m^2 brown mulching material; SL0, stems plus leaves—control (cultivated without mulching material); SL1, plant stems plus leaves cultivated with 100 g/m^2 black mulching material; SL2, plant stems plus leaves cultivated with 100 g/m^2 green mulching material, SL3, plant stems plus leaves cultivated with 100 g/m^{2}brown mulching material; SL4, plant stems plus leaves cultivated with 80 g/m^2 black mulching material; SL5, plant stems plus leaves planted with 80 g/m^2 green mulching material; SL6, plant stems plus leaves planted with 80 g/m^2 brown mulching material; * chlorogenic acid equivalent.

The results of the DPPH analysis revealed that flowers were characterised by higher antioxidant activity than the plant stems plus leaves (Table 3). Among the flower samples, the highest antioxidant properties were obtained in sample F4 compared to other flower samples. Variation was also observed in samples of plant stems plus leaves. In this group, the highest antioxidant activity was measured in sample SL4. However, samples SL1 and SL3 showed the lowest.

Among the flower samples, the highest antioxidant capacity, measured with the ABTS method, was obtained in sample F6, while the lowest was observed in sample F3. The result of an ABTS test carried out on the plant stems plus leaves indicated that samples SL0 and

SL4 were characterised by significantly higher antioxidant activity; therefore, the samples SL1 and SL3 were characterised by the lowest (Table 2). Different mulching materials with respect to thickness significantly affected the antioxidant properties in *E. purpurea* flowers. The flowers of samples planted with 80 g/m^2 of mulch material were characterised by significantly higher antioxidant activity compared to the flowers planted with 100 g/m^2 of material.

The obtained results of a FRAP test indicated that flowers had higher antioxidant properties than plant stems plus leaves (Table 2). In the group of flowers, significantly, the highest antioxidant activity was assessed in samples F2 and F6. However, the lowest was observed in sample F1. Therefore, in the group of plant stems plus leaves, samples SL0 and SL4 were characterised by the highest antioxidant capacity; therefore, the samples SL1 and SL3 recorded as the lowest.

2.3. Polyphenolic Profile

The 25 types of polyphenols were detected by the HPLC method in the analysed samples, and their concentrations are shown in Tables 3 and 4.

P-coumaric acid, chlorogenic acid and rutin were the dominant polyphenolic compounds in all samples, both in the flower as well as the plant stem plus leaves. Additionally, in the group of plant stems plus leaves, a high content of naringin was found. The obtained results showed that the flower samples had higher concentrations of almost all the polyphenol types in comparison with the plant stem plus leaf samples, regardless of the colour or thickness of the mulching material. The exceptions were hispidulin, which was identified only in plant stem plus leaf samples, as well as gallic acid, which was found in all samples in the group of plant stems plus leaves and only in samples F1 and F3 in the group of flower. From the result of the analysis of variance, a high significant variation in the concentration of polyphenol types was recorded within the different sample colours and weight of mulch applied at $p < 0.05$.

Furthermore, in flower samples that were planted with 100 g/m^2 of mulching material (regardless of colour), a higher content of polyphenols was observed, such as sinapinic acid, catechin, apigenin and carnosol, in comparison with those planted with 80 g/m^2 of mulching material. Therefore, a higher concentration of acacetin in flower samples as well as rutin and hesperidin in plant stems plus leaves samples was found in samples planted with 80 g/m^2 of mulching material, regardless of colour, compared to those planted with 100 g/m^2 of mulching material.

Additionally, the application of mulching material with a thickness of 80 g/m^2 and a black and or brown colour resulted in an increase in the content of some polyphenolic compounds in flower samples (chlorogenic acid, 4-hydroxybenzoic acid, caffeic acid, p-coumaric acid and rutin) in comparison with using mulching material with a thickness of 100 g/m^2 in the corresponding colours.

Table 3. The profile of phenolic acids in flowers and a mixture of stems plus leaves of coneflower [mg/100 g D.M.].

Sample	Gallic Acid	Chlorogenic Acid	4 Hydroxybenzoic Acid	Caffeic Acid	Vanillic Acid	Syringic Acid	p-Coumaric Acid	Ferulic Acid	Sinapinic Acid mg/100 g	Rosmarinic Acid mg/100 g	Carnosic Acid * mg/100 g
F0	nd	276 ± 0.03 ^{i}	54.46 ± 0.04 ^{h}	14.42 ± 1.66 ^{e}	3.03 ± 0.01 ^{cd}	13.30 ± 0.09 ^{h}	1719 ± 1.58 ^{j}	2.83 ± 0.10 ^{a}	51.64 ± 1.36 ^{c}	3.91 ± 0.09 ^{c}	91.05 ± 0.03 ^{f}
F1	2.53 ± 0.3 ^{b}	195 ± 0.25 ^{e}	40.79 ± 0.11 ^{e}	16.68 ± 0.07 ^{f}	9.46 ± 1.36 ^{i}	7.96 ± 0.04 ^{g}	1354 ± 2.17 ^{h}	2.63 ± 0.03 ^{a}	50.20 ± 1.46 ^{c}	2.10 ± 0.03 ^{ab}	43.61 ± 0.69 ^{c}
F2	nd	331 ± 0.07 ^{k}	63.64 ± 1.06 ^{j}	31.22 ± 0.04 ^{j}	13.83 ± 0.09 ^{j}	4.57 ± 0.17 ^{bc}	2053 ± 7.38 ^{m}	3.72 ± 0.15 ^{b}	52.71 ± 1.16 ^{c}	2.60 ± 0.06 ^{b}	67.30 ± 0.73 ^{de}
F3	5.27 ± 0.01 ^{f}	187 ± 0.06 ^{d}	37.45 ± 0.04 ^{d}	14.26 ± 0.91 ^{e}	6.07 ± 0.01 ^{g}	3.76 ± 0.20 ^{ab}	1385 ± 4.08 ^{i}	nd	40.59 ± 0.61 ^{ab}	1.95 ± 0.03 ^{a}	70.89 ± 0.22 ^{e}
F4	nd	358 ± 0.24 ^{l}	58.78 ± 0.03 ^{i}	21.00 ± 0.00 ^{g}	5.28 ± 0.03 ^{f}	2.83 ± 1.06 ^{a}	1934 ± 7.53 ^{k}	nd	42.61 ± 0.24 ^{b}	20.90 ± 0.56 ^{g}	38.63 ± 0.26 ^{c}
F5	nd	230 ± 6.56 ^{h}	47.33 ± 1.39 ^{f}	25.33 ± 0.00 ^{i}	8.01 ± 0.02 ^{h}	13.56 ± 1.28 ^{h}	2035 ± 1.31 ^{l}	nd	38.28 ± 0.54 ^{a}	14.30 ± 0.09 ^{e}	61.31 ± 14.73 ^{d}
F6	nd	221 ± 0.94 ^{g}	41.40 ± 0.05 ^{e}	22.29 ± 0.45 ^{h}	6.59 ± 0.02 ^{g}	16.43 ± 0.29 ^{i}	2107 ± 4.45 ^{n}	nd	39.58 ± 0.29 ^{a}	1.60 ± 0.03 ^{a}	106.44 ± 0.17 ^{g}
SL0	2.50 ± 0.01 ^{b}	282 ± 0.33 ^{j}	54.34 ± 0.09 ^{h}	8.31 ± 0.04 ^{d}	3.00 ± 0.01 ^{cd}	14.10 ± 0.15 ^{h}	470 ± 0.86 ^{f}	nd	100.35 ± 0.65 ^{g}	13.90 ± 0.46 ^{e}	2.75 ± 0.16 ^{a}
SL1	2.54 ± 0.01 ^{b}	86 ± 0.09 ^{b}	17.04 ± 0.03 ^{b}	5.58 ± 0.03 ^{ab}	1.64 ± 0.01 ^{ab}	5.23 ± 0.10 ^{cd}	318 ± 2.85 ^{c}	nd	69.39 ± 1.54 ^{d}	5.66 ± 0.09 ^{d}	7.30 ± 0.63 ^{ab}
SL2	3.00 ± 0.07 ^{d}	202 ± 0.05 ^{f}	40.67 ± 0.11 ^{e}	7.27 ± 0.02 ^{cd}	4.30 ± 0.10 ^{e}	5.62 ± 0.03 ^{de}	422 ± 3.55 ^{e}	67.96 ± 0.16 ^{d}	99.81 ± 1.99 ^{g}	5.34 ± 0.07 ^{d}	7.13 ± 0.11 ^{ab}
SL3	2.87 ± 0.02 ^{c}	197.66 ± 1.09 ^{e}	37.35 ± 0.02 ^{d}	4.47 ± 0.00 ^{a}	1.41 ± 0.02 ^{a}	6.77 ± 0.00 ^{f}	203 ± 2.01 ^{a}	41.78 ± 0.27 ^{c}	51.41 ± 0.88 ^{c}	3.89 ± 0.06 ^{c}	8.69 ± 0.38 ^{ab}
SL4	3.60 ± 0.11 ^{e}	274.60 ± 0.14 ^{i}	49.76 ± 0.10 ^{g}	6.52 ± 0.03 ^{bc}	4.65 ± 0.08 ^{ef}	6.46 ± 0.02 ^{ef}	548 ± 3.93 ^{g}	96.14 ± 0.14 ^{e}	120.44 ± 1.08 ^{h}	22.83 ± 0.51 ^{h}	12.86 ± 0.89 ^{b}
SL5	5.53 ± 0.09 ^{g}	70.19 ± 0.16 ^{a}	13.27 ± 0.02 ^{a}	5.24 ± 0.03 ^{a}	3.24 ± 0.17 ^{d}	4.53 ± 0.02 ^{bc}	309 ± 2.04 ^{b}	nd	76.15 ± 0.03 ^{e}	1.90 ± 0.03 ^{a}	6.30 ± 0.62 ^{ab}
SL6	2.10 ± 0.00 ^{a}	109.20 ± 0.16 ^{c}	19.84 ± 0.01 ^{c}	6.82 ± 0.06 ^{c}	2.30 ± 0.00 ^{bc}	7.49 ± 0.16 ^{fg}	360 ± 2.94 ^{d}	nd	83.93 ± 1.51 ^{f}	18.41 ± 0.33 ^{f}	8.06 ± 1.08 ^{ab}

Mean values in the same column with different letters (a–n) are statistically different ($p < 0.05$); results are expressed as mean ± SD; D.M., dry matter; * phenolic diterpene; F0, flower control (cultivated without mulching material); F1, flower cultivated with 100 g/m^2 black mulching material; F2, flower cultivated with 100 g/m^2 green mulching material; F3, flower cultivated with 100 g/m^2 brown mulching material; F4, flower cultivated with 80 g/m^2 black mulching material; F5, flower cultivated with 80 g/m^2 green mulching material; F6, flower cultivated with 80 g/m^2 brown mulching material; SL0, plant stems plus leaves—control (cultivated without mulching material); SL1, plant stems plus leaves cultivated with 100 g/m^2 black mulching material, SL2, plant stems plus leaves cultivated with 100 g/m^2 green mulching material; SL3, plant stems plus leaves cultivated with 100 g/m^2 brown mulching material; SL4, plant stems plus leaves cultivated with 80 g/m^2 black mulching material; SL5, plant stems plus leaves planted with 80 g/m^2 green mulching material; SL6, plant stems plus leaves planted with 80 g/m^2 brown mulching material; nd—not identified.

Table 4. The profile of polyphenols in flowers and a mixture of stems plus leaves of coneflower [mg/100 g D.M.].

Sample Name	Catechin	Epicatechin	Naringin	Rutin	Hesperidin	Myricetin	Luteolin	Kaempferol	Apigenin	Hispidulin	Acacetin	Carnosol *
F0	20.10 ± 4.65 ^{b}	29.10 ± 5.56 ^{ef}	3.79 ± 0.09 ^{a}	119 ± 0.9 ^{g}	53.58 ± 0.11 ^{e}	15.57 ± 0.00 ^{c}	1.89 ± 0.00 ^{c}	2.11 ± 0.00 ^{a}	3.05 ± 0.01 ^{b}	nd	11.17 ± 0.03 ^{i}	11.26 ± 0.19 ^{ab}
F1	24.06 ± 7.16 ^{bc}	32.34 ± 10.41 ^{f}	35.58 ± 1.33 ^{b}	203 ± 0.01 ^{i}	86.24 ± 0.17 ^{h}	28.50 ± 0.07 ^{f}	nd	2.68 ± 0.01 ^{e}	3.98 ± 0.24 ^{e}	nd	4.32 ± 0.00 ^{d}	24.23 ± 24.74 ^{abc}
F2	30.18 ± 0.13 ^{c}	23.92 ± 0.07 ^{bcde}	115 ± 1.83 ^{e}	257 ± 1.14 ^{m}	107. ± 0.21 ^{l}	31.69 ± 0.11 ^{h}	2.44 ± 0.00 ^{f}	2.60 ± 0.02 ^{d}	3.67 ± 0.00 ^{d}	nd	6.06 ± 0.04 ^{f}	31.47 ± 31.71 ^{abc}
F3	25.34 ± 0.10 ^{bc}	28.88 ± 0.00 ^{def}	154 ± 4.60 ^{g}	177 ± 0.64 ^{h}	103 ± 1.83 ^{k}	22.57 ± 0.03 ^{e}	nd	3.41 ± 0.03 ^{f}	5.68 ± 0.01 ^{f}	nd	6.61 ± 0.03 ^{g}	8.64 ± 0.16 ^{a}
F4	6.23 ± 0.26 ^{a}	20.71 ± 0.48 ^{abcd}	197 ± 2.74 ^{h}	217 ± 0.89 ^{l}	98.16 ± 0.70 ^{j}	38.77 ± 1.68 ^{i}	1.97 ± 0.02 ^{d}	2.23 ± 0.00 ^{b}	3.15 ± 0.00 ^{bc}	nd	4.72 ± 0.00 ^{e}	8.37 ± 0.17 ^{a}
F5	3.53 ± 4.99 ^{a}	23.88 ± 5.31 ^{bcde}	14.83 ± 19.19 ^{a}	215 ± 0.54 ^{k}	87.87 ± 0.06 ^{i}	30.09 ± 0.12 ^{g}	2.02 ± 0.05 ^{d}	2.62 ± 0.00 ^{d}	1.45 ± 0.03 ^{a}	nd	10.55 ± 0.05 ^{h}	7.22 ± 0.14 ^{a}
F6	3.84 ± 0.39 ^{a}	14.36 ± 0.19 ^{a}	3.05 ± 0.08 ^{a}	210 ± 1.11 ^{j}	76.98 ± 0.03 ^{g}	21.37 ± 0.02 ^{d}	1.69 ± 0.02 ^{a}	2.44 ± 0.02 ^{c}	3.33 ± 0.03 ^{c}	nd	14.68 ± 0.05 ^{j}	7.96 ± 0.03 ^{a}
SL0	nd	27.01 ± 0.18 ^{cdef}	134 ± 0.95 ^{f}	94.36 ± 0.49 ^{f}	66.56 ± 0.64 ^{f}	2.93 ± 0.06 ^{b}	1.79 ± 0.07 ^{b}	nd	nd	1.67 ± 0.00 ^{a}	nd	26.07 ± 23.96 ^{abc}
SL1	7.78 ± 0.49 ^{a}	19.91 ± 0.25 ^{abc}	56 ± 0.43 ^{c}	42.95 ± 0.06 ^{a}	26.21 ± 0.12 ^{a}	nd	nd	nd	nd	1.71 ± 0.00 ^{a}	nd	43.38 ± 0.13 ^{c}
SL2	10.11 ± 0.70 ^{a}	13.94 ± 0.03 ^{a}	129 ± 1.70 ^{f}	58.78 ± 0.36 ^{c}	42.58 ± 0.46 ^{c}	1.63 ± 0.02 ^{a}	2.16 ± 0.02 ^{e}	nd	nd	nd	nd	46.30 ± 0.00 ^{c}
SL3	48.39 ± 5.85 ^{d}	15.77 ± 0.39 ^{ab}	89 ± 0.33 ^{d}	54.40 ± 0.02 ^{b}	32.28 ± 0.56 ^{b}	1.86 ± 0.02 ^{ab}	nd	nd	nd	1.93 ± 0.00 ^{b}	1.64 ± 0.02 ^{a}	43.27 ± 0.43 ^{c}
SL4	nd	19.60 ± 0.06 ^{abc}	187 ± 15.97 ^{h}	83.73 ± 1.12 ^{e}	53.98 ± 0.11 ^{e}	1.98 ± 0.02 ^{ab}	nd	nd	nd	2.49 ± 0.00 ^{c}	1.94 ± 0.02 ^{b}	45.79 ± 0.44 ^{c}
SL5	nd	19.47 ± 0.33 ^{abc}	63. ± 0.21 ^{c}	84.13 ± 0.21 ^{e}	47.43 ± 0.33 ^{d}	2.21 ± 0.05 ^{ab}	nd	2.45 ± 0.02 ^{c}	nd	1.94 ± 0.05 ^{b}	nd	45.68 ± 0.28 ^{c}
SL6	nd	31.57 ± 0.65 ^{ef}	89. ± 0.08 ^{d}	78.94 ± 0.25 ^{d}	48.44 ± 0.33 ^{d}	2.11 ± 0.13 ^{ab}	nd	nd	1.56 ± 0.00 ^{a}	nd	2.33 ± 0.01 ^{c}	39.33 ± 0.25 ^{bc}

Mean values in the same column with different letters (a–m) are statistically different ($p < 0.05$); results are expressed as mean ± SD; D.M., dry matter; * phenolic diterpene; F0, flower control (cultivated without mulching material); F1, flower cultivated with 100 g/m^2 black mulching material; F2, flower cultivated with 100 g/m^2 green mulching material; F3, flower cultivated with 100 g/m^2 brown mulching material; F4, flower cultivated with 80 g/m^2 black mulching material; F5, flower cultivated with 80 g/m^2 green mulching material; F6, flower cultivated with 80 g/m^2 brown mulching material; SL0, plant stems plus leaves—control (cultivated with no mulching material); SL1, plant stems plus brown mulching material; SL4, plant stems plus leaves cultivated with 80 g/m^2 black mulching material; SL5, plant stems plus leaves planted with 80 g/m^2 green mulching material; SL6, plant stems plus leaves planted with 80 g/m^{2}brown mulching material; nd—not identified.

3. Discussion

The flowers of *E. purpurea* generally had higher protein content than the plant stems with leaves (Table 1). It can be also be suggested that mulch material used for cultivation had an effect on the content of protein and ash, especially in flowers. The use of black and green material during the cultivation of the plants resulted in a higher content of protein in the flowers and the mixture of the leaves and plant stems as compared to their controls. These changes can be explained by the protection of the soil and roots from dryness as well as changes in the temperature. These factors could synthesise protein. It was reported that the use of mulch material in the cultivation of various plants protects the evaporation of water from the soil; this increases water use efficiency, increases the soil temperature and nitrogen balance, as well as the yield of plants [15,18–21]. To the best of the authors' knowledge of the currently available literature, there are no data concerning the content of crude protein, fat and ash in *E. purpurea's* morphological parts cultivated in the traditional way or with mulching material. The composition of various types of extract from roots was evaluated for the bioactive substance content including phenolic acids, alkamides, polysaccharides and others [5,22,23].

In some studies, the proximate analysis was reported in various flowers that can be used as the herbs or ingredients of beverages. Rachkeeree et al. [24] reported that, in edible flowers of *Alpinia galanga* and *Hedychium forrestii*, the protein content was 33 g/100 g D.M. and 12.88 g/100 g D.M.. In the same flowers, the fat content was 11.9 g/100 g D.M. and 20.5 g/100 g D.M.. Li et al. [25] reported that the content of soluble protein and carbohydrates in peony flowers depended on the stage of the blooming flower. In chive flowers, the content of crude protein and crude fat was 18.5 g/100 g D.M. and 3.45 g/100 g D.M., respectively [26]. In our study, the variation observed in protein and ash content could be explained by many factors, involving the environment, abiotic stress, the genetic heritance maturity of the plants and the type of mulching material applied during the cultivation of the plants.

An important finding of this study is that the total polyphenol content and antioxidant activity of *E. Purpurea* flowers and plant stems plus leaves were affected by the mulch material used. It can be suggested that the mulch material could reduce soil aeration. Additionally, no mechanical weeding of the crop was performed, which also influenced the contact of air with soil. It could increase free radical production and increase the synthesis of phenolic compounds. According to data in the literature, polyphenols are involved in the stress protection of plants under harmful environmental conditions—not only drought, flooding and high or low temperature, but also ultraviolet radiation, salinity and heavy metal pollution. Their bioactivity and role in stress defence are generally attributed to their antioxidant activity. In addition, the different colours of the mulch determine changes in the intensity and absorption of radiation. Polyphenols also protect the plant against biotic stress. They are characterised by insecticidal/insect repellent activity and are a potential pest control strategy [27–29]. In this study, the content of total polyphenols in flower samples ranged from 9728.05 mg/100 g D.M. to 14,222 mg/100 g D.M. and in the combination of plant stems plus leaves ranged from 3256.16 mg/100g D.M. to 6395.19 mg/100 g D.M. The results of our study are in the range of the data published by Tsai et al. [30]. These authors measured the total phenolic content in flowers of *E. purpurea* in the range 9707 mg/100 g–39,982 mg/100 g. Chen et al. [31] reported that the level of phenolic compounds was 18,208 mg/100 g. Lower levels of total phenolic compounds than in our study were measured by Pellati et al. [32] in the roots of *E. purpurea* from Italy (2332 mg/100 g D.M.). These differences can be explained by the various morphological parts of the plant, climate condition and drying method. These authors used 40 °C for drying in the laboratory oven. In our study, we were using the freeze-dried method. In another study, the content of phenolic compound content in roots was measured in the range of 22.79 ± 0.37 mg/1 g [4,33].

Recent studies have shown that there is no universal method to evaluate antioxidant activity quantitatively and accurately [34]; therefore, the antioxidant activity of plants should be evaluated using several methods. Previous studies by Schlesier et al. [35] showed that when analysing antioxidant activity, it is preferable to use at least two methods. In this study, however, the analysis of the antioxidant activity of flowers and the combination of plant stems plus leaves of *E. purpurea* was performed using three methods: DPPH, ABTS and FRAP.

The antioxidant activity of flower samples, measured by the DPPH method in the present study for control samples, was 690.27 ± 40.75 µmol Trolox/1 g D.M. and that of the combination of plant stems with leaves was 351.93 ± 12.60 µmol Trolox/1 g D.M. It was reported that the flowers of *E. purpurea* have strong antioxidant activity [22,30,31].

Lower antioxidant activity, measured by the DPPH method, than in our study was reported in the leaves of *E. purpurea* (75.0 µmol Trolox/100 g D.M.) [36]. Hu and Kitts [22] reported that 0.8 mg/mL of the extract of *E. purpurea* scavenged in 20% DPPH radicals. Wojdyło et al. [36] reported lower antioxidant capacity, determined by the FRAP test, in the leaves of *E purpurea* (94.6 and 191.0 µM Trolox/100 g D.M., respectively), compared to our results.

The flower samples showed higher values of antioxidant activity compared to the combination of plant stems with leaves. Our data are different from the results reported by Ramazan et al. [1].

We have found that the major phenolic compounds in flowers and the mixture of plant stems with leaves were p-coumaric acid, chlorogenic acid, rutin, 4-hydroxybenzoic acid and hesperidin. Our results are different from data published by Tsali et al. [30], who reported that in flowers of *E. purpurea*, caffeic acid, chicoric acid and their derivatives were measured in higher amounts. A study carried out by Chen et al. [31] and Ramazan et al. [1] showed that cichoric acid was in the highest concentration in the leaves and flowers of *E. purpurea* extracted with methanol. Pellati et al. [33] reported that in roots, caftaric acid, chlorogenic acid and caffeic acid are the major polyphenolic compounds. In our study, p-coumaric acid was found in higher amounts in flowers and leaves. We have found that rutin was the major flavone in the evaluated plants (Table 4). What is more, the rutin content was higher in all the flower samples in comparison with the plant stem plus leaf samples. It has also been reported to be one of the polyphenolic compounds present in high concentration in medicinal plants and is responsible for the antioxidant capacity of plant extracts [37].

In this study, rutin concentration was the third highest concentration found in all the samples that were analysed, compared to other compounds. In general, flower samples showed more concentrations of almost all the polyphenolic compounds except for gallic acid, ferulic acid and sinapinic acid that showed more concentration in the plant stems plus leaves samples.

Based on the results of the antioxidant activity of *E. purpurea*, it is suggested that this plant can be used as a functional food (in the form of beverages or food additives). The high content of polyphenolic compounds means that *E. purpurea* extract may be helpful in the prevention and treatment of chronic non-communicable diseases such as obesity, atherosclerosis and other cardiovascular diseases, diabetes and cancer [38]. In addition, it can be suggested to use the whole extract due to the rich polyphenolic composition and the pleiotropic effects of polyphenols [39]. A single compound may not have as strong an effect as a mixture of them.

It can be suggested to producers to separate the flowers from the plan stems with leaves when harvesting the herb. The introduction of mulching material can also improve the yield of crops. Extracts obtained especially from the flowers can be used as a good source of antioxidants in food, used in the pharmaceutical or cosmetic industry. The dried flowers or/and plant stems plus leaves can also be used to prepare tea mixtures.

4. Material and Methods

4.1. Plant Material

Seedlings of *E. purpurea* were grown at the Experimental Station of the Agricultural University in Krakow, Poland, located in Mydlniki Kraków, (Poland). Synthetic mulching material was used to protect growing seedlings and avoid the use of herbicides. Synthetic mulching material was purchased in a local market. During cultivation, plants of *E. purpurea* and various types of mulch materials different colours (black, green and brown; thickness 100 g/m^2 and 80 g/m^2, respectively) were used (Table 5; Figure 1). The research materials were as follows. Samples of flower and combination of plant stems plus leaves were harvested two years after seedlings' planting according to Farmacopea recommendations [40]. Samples were collected at the end of July during full blooming (Figure 2). In the laboratory, the flowers were separated from the plant stems plus leaves. The plants were washed to remove all impurities, e.g., soil and insects. Next, the samples were frozen and then lyophilised (Alpha 1–4 LS Cplus, Martin Christ Gefriertrocknungsanlagen GmbH, Osterodeam, Germany) for 24 h (capacitor temperature: −55 °C, vacuum: 1000 mbar). Lyophilised samples were ground and stored until the analyses.

Figure 1. Mulching material. (**A**) Field covered with various types of mulching material prepared for cultivation. (**B**) *E purpureae* seedlings. (**C**) Example of black mulching material. (**D**) Example of green mulching material. (**E**) Example of brown mulching material (own source).

Table 5. *E. purpurea* treatment codes.

Code	Treatment
F0	Flower/control
F1	Flower/black mulching; density 100 g/m^2
F2	Flower/green mulching; density 100 g/m^2
F3	Flower/brown mulching; density 100 g/m^2
F4	Flower/black mulching; density 80 g/m^2
F5	Flower/green mulching; density 80 g/m^2
F6	Flower brown mulching; density 80 g/m^2
SL0	Stem plus leaves control
SL1	Stem plus leaves/black mulching; density 100 g/m^2
SL2	Stem plus leaves/green mulching; density 100 g/m^2
SL3	Stem plus leaves/brown mulching; density 100 g/m^2
SL4	Stem plus leaves/black mulching; density 80 g/m^2
SL5	Stem plus leaves/green mulching; density 80 g/m^2
SL6	Stem plus leaves/brown mulching; density 80 g/m^2

Figure 2. Mature *E. purpurea*. (**A**) Mature flowers. (**B**) Example of plants with the mulching material (own source).

4.2. Concentration of Protein, Ash and Crude Fat

In freeze-dried samples, the concentration of nitrogen was measured in accordance to the Kjeldahl method (AOAC no. 978.04) [41] as was previously described [42]. The percentage of total nitrogen was calculated following the formula $N = (V \times M \times 14.007 \times 100)/m$, where N—total nitrogen content in g/100 g D.M. of the test research material [g/100 g D.M.]; V—volume of HCl used for titration of the sample [cm^3]; M—molar concentration of HCl [mol/dm^3]; 14.007—amount of nitrogen, which corresponds to 1 cm^3 of HCl with a concentration of 1 mol/dm^3; m—sample mass [g]. For the calculation of the protein, the conversion factor 6.25 was used. Crude fat content was determined with the Soxlet method in Soxtec Avanti's 2050 Auto Extraction Unit (Tecator Foss, Hillerød, Sweden). For the extraction of crude fat petroleum, ether (POCh Gliwice, Poland) was used. Ash was determined in an electric muffle furnace (525 °C) (SNOL82.110, Utena, Lithuania).

4.3. Methanolic Extract Preparation

About 0.5 g of the ground plant material was weighed and mixed with an 80 cm^3 solution of 70% methanol acidified with 0.1% of formic acid (*v*/*v*; POCh Giwice, Poland) and then placed in a shaker at room temperature for 2 h and protected from light. The mixture was centrifuged for 15 min (402× *g*, room temperature), and the supernatant was moved into a plastic container and stored at −22 °C [42].

4.4. Total Polyphenols Content and Antioxidant Activity

The content of total polyphenols in extracts of samples was determined using Folin–Ciocalteu reagent (Sigma-Aldrich, Saint Louis, MO, USA) [43]. The results are presented as the chlorogenic acid equivalent (CGA) in mg per 100 g of dry matter (D.M.) as was previously reported [43].

The antioxidant activity of methanolic extracts of coneflowers and leaves plus stems was determined as follows: ABTS (2.2′-Azino-bis (3-ethylbenzthiazoline-6-sulfonic acid)), FRAP (ferric-reducing antioxidant power) and DPPH (2,2-diphenyl-1-picrylhydrazyl). The determination of antioxidant activity using $ABTS^{•+}$ free radicals was carried out according to the method described by Re et al. [44]. The stock solution was prepared by dissolving ABTS in water to a concentration of 7 mmol. The ABTS radical cation ($ABTS^{•+}$) was prepared by reacting the ABTS solution with 2.45 mM potassium persulfate (final concentration) and allowing the mixture to stand for 16 h in the dark at room temperature before use. An $ABTS^{•+}$ working solution was prepared by diluting the stock solution with 70% methanol to the absorbance of 0.740–0.750 at 734 nm. Next, 30 µL of the flower extract or 80 µL of the plant stem plus leaves extract, 970 µL or 920 µL of 70% methanol and then 2 mL of the $ABTS^{•+}$ solution were added to a test tube. The mixture was stored in the dark at 30 °C for 6 min. The absorbance of the samples was determined at 734 nm (UV-1800, RayLeigh, Beijing Beifen Ruili Analytical Instrument Co., Ltd., Beijing, China). The FRAP assay was performed according to the Benzie and Strain [45] method. The working solution was prepared by mixing 100 mL of acetate buffer (pH 3.6), 10 mL of TPTZ solution (10 mmol/L TPTZ in 40 mmol/L HCl) and 10 mL of 20 mmol/L $FeCl_3{\cdot}6H_2O$ solution. Then, 30 µL of the flower extract or 80 µL of the plant stem plus leaf extract, 970 µL or 920 µL of 70% methanol and 3 mL of working solution of the FRAP reagent were transferred into a test tube. The mixture was stored in a dark at room temperature for 10 min. The absorbance of the samples was assessed at 593 nm (UV-1800, RayLeigh, Beijing Beifen Ruili Analytical Instrument Co., Ltd., Beijing, China). The determination of antioxidant activity with the DPPH method was conducted in accordance to Miliauskas et al. [46]. The stock solution was prepared by dissolving 5 mg of DPPH in 100 mL of methanol. The $DPPH^{•}$ working solution was obtained by diluting the stock solution with methanol to the absorbance of 0.900–1.000 at 515 nm. Next, 50 µL of the flower or the plant stem plus leaf extract was added to 1450 µL of methanol and 3 mL of FRAP working solution; all were added into a test tube. The mixture was stored in the dark at room temperature for 10 min. All results were compared to the concentration–response curve of the standard Trolox solution and presented as µmol of Trolox equivalent per g of D.M. (TEAC) of samples. For the ABTS and DPPH assays, the range of Trolox concentrations was 1.95–62.5 µmol Trolox/L; however, for the FRAP assay, it was 31.25–250 µmol Trolox/L.

4.5. Polyphenols Profile

For HPLC polyphenolic compound analysis, acidified methanolic extracts were used to determine polyphenolic compounds with the HPLC method using the Prominence-i LC 2030C D3 Plus system (Shimadzu, Kyoto, Japan) along with a DAD detector and a Luna Omega 5 µm Polar C18, 100 A, 250 × 10 mm Phenomenex column (Torrance, CA, USA). The separation of phenolic compounds was performed at 40 °C. The mobile phase was a mixture of eluents: A, 0.1% formic acid in water (*v*/*v*), and B, 0.1% formic acid in methanol (*v*/*v*). The mobile phase flow rate was 1.2 mL/min. The analysis was carried out with the following gradient conditions: from 20% to 40% B in 10 min, 40% B for 10 min, from 40% to 50% B in 10 min, from 50% to 60% B in 5 min, 60% B for 5 min, from 60 to 70% B in 5 min, from 70% to 90% B in 5 min, 90% B for 5 min, from 90% to 20% B (the initial condition) in 1 min, and 20% B for 4 min. This resulted in a total run time of 60 min. The injection volume of tested samples was 20 µL. A stock standard solution (100 mg/L) of each polyphenolic compound was prepared in 70% methanol acidified methanol (0.1% formic acid (POCH, Gliwice, Poland). The identification of compounds was made on the basis of retention times. The calibration curves of the polyphenol standards were made by the dilution of stock standard solutions in 0.1% formic acid in 70% methanol (*v*/*v*). The range of the calibration curve was 0.25–4.0 mg/L. All the solutions were filtered through a 0.22 μm filter as was previously reported [47].

The following polyphenolic compounds were determined based on standards: gallic acid, catechin, chlorogenic acid, 4-hydroxybenzoic acid, epicatechin, caffeic acid, vanillic acid, syringic acid, p-coumaric acid, ferulic acid, sinapinic acid, naringin, rutin, hesperidin, rosmarinic acid, myricetin, quercetin, luteolin, kaempferol, apigenin, isorhamnetin, hispidulin, acacetin, carnosol and carnosic acid (Sigma, Saint Louis, MO, USA). Measurements were based on LabSolution ver. 5.93 from Shimadzu Corporation (Kyoto, Japan). The levels of quercetin, luteolin, apigenin and myricetin were determined only in their free aglycone form.

4.6. Statistical Analysis

The analyses were performed in two or three parallel replications. Multiple-way analysis of variance was also carried out using the program Statistica version 13.1, Dell Inc., Tulsa, OK, USA, 2016. Significant differences were assessed using Duncan's test ($p \leq 0.05$).

5. Conclusions

Worldwide, edible flowers are gaining increased interest among consumers. The results of this study indicate that the flower and plant stem plus leaves of *E. purpurea* are rich sources of polyphenols and exhibit high antioxidant activity, making it a potential functional food. The extract of *E. purpurea*, especially the flower extract, can be helpful in the prevention and treatment of chronic non-communicable diseases. Additionally, under favourable metrological conditions, it is possible to harvest them twice in one year. The introduction of mulching material can also improve the yield of crops. As a source of phytochemicals, it opens up new possibilities for the creation of new functional foods, such as beverages or food additives (lyophilised plant or powdered extract). On the other hand, there is a need for further research on *E. purpurea*, in particular, in vivo studies to confirm beneficial effects in the organism, determine the dose and check the safety of use and possible side effects.

Author Contributions: Conceptualisation, U.S. and A.K.; methodology, U.S., A.K., K.D. and J.S.; formal analysis, U.S. and A.K.; investigation C.A.O., K.D., J.S. and U.S.; resources, U.S. and A.K.; data curation C.A.O., K.D., J.S. and U.S.; Writing—original draft preparation, C.A.O. and K.D.; Writing—review and editing, K.D., A.K. and U.S.; funding acquisition, U.S. and A.K. All authors have read and agreed to the published version of the manuscript.

Funding: The study was financed by the Ministry of Science and Higher Education of the Republic of Poland for University of Agriculture in Kraków, Poland.

Institutional Review Board Statement: Not applicable.

Informed Consent Statement: Not applicable.

Data Availability Statement: The data presented in this study are available on request from the corresponding author.

Conflicts of Interest: The authors declare no conflicts of interest.

References

1. Ramazan, E.; Isa, T.; Musa, U.; Ibrahim, D.; Fatih, G.; Mahfuz, E.; Omer, K. Chemical constituents, quantitative analysis and antioxidant activities of *Echinacea purpurea* (L.) Moench and *Echinacea pallida* (nutt.) Nutt. *J. Food Biochem.* **2015**, *39*, 622–630. [CrossRef]

2. Dosoky, N.S.; Kirpotina, L.N.; Schepetkin, I.A.; Khlebnikov, A.I.; Lisonbee, B.L.; Black, J.L.; Woolf, H.; Thurgood, T.L.; Graf, B.L.; Satyal, P.; et al. Volatile Composition, Antimicrobial Activity, and In Vitro Innate Immunomodulatory Activity of *Echinacea purpurea* (L.) Moench Essential Oils. *Molecules* **2023**, *28*, 7330. [CrossRef]
3. Osińska, E.; Pióro-Jabrucka, E. *Uprawa i Przetwórstwo Roślin Zielarskich (Cultivation and Processing of Herbal Plants)*; Centrum Doradztwa Rolniczego w Brwinowie: Brwinów, Poland, 2022; ISBN 978-83-88082-64-1. (In Polish)
4. Newerli-Guz, J. Uprawa roślin zielarskich w Polsce. *Rocz. Nauk. Stowarzyszenia Ekon. Rol. I Agrobiznesu* **2016**, *18*, 268–274.
5. Barnes, J.; Anderson, L.A.; Gibbons, S.; Phillipson, J.D. Echinacea species (*Echinacea angustifolia* (DC.) Hell., *Echinacea pallida* (Nutt.) Nutt., *Echinacea purpurea* (L.) Moench): A review of their chemistry, pharmacology and clinical properties. *J. Pharm. Pharmacol.* **2005**, *57*, 929–954. [CrossRef]
6. Bergeron, C.; Livesey, J.F.; Awang, D.V.C.; Arnason, J.T.; Rana, J.; Baum, B.R.; Letchamo, W.A. Quantitative HPLC Method for the Quality Assurance of *Echinacea* Products on the North American Market. *Phytochem. Anal.* **2000**, *11*, 207–215. [CrossRef]
7. Sharma, S.M.; Anderson, M.; Schoop, S.R.; Hudson, J.B. Bactericidal and anti-inflammatory properties of a standardized *Echinacea* extract (Echinaforce): Dual actions against respiratory bacteria. *Phytomedicine* **2010**, *17*, 563–568. [CrossRef] [PubMed]
8. Manayi, A.; Vazirian, M.; Saeidnia, S. *Echinacea purpurea*: Pharmacology, phytochemistry and analysis methods. *Pharmacogn. Rev.* **2015**, *9*, 63–72. [CrossRef]
9. Stojicevic, S.; Stanisavljevic, A.; Velickovic, D.; Veljkovic, V.; Lazic, M. Antioxidant and antimicrobial activities of *Echinacea* (*Echinacea purpurea* L.). extracts obtained by classical and ulterasound extraction. *Chin. J. Chem. Eng.* **2009**, *17*, 478–483.
10. Binns, S.E.; Purgina, B.; Bergeron, C.; Smith, M.L.; Ball, L.; Baum, B. Light-mediated antifungal activity of *Echinacea* extracts. *Planta Med.* **2000**, *66*, 241–244. [CrossRef]
11. Hudson, J.B. Applications of the phytomedicine *Echinacea purpurea* (Purple Coneflower) in infectious diseases. *J. Biomed. Biotechnol.* **2012**, *2012*, 769896. [CrossRef]
12. Gurley, B.J.; Gardner, S.F.; Hubbard, M.A.; Williams, D.K.; Gentry, W.B.; Carrier, J.; Khan, I.A.; Edwards, D.J.; Shah, A. In vivo assessment of botanical supplementation on human cytochrome P450 phenotypes: *Citrus aurantium*, *Echinacea purpurea*, milk thistle, and saw palmetto. *Clin. Pharmacol. Ther.* **2004**, *76*, 428–440. [CrossRef] [PubMed]
13. Kołodziej, B. *Uprawa Ziół*; PWRiL: Warszawa, Poland, 2010; ISBN 978-83-09-99021-3. (In Polish)
14. Zheng, Y.; Dixon, M.; Saxena, P.K. Growing environment and nutrient availability affect the content of some phenolic compounds in *Echinacea purpurea* and *Echinacea angustifolia*. *Planta Med.* **2006**, *72*, 1407–1414. [CrossRef] [PubMed]
15. Kader, M.A.; Senge, M.; Mojid, M.A.; Ito, K. Recent advances in mulching materials and methods for modifying soil environment. *Soil Tillage Res.* **2017**, *168*, 155–166. [CrossRef]
16. Lamont, W.J. Plastics: Modifying the Microclimate for the Production of Vegetable Crops. *Hort. Technol.* **2005**, *15*, 477–481. [CrossRef]
17. Ibarra-Jiménez, L.; Hugolira-Saldivar, R.; Valdez-Aguilar, L.A.; Lozano-Del Río, J. Colored plastic mulches affect soil temperature and tuber production of potato. *Acta Agric. Scand.* **2011**, *61*, 365–371. [CrossRef]
18. Yin, T.; Yao, Z.; Yan, C.; Liu, Q.; Ding, X.; He, W. Maize yield reduction is more strongly related to soil moisture fluctuation than soil temperature change under biodegradable film vs plastic film mulching in a semi-arid region of northern China. *Agric. Water Manag.* **2023**, *287*, 108351. [CrossRef]
19. Xiukang, W.; Zhanbin, L.; Yingying, X. Effects of mulching and nitrogen on soil temperature water content, nitrate-N content and maize yield in the Loess Plateau of China. *Agric. Water Manag.* **2015**, *161*, 53–64. [CrossRef]
20. Almeida, W.F.D.; Lima, L.A.; Pereira, G.M. Drip pulses and soil mulching effect on American crisp head lettuce yield. *Eng. Agrícola* **2015**, *35*, 1009–1018. [CrossRef]
21. Filipovic, V.; Romic, D.; Romic, M.; Borošic, J.; Filipovic, L.; Mallmann, F.J.K.; Robinson, D. A Plastic mulch and nitrogen fertigation in growing vegetables modify soil temperature, water and nitrate dynamics: Experimental results and a modeling study. *Agric. Water Manag.* **2016**, *176*, 100–110. [CrossRef]
22. Hu, C.; Kitts, D.D. Studies on the antioxidant activity of *Echinacea* root extract. *J. Agric. Food Chem.* **2000**, *48*, 1466–1472. [CrossRef]
23. Sharifi-Rad, M.; Mnayer, D.; Morais-Braga, M.F.B.; Carneiro, J.N.P.; Bezerra, C.F.; Coutinho, H.D.M.; Salehi, B.; Martorell, M.; Del Mar Contreras, M.; Soltani-Nejad, A.; et al. Echinacea plants as antioxidant and antibacterial agents: From traditional medicine to biotechnological applications. *Phytother. Res.* **2018**, *32*, 1653–1663. [CrossRef] [PubMed]
24. Rachkeeree, A.; Kantadoung, K.; Suksathan, R.; Puangpradab, R.; Page, P.A.; Sommano, S.R. Nutritional Compositions and Phytochemical Properties of the Edible Flowers from Selected Zingiberaceae Found in Thailand. *Front Nutr.* **2018**, *5*, 3. [CrossRef] [PubMed]
25. Li, W.; Song, X.; Hua, Y.; Tao, J.; Zhou, C. Effects of Different Harvest Times on Nutritional Component of Herbaceous Peony Flower Petals. *J. Chem.* **2020**, *2020*, 4942805. [CrossRef]
26. Grzeszczuk, M.; Wesołowska, A.; Jadczak, D.; Jakubowska, B. Nutritional value of chive edible flowers. *Acta Sci. Pol. Hortorum Cultus* **2011**, *10*, 85–94.
27. Singh, S.; Kaur, I.; Kariyat, R. The Multifunctional Roles of Polyphenols in Plant-Herbivore Interactions. *Int. J. Mol. Sci.* **2021**, *22*, 1442. [CrossRef] [PubMed]
28. Zagoskina, N.V.; Zubova, M.Y.; Nechaeva, T.L.; Kazantseva, V.V.; Goncharuk, E.A.; Katanskaya, V.M.; Baranova, E.N.; Aksenova, M.A. Polyphenols in Plants: Structure, Biosynthesis, Abiotic Stress Regulation, and Practical Applications (Review). *Int. J. Mol. Sci.* **2023**, *24*, 13874. [CrossRef] [PubMed]

29. El-Beltagi, H.S.; Basit, A.; Mohamed, H.I.; Ali, I.; Ullah, S.; Kamel, E.A.R.; Shalaby, T.A.; Ramadan, K.M.A.; Alkhateeb, A.A.; Ghazzawy, H.S. Mulching as a Sustainable Water and Soil Saving Practice in Agriculture: A Review. *Agronomy* **2022**, *12*, 1881. [CrossRef]
30. Tsai, Y.L.; Chiou, S.Y.; Chan, K.C.; Sung, J.M.; Lin, S.D. Caffeic acid derivatives, total phenols, antioxidant and antimutagenic activities of *Echinacea purpurea* flower extracts. *LWT* **2012**, *46*, 169–176. [CrossRef]
31. Chen, Y.L.; Sung, J.M.; Lin, S.D. Effect of Extraction Methods on the Active Compounds and Antioxidant Properties of Ethanolic Extracts of *Echinacea purpurea* Flower. *Am. J. Plant Sci.* **2015**, *6*, 201–212. [CrossRef]
32. Pellati, F.; Benvenuti, S.; Magro, L.; Melegari, M.; Soragni, F. Analysis of phenolic compounds and radical scavenging activity of *Echinacea* spp. *J. Pharm. Biomed. Anal.* **2004**, *35*, 289–301. [CrossRef]
33. Pellati, F.; Benvenuti, S.; Melegari, M.; Lasseigne, T. Variability in the Composition of Antioxidant Compounds in *Echinacea* Species by HPLC. *Phytochem. Anal.* **2005**, *16*, 77–85. [CrossRef] [PubMed]
34. Prior, R.L.; Wu, X.; Schaich, K. Standardized Methods for the Determination of Antioxidant Capacity and Phenolics in Foods and Dietary Supplements. *J. Agric. Food Chem.* **2005**, *53*, 4290–4302. [CrossRef] [PubMed]
35. Schlesier, K.; Harwat, M.; Böhm, V.; Bitsch, R. Assessment of Antioxidant Activity by Using Different In Vitro Methods. *Free Radic. Res.* **2002**, *36*, 177–187. [CrossRef] [PubMed]
36. Wojdyło, A.; Oszmiański, J.; Czemerys, R. Antioxidant activity and phenolic compounds in 32 selected herbs. *Food Chem.* **2007**, *105*, 940–949. [CrossRef]
37. Skrzypczak-Pietraszek, E.; Piska, K.; Pietraszek, J. Enhanced production of the pharmaceutically important polyphenolic compounds in *Vitex agnus castus* L. shoot cultures by precursor feeding strategy. *Eng. Life Sci.* **2018**, *18*, 287–297. [CrossRef] [PubMed]
38. Costa, C.; Tsatsakis, A.; Mamoulakis, C.; Teodoro, M.; Briguglio, G.; Caruso, E.; Fenga, C. Current evidence on the effect of dietary polyphenols intake on chronic diseases. *Food Chem. Toxicol.* **2017**, *110*, 286–299. [CrossRef] [PubMed]
39. Kishimoto, Y.; Tani, M.; Kondo, K. Pleiotropic preventive effects of dietary polyphenols in cardiovascular diseases. *Eur. J. Clin. Nutr.* **2013**, *67*, 532–535. [CrossRef] [PubMed]
40. Germosén Robineau, L.; García González, M.; Morón, F.; Costaguta, M.; Delens, M.; Olmedo, D.; Méndez, M.; Boulogne, I.; García, R.; Durán, R.; et al. *Farmacopea Vegetal Caribeña*; Universidad de Cartagena: Cartagena, Colombia, 2017; ISBN 978-958-5439-06-1.
41. AOAC. *Official Methods of Analysis of the AOAC*, 21st ed.; AOAC: Gaithersburg, MD, USA, 2019.
42. Skoczylas, J.; Jędrszczyk, E.; Dziadek, K.; Dacewicz, E.; Kopeć, A. Basic Chemical Composition, Antioxidant Activity and Selected Polyphenolic Compounds Profile in Garlic Leaves and Bulbs Collected at Various Stages of Development. *Molecules* **2023**, *28*, 6653. [CrossRef]
43. Swain, P.; Hillis, W.E. The phenolic constituents of *Prunus domestica* (L.). The quantity of analisys of phenolic constituents. *J. Sci. Food Agric.* **1959**, *10*, 63–68. [CrossRef]
44. Re, R.; Pellegrini, N.; Proteggente, A.; Pannala, A.; Yang, M.; Rice-Evans, C. Antioxidant activity applying an improved ABTS radical cation decolorization assay. *Free Radic. Biol. Med.* **1999**, *26*, 1231–1237. [CrossRef]
45. Benzie, I.F.F.; Strain, J.J. The Ferric Reducing Ability of Plasma (FRAP) as a Measure of "Antioxidant Power": The FRAP Assay. *Anal. Biochem.* **1996**, *239*, 70–76. [CrossRef]
46. Miliauskas, G.; Venskutonis, P.R.; Van Beek, T.A. Screening of radical scavenging activity of some medicinal and aromatic plant extracts. *Food Chem.* **2004**, *85*, 231–237. [CrossRef]
47. Dziadek, K.; Kopeć, A.; Dziadek, M.; Sadowska, U.; Cholewa-Kowalska, K. The Changes in Bioactive Compounds and Antioxidant Activity of Chia (*Salvia hispanica* L.) Herb under Storage and Different Drying Conditions: A Comparison with Other Species of Sage. *Molecules* **2022**, *27*, 1569. [CrossRef]

molecules

MDPI

Article

Comparative Antibacterial and Efflux Pump Inhibitory Activity of Isolated Nerolidol, Farnesol, and α-Bisabolol Sesquiterpenes and Their Liposomal Nanoformulations

Jorge Ederson Gonçalves Santana [1], Cícera Datiane de Morais Oliveira-Tintino [2], Gabriel Gonçalves Alencar [2], Gustavo Miguel Siqueira [2], Daniel Sampaio Alves [2], Talysson Felismino Moura [2], Saulo Relison Tintino [2], Irwin Rose Alencar de Menezes [2], João Pedro Viana Rodrigues [3], Vanessa Barbosa Pinheiro Gonçalves [3], Roberto Nicolete [3], Talha Bin Emran [4,5,6], Clara Mariana Gonçalves Lima [7,*], Sheikh F. Ahmad [8], Henrique Douglas Melo Coutinho [2,*] and Teresinha Gonçalves da Silva [1]

1 Departamento de Antibióticos, Universidade Federal de Pernambuco (UFPE), Recife 50670-901, Brazil; edersantana22@hotmail.com (J.E.G.S.); teresinha.goncalves@ufpe.br (T.G.d.S.)
2 Departament of Biological Chemistry, Universidade Regional do Cariri (URCA), Crato 63105-010, Brazil; datianemorais@hotmail.com (C.D.d.M.O.-T.); gabriel.goncalves101@urca.br (G.G.A.); gustavo.miguelsiqueira@urca.br (G.M.S.); daniel.sampaio10@urca.br (D.S.A.); talysson.f.moura@urca.br (T.F.M.); saulorelison@gmail.com (S.R.T.); irwin.alencar@urca.br (I.R.A.d.M.)
3 Oswaldo Cruz Foundation (Fiocruz Ceará), Eusebio 61773-270, Brazil; jpedroviana@alu.ufc.br (J.P.V.R.); pinheiro.vanessaf@gmail.com (V.B.P.G.); rnicolete@gmail.com (R.N.)
4 Department of Pathology and Laboratory Medicine, Warren Alpert Medical School, Brown University, Providence, RI 02912, USA; talhabmb@bgctub.ac.bd
5 Legorreta Cancer Center, Brown University, Providence, RI 02912, USA
6 Department of Pharmacy, Faculty of Allied Health Sciences, Daffodil International University, Dhaka 1207, Bangladesh
7 Department of Food Science, Federal University of Lavras, Lavras 37203-202, Brazil
8 Department of Pharmacology and Toxicology, College of Pharmacy, King Saud University, Riyadh 11451, Saudi Arabia
* Correspondence: claramarianalima@gmail.com (C.M.G.L.); hdmcoutinho@gmail.com (H.D.M.C.)

Abstract: The efflux systems are considered important mechanisms of bacterial resistance due to their ability to extrude various antibiotics. Several naturally occurring compounds, such as sesquiterpenes, have demonstrated antibacterial activity and the ability to inhibit efflux pumps in resistant strains. Therefore, the objective of this research was to analyze the antibacterial and inhibitory activity of the efflux systems NorA, Tet(K), MsrA, and MepA by sesquiterpenes nerolidol, farnesol, and α-bisabolol, used either individually or in liposomal nanoformulation, against multi-resistant *Staphylococcus aureus* strains. The methodology consisted of in vitro testing of the ability of sesquiterpenes to reduce the Minimum Inhibitory Concentration (MIC) and enhance the action of antibiotics and ethidium bromide (EtBr) in broth microdilution assays. The following strains were used: *S. aureus* 1199B carrying the NorA efflux pump, resistant to norfloxacin; IS-58 strain carrying Tet(K), resistant to tetracyclines; RN4220 carrying MsrA, conferring resistance to erythromycin. For the EtBr fluorescence measurement test, K2068 carrying MepA was used. It was observed the individual sesquiterpenes exhibited better antibacterial activity as well as efflux pump inhibition. Farnesol showed the lowest MIC of 16.5 μg/mL against the *S. aureus* RN4220 strain. Isolated nerolidol stood out for reducing the MIC of EtBr to 5 μg/mL in the 1199B strain, yielding better results than the positive control CCCP, indicating strong evidence of NorA inhibition. The liposome formulations did not show promising results, except for liposome/farnesol, which reduced the MIC of EtBr against 1199B and RN4220. Further research is needed to evaluate the mechanisms of action involved in the inhibition of resistance mechanisms by the tested compounds.

Keywords: efflux pump; fluorescence; liposome; nanoformulation; sesquiterpenes; *Staphylococcus aureus*

1. Introduction

The indiscriminate use of antibiotics has influenced the alarming levels of bacterial resistance to multiple drugs. Bacterial resistance to antibiotics is a natural, evolutionary, and adaptive phenomenon of these microorganisms, causing inactivation or reducing the action of antibiotics and biocides. Among the existing resistance mechanisms, active efflux systems stand out, which reduce the intracellular concentration of the antibiotic in the bacterial cell [1–3]

Efflux pumps are considered one of the most important mechanisms of bacterial resistance due to their broad range of substrates. They can be found in Gram-positive and Gram-negative bacteria, facilitating the extrusion of almost all existing classes of conventional antibiotics [4–9].

Research on efflux pumps is growing, aiming to develop or enhance effective drugs. In this perspective, many studies have evaluated medicinal plants that show high potential for inhibiting bacterial infections [10]. Many bioactive compounds present in medicinal plants demonstrate direct antimicrobial activity, synergistic action, and potentiation of drugs, as well as inhibition of bacterial resistance mechanisms. Among the studied classes, sesquiterpenes are active metabolites of essential oils from medicinal plants that possess important antimicrobial characteristics [11,12].

Studies have shown sesquiterpenes exhibit activity in inhibiting efflux pumps in resistant strains of *Staphylococcus aureus*. Furthermore, formulation and encapsulation studies have demonstrated liposomal nanoformulations can enhance absorption, improve distribution, and prolong the plasma half-life of compounds such as sesquiterpenes, which have limitations in their pharmacological potential due to their low solubility in biological fluids [13–16].

Liposomes are artificial vesicles composed of one or more concentric phospholipid bilayers. They are formed from phospholipids and cholesterol, which are biocompatible and non-toxic materials. These phospholipids have a hydrophilic head and a hydrophobic tail made of fatty acids, providing compartments of different polarities and compatibility for encapsulating hydrophilic or hydrophobic compounds. They have the ability to trap lipophilic agents in the lipid membrane and hydrophilic compounds in the aqueous core. The physicochemical properties of liposomes, such as permeability, membrane fluidity, charge density, determine the interaction of liposomes with the body's targets after systemic administration, making them an efficient drug carrier system [17–19].

Research has shown the significant bioactivity of naturally sourced compounds encapsulated in liposomes. Among these biological activities, one can mention antibacterial action [20], antioxidant [21–23], antitumoral [24,25], and protection against neurodegenerative diseases [26]. Sesquiterpenes are organic compounds with lipophilic characteristics. Many substances, such as proteins, lipid-polymer conjugates, and fats, are frequently used as vehicles for lipophilic substances. The use of nanoformulation technology facilitates the delivery of sesquiterpenes without altering their organoleptic characteristics and physicochemical properties [27,28].

Innovations using vesicular carriers are on the rise due to their proven effectiveness in drug delivery, enhancing the bioactivity of compounds for topical or systemic action [29]. Various vesicular carrier systems have demonstrated in vitro antimicrobial action, leading to the optimization of the drug [30].

Currently, research into new targets for antibiotic therapy is on the rise. In this regard, the search for different strategies to inhibit efflux pumps is essential to restore the effectiveness of antibiotics. Therefore, the discovery of natural compounds that can act as antibiotic adjuvants or inhibit resistance mechanisms becomes relevant. Furthermore, the liposomal encapsulation of bioactive molecules has shown promise in antimicrobial treatment because liposomes enhance the delivery and distribution of the drug within biological systems [20].

In light of this, the present study aims to analyze the inhibitory activity of the efflux systems NorA, Tet(K), MsrA, and MepA by sesquiterpenes nerolidol, farnesol, and α-bisabolol, used individually and in liposomal nanoformulation, against multi-resistant *S. aureus* strains.

2. Results and Discussion

2.1. *Physical-Chemical Profile of Liposomes*

The physical-chemical characterization of the nanoformulations consisted of determining the average size of the liposomes, intensity, Zeta potential, concentration, polydispersity index (PDI), and encapsulation efficiency (Table 1).

Table 1. Physical-chemical characteristics of liposomal nanoformulations control and containing nerolidol, farnesol, and α-bisabolol.

	Liposome Control	Liposome/ Nerolidol	Liposome/ Farnesol	Liposome/ α-Bisabolol
Size	218.9 nm ± 45.1	241.8 nm ± 73.1	201.4 nm ± 66.6	183.5 nm ± 58
Intensity	2.5 a.u.	3 a.u.	5 a.u.	5 a.u.
Zeta potential	−18.8 mV	−24.1 mV	−13 mV	28.2 mV
pH	7.4	7.4	7.4	7.4
Concentration	6.99×10^8 particles/mL	6.82×10^8 particles/mL	4.42×10^8 particles/mL	4.02×10^8 particles/mL
Polydispersity index (PDI)	0.50	0.61	0.92	0.755
Encapsulation efficiency	-	85.4%	79%	87%

The nanoformulations presented average sizes of 218.9 nm, 241.8 nm, 201.4 nm, and 183.5 nm, respectively, for the control liposomes and those containing nerolidol, farnesol, and α-bisabolol. The average size of nanoparticles is a parameter that can influence their biological activity. Different particle sizes can exhibit distinct degrees of biodistribution, absorption, and therapeutic efficacy in different targets. Approaching liposome particle sizes around 100 nm, as seen in the α-bisabolol formulation, is particularly relevant because this size is associated with enhanced biological activity. This implies the formulation can be more effective in delivering its active compounds, leading to superior therapeutic outcomes. However, it is important to emphasize the final biological activity is also influenced by factors such as liposome composition, surface charge, morphology, and release properties of the encapsulated active compound [31,32].

Intensity is a widely used technique for determining the concentration and size of particles in suspension. Intensity is related to the number of particles in the sample and the encapsulation efficiency of the encapsulated sesquiterpenes. This parameter is used as an indirect measure to estimate the amount of encapsulated agent relative to the total quantity of particles present [33]. The present study showed satisfactory signal intensity, which was 2.5 a.u., 3 a.u., 5 a.u., and 5 a.u., respectively.

The Zeta potential, ranging from −24.1 mV to 28.2 mV in the formulations, serves as an indicator of particle charge and colloidal stability. The Zeta potential is a measure of the electric charge of nanoparticles and is indicative of their colloidal stability. This value provides information about the electrostatic repulsion between particles, the tendency of aggregation, and the interaction with cells and tissues. The potential is determined by the potential difference between the surface of the particles and the surrounding dispersing liquid. It is influenced by various factors such as ionic strength, suspension composition, pH, and interfacial interactions [34,35]. Values further from zero indicate more suspension stability. Therefore, the liposomes/sesquiterpenes studied show significant Zeta potential values, which are −18.8 mV, −24.1 mV, −13 mV, and 28.2 mV. Significant Zeta potential values, as observed in the nerolidol and α-bisabolol formulations, suggest good colloidal stability.

Concentration refers to the quantity of nanoparticles present per ml of the analyzed sample. For instance, the nanoformulation containing nerolidol has a concentration of 6.82×10^8 particles/mL, the farnesol formulation contains 4.42×10^8 particles/mL, and the nanoformulation with α-bisabolol contains 4.02 particles/mL. In addition to implications related to safety and efficacy, nanoparticle concentration can also impact the physical and chemical properties of the particles. For example, the colloidal stability of nanoparticles can be influenced by the concentration. A study conducted by Hufschmid et al. [36] investigated the stability of iron oxide nanoparticles at different concentrations. The results revealed high concentrations of nanoparticles led to a higher rate of agglomeration and sedimentation, which could compromise the colloidal stability of the particles.

The concentration of nanoparticles in the solution can influence the dosage and therapeutic efficacy. Although higher concentrations of nanoparticles result in greater bioactivity and faster biodisponibility, it can also potentiate their toxicity [37]. High concentrations of nanoparticles can accumulate in tissues and organs, leading to oxidative stress, cellular damage, and consequently, adverse effects. Therefore, it is necessary to find the ideal concentration for pharmacological action [38].

The Polydispersity Index is a measure of particle size uniformity within the sample. PDI values range from 0.50 (control group) to 0.92 (farnesol formulation). A PDI close to 1 indicates particles have similar sizes, whereas a higher PDI suggests a broader size distribution. The farnesol formulation exhibits a higher PDI, which may indicate greater variation in particle size. The elevated PDI observed in some liposome formulations (up to 0.92) suggests a broader size distribution of suspended particles. However, it is worth highlighting the suitability of the microfluidic technique employed in this study. Microfluidics is renowned for its precision in liposome formation, allowing for precise control of particle size and uniformity. The high mixing efficiency and shear within microchannels result in consistently sized liposomes. Microfluidics can be more lipid-efficient compared to other methods, thereby reducing costs, and optimizing formulation [39,40].

The encapsulation efficiency refers to the amount of sesquiterpene encapsulated In the lipid nanoparticle. Higher values indicate greater encapsulation efficiency. For example, the liposomes studied here showed an encapsulation efficiency of 85%, 79%, and 87%, respectively, for nerolidol, farnesol, and bisabolol, indicating a high percentage of the compound inside the liposome.

A study conducted by Minelli et al. [41] investigated the efficacy of solid lipid nanoparticles carrying farnesol in inhibiting the growth of colon cancer cells. The results showed farnesol-loaded nanoparticles exhibited a higher internalization rate in cancer cells and induced greater apoptosis compared to free farnesol. Additionally, the nanoparticles demonstrated lower toxicity to healthy cells. These studies highlight the importance of nanoparticles in enhancing the delivery efficiency of sesquiterpenes and improving their therapeutic activities. Nanoparticles allow for the protection of sesquiterpenes against degradation and enable their controlled release at the target site. Moreover, nanoparticle formulation can enhance the water solubility of sesquiterpenes, allowing for more effective administration [42].

2.2. Antibacterial Activity and Efflux Pump Inhibition Assessed through MIC Reduction

The isolated sesquiterpenes showed direct antibacterial activity, with nerolidol having MIC values of 32 μg/mL and 128 μg/mL against IS-58 and RN4220 strains, respectively. Farnesol presented MIC values of 25.4 μg/mL, 32 μg/mL, and 16 μg/mL against 1199B, IS-58, and RN4220 strains, respectively. α-bisabolol exhibited MIC values of 128 μg/mL, 64 μg/mL, and 161.3 μg/mL against 1199B, IS-58, and RN4220 strains, respectively (Table 2).

Given sesquiterpenes are organic compounds with lipophilic characteristics, they were incorporated into the lipophilic layer of the liposomes, becoming interspersed within their membrane. With this, the sesquiterpene can be gradually released upon contact with the bacterial membrane or the surrounding environment. The non-promising results observed with the liposome/sesquiterpene complex may have occurred due to the difference between the lipids present in the liposome and the phospholipids in the bacterial membrane of *S. aureus* [43,44].

Table 2. The minimum inhibitory concentration of isolated and encapsulated nerolidol, farnesol, and α-bisabolol against *S. aureus* strains 1199B, IS-58, and RN4220. Concentrations in (μg/mL).

Compound	1199B	IS-58	RN4220
Nerolidol	1024 μg/mL ± 0.5	32 μg/mL ± 0.5 *	128 μg/mL ± 0.5 *
Farnesol	25.4 μg/mL ± 0.8 *	32 μg/mL ± 0.5 *	16 μg/mL ± 0.5 *
α-bisabolol	128 μg/mL ± 0.5 *	64 μg/mL ± 0.5 *	161.3 μg/mL ± 0.8 *
Liposome/Nerolidol	≥1024 μg/mL ± 0.5	≥1024 μg/mL ± 0.5	≥1024 μg/mL ± 0.5
Liposome/Farnesol	≥1024 μg/mL ± 0.5	≥1024 μg/mL ± 0.5	≥1024 μg/mL ± 0.5
Liposome/α-bisabolol	≥1024 μg/mL ± 0.5	≥1024 μg/mL ± 0.5	≥1024 μg/mL ± 0.5

* Clinically relevant antibacterial activity.

Figure 1 and Table 3 show the inhibitory action results of the sesquiterpenes nerolidol, farnesol, and α-bisabolol against *S. aureus* 1199B strains carrying the NorA efflux pump. In Figure 1A, it can be observed the combination of nerolidol with norfloxacin significantly reduced the MIC of this antibiotic from 50.6 μg/mL to 12.7 μg/mL, compared to the antibiotic control alone, indicating potentiation of the antibacterial activity. The combination of nerolidol with EtBr also reduced the EtBr MIC from 64 μg/mL to 5 μg/mL. Both in combination with the antibiotic and in combination with EtBr, nerolidol showed significantly better results than the positive control CCCP. However, nerolidol encapsulated in a liposomal nanoformulation did not show a potentiating effect on norfloxacin or EtBr.

Table 3. Minimum inhibitory concentration of nerolidol, farnesol, and α-bisabolol sesquiterpenes, whether isolated or encapsulated in liposomes, against the *S. aureus* 1199B strain, associated with norfloxacin and ethidium bromide. * Statistically significant compared to the norfloxacin or EtBr control.

Group with Antibiotic	MIC	Group with EtBr	MIC
Norfloxacin	50.8 μg/mL ± 0.8	EtBr	64 μg/mL ± 0.5
Norfloxacin + CCCP	32 μg/mL ± 0.5 *	EtBr + CCCP	32 μg/mL ± 0.5 *
Norfloxacin + Nerolidol	12.7 μg/mL ± 0.8 *	EtBr + Nerolidol	5 μg/mL ± 0.5 *
Norfloxacin + Farnesol	64 μg/mL ± 0.5	EtBr + Farnesol	50.8 μg/mL ± 0.8 *
Norfloxacin + α-bisabolol	64 μg/mL ± 0.5	EtBr + α-bisabolol	80.6 μg/mL ± 0.8
Norfloxacin + Liposome control	101.6 μg/mL ± 0.8	EtBr + Liposome control	64 μg/mL ± 0.5
Norfloxacin + Liposome/Nerolidol	64 μg/mL ± 0.5	EtBr + Liposome/Nerolidol	64 μg/mL ± 0.5
Norfloxacin + Liposome/Farnesol	101.6 μg/mL ± 0.8	EtBr + Liposome/Farnesol	40.3 μg/mL ± 0.8 *
Norfloxacin + Liposome/α-bisabolol	64 μg/mL ± 0.5	EtBr + Liposome/α-bisabolol	50.8 μg/mL ± 0.8 *

Isolated or encapsulated farnesol, when associated with norfloxacin, did not show significant results in reducing the MIC. However, when associated with EtBr, both isolated and encapsulated farnesol reduced the EtBr MIC to 50 μg/mL and 40.3 μg/mL, respectively, compared to the control (64 μg/mL) (Figure 1B).

The α-bisabolol in association with norfloxacin did not potentiate the antibacterial action of the antibiotic. When associated with EtBr, isolated α-bisabolol did not have a synergistic effect. However, α-bisabolol in the liposomal nanoformulation showed a synergistic effect with a reduced MIC of 50.8 μg/mL, compared to the control of 64 μg/mL (Figure 1C). These results indicate the isolated sesquiterpenes nerolidol and farnesol possibly act on the inhibition of NorA, while the encapsulated forms of farnesol and α-bisabolol may act on the inhibition of NorA.

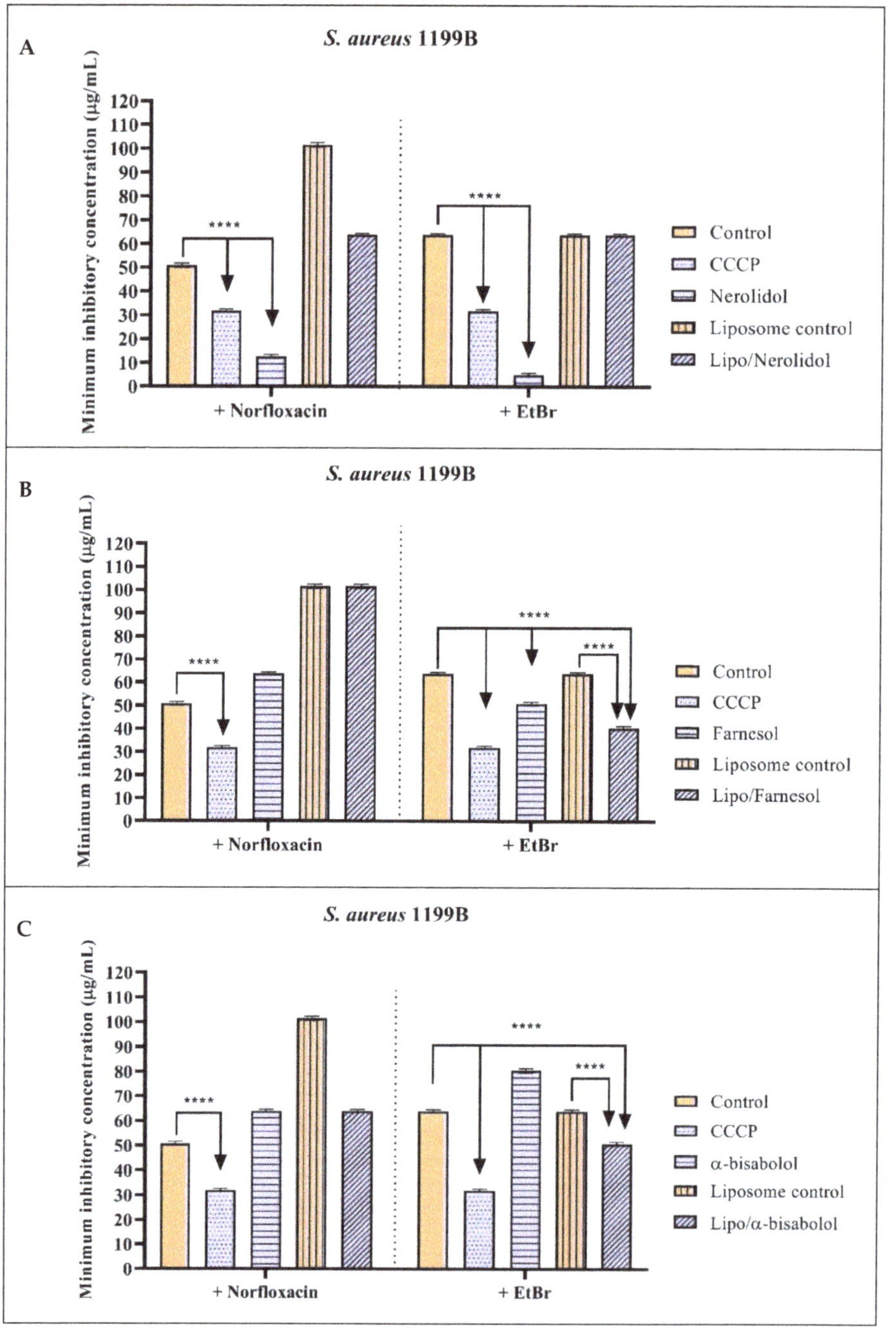

Figure 1. Evaluation of the NorA efflux pump inhibitory activity by nerolidol (**A**), farnesol (**B**), and α-bisabolol (**C**) sesquiterpenes isolated and encapsulated in liposomes, against the *S. aures* 1199B strain. Associated with norfloxacin and ethidium bromide. Two-way ANOVA followed by Bonferroni post hoc. CCCP = carbonyl cyanide 3-chlorophenylhydrazone; EtBr = ethidium bromide; **** = $p < 0.0001$ vs. control.

Figure 2 shows the results of the action of the sesquiterpenes nerolidol, farnesol, and α-bisabolol against the *S. aureus* IS-58 strain carrying the Tet(K) efflux pump. Among all substances tested, only α-bisabolol exhibited a potentiating effect, reducing the EtBr MIC to 8 μg/mL. This result indicates that α-bisabolol may inhibit the Tet(K) efflux mechanism in the *S. aureus* IS-58 strain (Table 4).

Table 4. Minimum inhibitory concentration of nerolidol, farnesol, and α-bisabolol sesquiterpenes, whether isolated or encapsulated in liposomes, against the *S. aureus* IS-58 strain, associated with norfloxacin and ethidium bromide. * Statistically significant compared to the tetracyclin or EtBr control.

Group with Antibiotic	MIC	Group with EtBr	MIC
Tetracycline	64 μg/mL ± 0.5	EtBr	10 μg/mL ± 0.5
Tetracycline + CCCP	25.4 μg/mL ± 0.8 *	EtBr + CCCP	2 μg/mL ± 0.5 *
Tetracycline + Nerolidol	64 μg/mL ± 0.5	EtBr + Nerolidol	25.4 μg/mL ± 0.8
Tetracycline + Farnesol	64 μg/mL ± 0.5	EtBr + Farnesol	10 μg/mL ± 0.5
Tetracycline + α-bisabolol	64 μg/mL ± 0.5	EtBr + α-bisabolol	8 μg/mL ± 0.5 *
Tetracycline + Liposome control	128 μg/mL ± 0.5	EtBr + Liposome control	36 μg/mL ± 0.5
Tetracycline + Liposome/Nerolidol	256 μg/mL ± 0.5	EtBr + Liposome/Nerolidol	36 μg/mL ± 0.5
Tetracycline + Liposome/Farnesol	128 μg/mL ± 0.5	EtBr + Liposome/Farnesol	36 μg/mL ± 0.5
Tetracycline + Liposome/α-bisabolol	64 μg/mL ± 0.5	EtBr + Liposome/α-bisabolol	12.7 μg/mL ± 0.8

Against the *S. aureus* strain expressing the MsrA efflux system, only farnesol showed significant effects when associated with erythromycin, reducing the erythromycin MIC to 256 μg/mL. In association with EtBr, encapsulated nerolidol and isolated and encapsulated farnesol showed significant effects in reducing the MIC to 8 μg/mL, 21 μg/mL, and 21.8 μg/mL, respectively, indicating the occurrence of MsrA efflux pump inhibition (Figure 3 and Table 5).

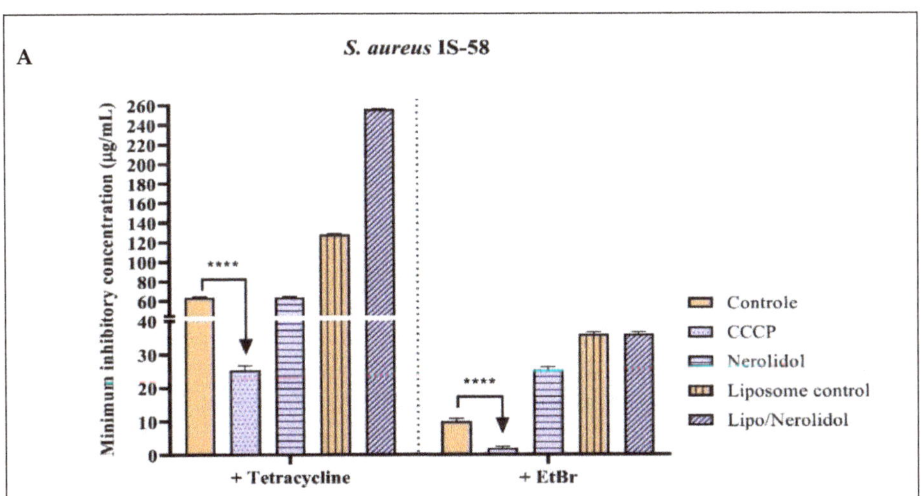

Figure 2. *Cont.*

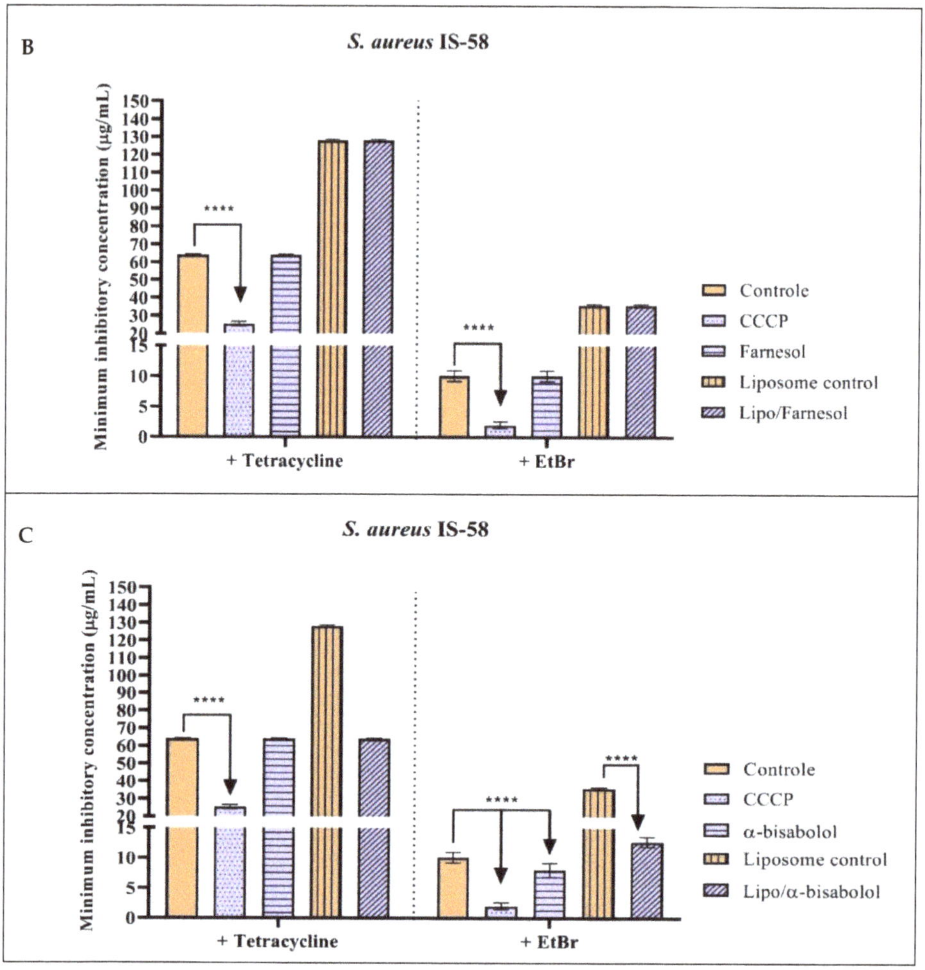

Figure 2. Evaluation of the Tet(K) efflux pump inhibitory activity by the nerolidol (**A**), farnesol (**B**), and α-bisabolol (**C**) sesquiterpenes isolated and encapsulated in liposomes, against the *S. aures* IS-58 strain. Associated with norfloxacin and ethidium bromide. Two-way ANOVA followed by Bonferroni post hoc. CCCP = Carbonyl cyanide 3-chlorophenylhydrazone; EtBr = ethidium bromide; **** = $p < 0.0001$ vs. control.

Table 5. Minimum inhibitory concentration of nerolidol, farnesol, and α-bisabolol sesquiterpenes, whether isolated or encapsulated in liposomes, against the *S. aureus* RN4220 strain, associated with norfloxacin and ethidium bromide. * Statistically significant compared to the tetracyclin or EtBr control.

Group with Antibiotic	MIC	Group with EtBr	MIC
Erythromycin	512 µg/mL ± 0.5	EtBr	36 µg/mL ± 0.5
Erythromycin + CCCP	0.5 µg/mL ± 0.5 *	EtBr + CCCP	2 µg/mL ± 0.5 *
Erythromycin + Nerolidol	512 µg/mL ± 0.5	EtBr + Nerolidol	36 µg/mL ± 0.5
Erythromycin + Farnesol	256 µg/mL ± 0.5 *	EtBr + Farnesol	21.8 µg/mL ± 0.8 *
Erythromycin + α-bisabolol	512 µg/mL ± 0.5	EtBr + α-bisabolol	36 µg/mL ± 0.5

Table 5. *Cont.*

Group with Antibiotic	MIC	Group with EtBr	MIC
Erythromycin + Liposome control	512 μg/mL ± 0.5	EtBr + Liposome control	36 μg/mL ± 0.5
Erythromycin + Liposome/Nerolidol	512 μg/mL ± 0.5	EtBr + Liposome/Nerolidol	8 μg/mL ± 0.5 *
Erythromycin + Liposome/Farnesol	512 μg/mL ± 0.5	EtBr + Liposome/Farnesol	21.8 μg/mL ± 0.8 *
Erythromycin + Liposome/α-bisabolol	512 μg/mL ± 0.5	EtBr + Liposome/α-bisabolol	36 μg/mL ± 0.5

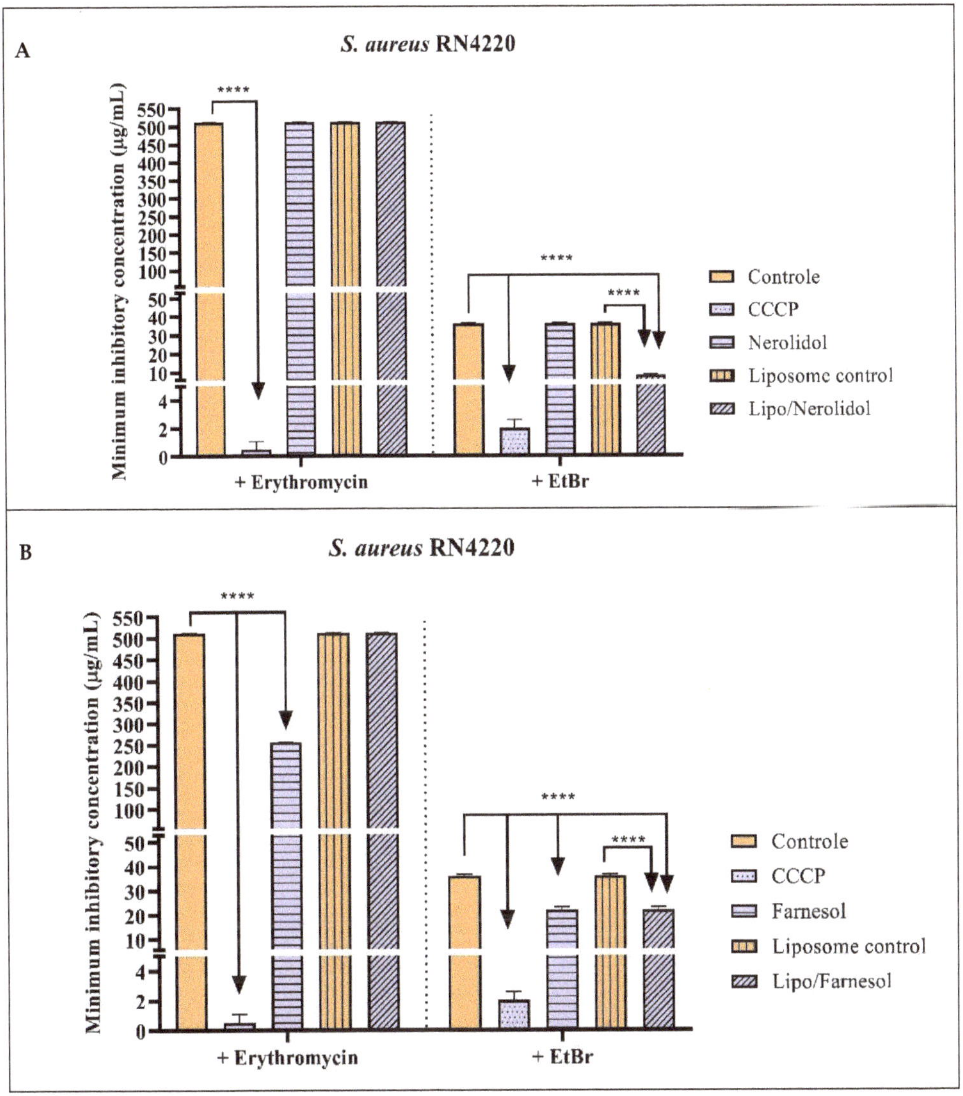

Figure 3. *Cont.*

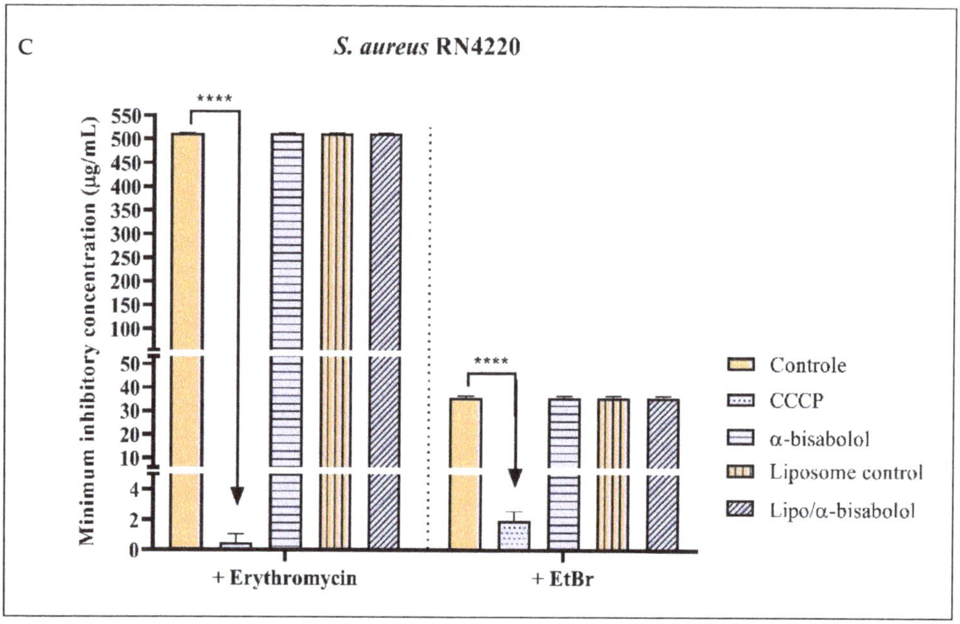

Figure 3. Evaluation of the MsrA efflux pump inhibitory activity by the nerolidol (**A**), farnesol (**B**), and α-bisabolol (**C**) sesquiterpenes isolated and encapsulated in liposomes, against the *S. aures* RN4220 strain. Associated with erythromycin and ethidium bromide. Two-way ANOVA followed by Bonferroni post hoc. CCCP = Carbonyl cyanide 3-chlorophenylhydrazone; EtBr = ethidium bromide; **** = $p < 0.0001$ vs. control.

The reduction of specific antibiotics' MIC and EtBr by sesquiterpenes is indicative of efflux pump inhibition or potentialization of antibiotic action [45–53]. Secondary metabolites are chemical compounds produced by plants as defense mechanisms against pathogens such as fungi and bacteria, as well as herbivorous animals. They can be useful in treating diseases and infections, exhibiting various proven bioactivities such as antioxidant, antidiabetic, antiproliferative, anti-inflammatory, and antimicrobial actions [54].

According to Alsheikh et al. [55], phytochemicals can exert antimicrobial action through mechanisms distinct from conventional antibiotics, such as inhibiting cell wall synthesis and interfering with bacterial physiology by reducing membrane potential and ATP synthesis. Additionally, they can modulate bacterial susceptibility to antibiotics.

Components of essential oils can act on efflux pumps, restoring the effectiveness of some antibiotics that are targets of resistance mechanisms. Sesquiterpenes exhibit broad antibacterial activity related to their lipophilic characteristics [56]. Oliveira et al. [57] emphasize that sesquiterpenes nerolidol, farnesol, and α-bisabolol have the potential to enhance the activity of conventional antimicrobials, such as gentamicin, oxacillin, and methicillin. Farnesol can potentiate the effect of conventional antimicrobials against *S. aureus* RN4220 strains that produce the MsrA efflux mechanism. In the study, researchers associated the sesquiterpene with fusidic acid, demonstrating the potentiation of this effect on the MIC.

According to Cruz et al. [58], α-bisabolol exhibited potentiating activity against antimicrobials in the presence of *S. aureus* strains expressing the Tet(K) and NorA efflux systems. In their studies, Moura et al. [11] demonstrated nerolidol is an effective sesquiterpene in infection treatment caused by multidrug-resistant bacteria.

In addition to the antimicrobial effects of sesquiterpenes on multidrug-resistant strains, studies reveal liposomal nanoformulations can enhance the therapeutic action of antimicrobials. The encapsulation of farnesol in liposomes resulted in significantly increased antifungal activity against strains of *Candida albicans*, *C. tropicalis*, and *C. krusei*, leading to a considerable reduction in IC50 [59].

Several in vivo and in vitro studies confirm the effectiveness of encapsulating compounds in liposomes, enhancing the antibacterial and anticancer action of these compounds [60–63]. These results are consistent with the data presented for farnesol in liposomes against the 1199B and RN4220 strains, where there was a reduction in the MIC of EtBr.

2.3. Evaluation of Efflux Pump Inhibition by Fluorescence Emission

When measuring fluorescence emission, it was observed nerolidol at 100 µg/mL and farnesol at 100 µg/mL increased fluorescence emission compared to the negative control, which consisted of inoculum plus EtBr. This increase was represented by 31.3% and 17.5%, respectively. The same result was observed with the efflux pump inhibitor CCCP, indicating the reproducibility of the experiment (Figure 4). The average increase in fluorescence suggested the possible inhibition of the MepA efflux pump, considering that inhibition of EtBr efflux led to an increase in its intracellular concentration and, consequently, enhanced the fluorescence of the sample [64–66].

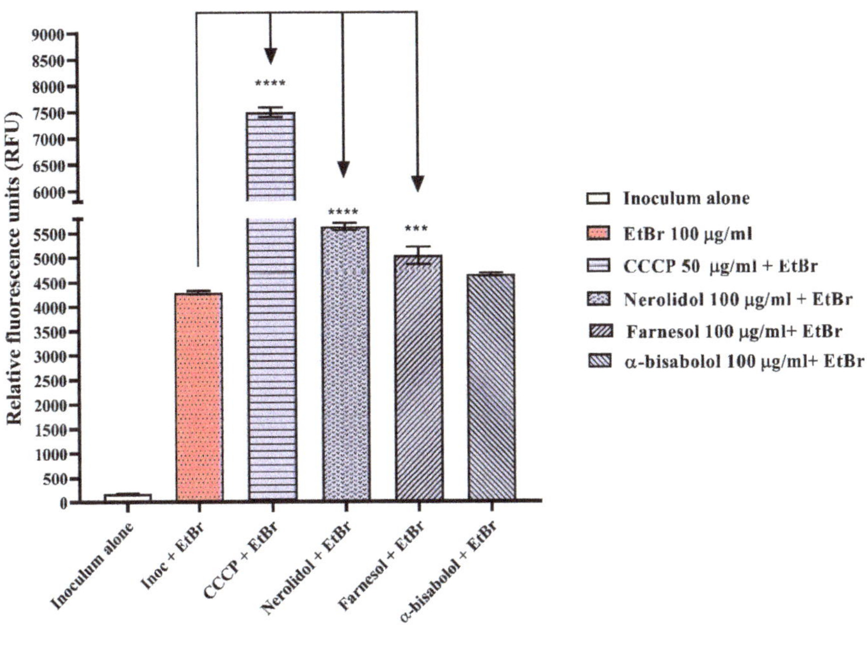

Figure 4. Evaluation of MepA efflux pump inhibition by measuring fluorescence emission in *S. aureus* K2068 strain, treated with nerolidol, farnesol, and α-bisabolol at 100 µg/mL. EtBr = ethidium bromide; Inoc = inoculum; **** = $p < 0.0001$ vs. inoc + EtBr; *** = $p < 0.001$ vs. inoc + EtBr.

3. Materials and Methods

3.1. Substances Used in Research

The sesquiterpenes nerolidol ($C_{15}H_{26}O$), farnesol ($C_{15}H_{26}O$), and α-bisabolol ($C_{15}H_{26}$) were used. Carbonyl cyanide m-chlorophenyl-hydrazone (CCCP) was used as the standard efflux pump inhibitor for positive control. The DNA intercalating agent used was ethidium bromide (EtBr). Specific antibiotics were used as substrates for each bacterial efflux pump: norfloxacin for the *S. aureus* 1199B strain carrying NorA; tetracycline for the *S. aureus* IS-58 strain carrying Tet(K); and erythromycin for the *S. aureus* RN4220 strain carrying the MsrA protein. The culture media used were solid medium Heart Infusion Agar (HIA, Difco, Forn

El Chebbak, Lebanon) and liquid medium Brain Heart Infusion (BHI). All products were purchased from Sigma-Aldrich Chemical Co. (St. Louis, MO, USA).

3.2. Synthesis of Liposomal Nanoformulations

Initially, a highly concentrated solution of 50 mg/mL of nerolidol, farnesol, and α-bisabolol was prepared for encapsulation. To create the organic phase that constituted the liposomal nanoparticles, a lipid solution was prepared, consisting of 1,2-dipalmitoyl-sn-glycero-3-phosphocholine (DPPC), cholesterol (CHOL), and distearoylphosphatidylcholine (DSPC) in a ratio of DPPC:CHOL:DSPC at 52:45:3 (mol/mol) to achieve a final lipid concentration of 35 mM. The microfluidics technique was employed using the NanoAssemblr Benchtop equipment (Precision Nanosystems^{TM}, Vancouver, BC, Cananda), For the nanoparticle fabrication. To optimize encapsulation efficiency, the following factors were used: a flow rate ratio of 2:1 and a total flow ratio of 12 mL/min for the left and right syringes, respectively. In the left syringe, the aqueous phase containing the diluted sesquiterpene solution was added until a total volume of 3 mL was reached. In the right syringe, the lipid solution was added with a total volume of 1 mL. For the preparation of control nanoformulations, or liposome controls, only phosphate-buffered saline (PBS) was added to the left syringe (aqueous solution). After the nanoparticle preparation, the resulting formulation was placed in Amicon® Ultra-15 3000 MWCO (Merck, Darmstadt, Germany) and centrifuged at 3000 rpm, 20 °C for 30 min, to remove the residual solvent used in lipid solubilization. The washing was performed with PBS buffer at pH 7.2. Finally, the formulations were stored in a refrigerated environment at 3 °C to 8 °C [59].

3.3. Microorganisms Used in the Assays

The following strains were used: *S. aureus* 1199B, resistant to hydrophilic fluoroquinolones via the NorA efflux protein; *S. aureus* IS-58, containing the PT181 plasmid carrying the Tet(K) gene that extrudes tetracyclines; *S. aureus* RN4220, carrying the pUL5054 plasmid that carries the gene for the MsrA protein which effluxes macrolides (Table 6). All strains were maintained on HIA medium at 4 °C and in glycerol in a freezer at −80 °C. The resistance gene-carrying strains were maintained in culture medium under subinhibitory antibiotic conditions to induce gene expression.

Table 6. Bacterial strains of *S. aureus* used in the microbiological assays.

Strain	Plasmid/Gene	Protein (Substrate Antibiotic)
1199B	*norA*	NorA (Norfloxacin)
IS-58	Plasmid PT181 (*tetK*)	Tet(K) (Tetracyclin)
RN4220	Plasmid Pul5054 (*msrA*)	MsrA (Erythromycin)

3.4. Antibacterial Activity Evaluated by Measuring the Minimum Inhibitory Concentration (MIC)

This test consisted of determining the minimum inhibitory concentration (MIC) of sesquiterpenes capable of inhibiting the growth of *S. aureus* strains 1199B, IS-58, and RN4220. The bacterial inoculum of the three strains was prepared in sterile saline solution, corresponding to a McFarland scale of 0.5, which corresponded to 1.5×10^8 Colony-Forming Units. Then, distribution media were prepared in Eppendorf tubes containing 900 μL of Brain Heart Infusion (BHI) culture medium and 100 μL of the inoculum. A total of 100 μL of the tube contents were transferred to a 96-well microdilution plate. Subsequently, a serial dilution (1:1) was performed with 100 μL of either isolated or encapsulated nerolidol, farnesol, or α-bisabolol sesquiterpenes. The microdilution was carried out until the penultimate well, leaving the last well as a growth control. The final concentrations of each sesquiterpene ranged from 512 μg/mL to 0.5 μg/mL. The plates were incubated in a bacteriological incubator for 24 h at 37 °C. The experiments were performed in triplicate. The reading was performed by adding 20 μL of resazurin (7-hydroxy-3H-phenoxazin-3-one 10-oxide), observing the change in color in each well.

Blue coloration indicated the absence of bacterial growth, while a color change to red indicated bacterial growth [67,68].

3.5. Evaluation of the Inhibition of Efflux Pumps NorA, Tet(K), and MsrA

The inhibition of efflux pumps was verified by the reduction of MIC of antibiotics and EtBr against *S. aureus* strains 1199B, IS-58, and RN4220. Bacterial inocula were prepared as described in the previous section. Test solutions were prepared in Eppendorf tubes containing 200 µL of inoculum, isolated or encapsulated sesquiterpene at subinhibitory concentration (MIC/8), and BHI culture medium, resulting in a final volume of 2 mL. The control solution contained only the inoculum and culture medium. Subsequently, the solutions were transferred to 96-well microtiter plates, with the addition of 100 µL of the content in each well. Then, 100 µL of norfloxacin, tetracycline, erythromycin, or EtBr antibiotics were serially diluted (1:1) until the penultimate well, resulting in concentrations ranging from 512 µg/mL to 0.5 µg/mL. The negative control contained only the antibiotic or EtBr alone. The positive control consisted of CCCP. The reading was performed as described in the previous section. The MIC was defined as the lowest concentration at which there was no bacterial growth in the well, characterized by the blue coloration of resazurin [67,68].

3.6. Efflux Pumps Inhibition Evaluated by the Increased Fluorescence Emission of EtBr

The strain *S. aureus* K2068 was seeded on a solid HIA culture medium and incubated in a bacteriological incubator at 37 °C for 24 h before conducting the experiments. The inoculum was prepared until obtaining 1.5×10^8 colony forming units (CFU), corresponding to the 0.5 value on the McFarland scale. The inoculum was prepared in PBS. For the test, sesquiterpenes were selected as they showed the best results in microdilution assays. Test solutions were prepared containing the K2068 inoculum and the sesquiterpenes nerolidol, farnesol, and α-bisabolol, all at 100 µg/mL. The positive control used was CCCP at 50 µg/mL. PBS buffer was added to each solution to reach a final volume of 1 mL. The solutions were incubated for 1 h and 30 min. Then, EtBr (ethidium bromide) at 100 µg/mL was added to all solutions except the inoculum alone group, which served as the growth control. The solutions were incubated for an additional 1 h. Subsequently, the solutions were centrifuged at 10,000 rpm for 2 min and washed with PBS to remove all EtBr and medium substances. The supernatant was discarded, and the resulting pellet was dissolved in PBS. The sample containing the dissolved pellet was distributed into microplates. The reading was performed using Cytation 1, BioTek® (Winooski, VT, USA) fluorescence microplate reader and Gen5™ 3.22 Software with excitation at 530 nm and emission wavelength at 590 nm. The reading was taken for the following groups: inoculum alone (growth control), inoculum + EtBr (negative control), inoculum + EtBr + CCCP (positive control), inoculum + EtBr + nerolidol 100 µg/mL, inoculum + EtBr + farnesol 100 µg/mL, and inoculum + EtBr + α-bisabolol 100 µg/mL. The assay was performed in triplicate, and the results were compared to the negative control group, EtBr [12].

3.7. Statistical Analysis

The assays were performed in triplicate. In microbiological tests, descriptive statistics were used to calculate the geometric mean and standard deviation, and the results were compared using Two-way ANOVA, followed by the Bonferroni post hoc test. Analysis of the fluorimetry assay and other tests was carried out using One-way ANOVA, followed by the Dunnett test. Results were considered significant when $p < 0.05$. GraphPad Prism 5.0 software was used.

4. Conclusions

Therefore, when isolated, nerolidol exhibited direct antibacterial activity against *S. aureus* IS-58 and RN4220 strains. Isolated farnesol and α-bisabolol showed direct antibacterial activity against *S. aureus* 1199B, IS-58, and RN4220. However, the liposomal formulation of these compounds did not show direct efficacy against *S. aureus* strains. In terms of efflux pump inhibition, these compounds demonstrated effectiveness. Liposome/farnesol and liposome/α -bisabolol acted as potential inhibitors of NorA present in *S. aureus* 1199B. Liposome/nerolidol and liposome/farnesol acted as potential inhibitors of MsrA in the *S. aureus* RN4220 strain. The isolated sesquiterpenes showed significant action: in their isolated form, nerolidol and farnesol acted as putative inhibitors of NorA and MepA; α-bisabolol acted as a putative inhibitor of Tet(K), and isolated farnesol acted as inhibitor of MsrA and MepA. Among all the substances tested, isolated nerolidol stood out for its potent inhibition of NorA, being even more effective than CCCP. Further studies are necessary to describe the molecular targets involved in these mechanisms.

Author Contributions: Conceptualization, C.M.G.L., T.G.d.S. and C.D.d.M.O.-T.; formal analysis, H.D.M.C., T.G.d.S., C.D.d.M.O.-T., I.R.A.d.M. and R.N.; investigation, J.E.G.S., C.D.d.M.O.-T., G.G.A., G.M.S., D.S.A., T.F.M., S.R.T., J.P.V.R. and V.B.P.G.; writing—original draft preparation, J.E.G.S., C.D.d.M.O.-T. and J.P.V.R.; writing—review and editing, C.D.d.M.O.-T. and T.G.d.S.; supervision, H.D.M.C., T.G.d.S., R.N. and C.D.d.M.O.-T.; project administration, H.D.M.C., T.B.E. and S.F.A.; funding acquisition, H.D.M.C., T.G.d.S., I.R.A.d.M. and R.N. All authors have read and agreed to the published version of the manuscript.

Funding: This research was funded by King Saud University, Riyadh, Saudi Arabia, Project Number (RSPD2023R709) and by Cearense Foundation for Supporting Scientific and Technological Development (FUNCAP), grant number DCT-0182-00104.02.00/21 and 04855159/2022.

Institutional Review Board Statement: Not applicable.

Informed Consent Statement: Not applicable.

Data Availability Statement: All experimental data generated or analyzed during this study are included in the article.

Acknowledgments: The authors acknowledge and extend their appreciation to the Researchers Supporting Project Number (RSPD2023R709), King Saud University, Riyadh, Saudi Arabia, for funding this study.

Conflicts of Interest: The authors declare no conflict of interest.

References

1. González-Bello, C. Antibiotic Adjuvants—A Strategy to Unlock Bacterial Resistance to Antibiotics. *Bioorg. Med. Chem. Lett.* **2017**, *27*, 4221–4228. [CrossRef]
2. Du, D.; Wang-Kan, X.; Neuberger, A.; van Veen, H.W.; Pos, K.M.; Piddock, L.J.V.; Luisi, B.F. Multidrug Efflux Pumps: Structure, Function and Regulation. *Nat. Rev. Microbiol.* **2018**, *16*, 523–539. [CrossRef] [PubMed]
3. Patini, R.; Mangino, G.; Martellacci, L.; Quaranta, G.; Masucci, L.; Gallenzi, P. The Effect of Different Antibiotic Regimens on Bacterial Resistance: A Systematic Review. *Antibiotics* **2020**, *9*, 22. [CrossRef] [PubMed]
4. Köhler, T.; Pechère, J.C.; Plésiat, P. Bacterial Antibiotic Efflux Systems of Medical Importance. *Cell. Mol. Life Sci.* **1999**, *56*, 771–778. [CrossRef]
5. Uddin, M.J.; Ahn, J. Associations between Resistance Phenotype and Gene Expression in Response to Serial Exposure to Oxacillin and Ciprofloxacin in *Staphylococcus aureus*. *Lett. Appl. Microbiol.* **2017**, *65*, 462–468. [CrossRef]
6. Li, X.Z.; Nikaido, H. Efflux-Mediated Drug Resistance in Bacteria. *Drugs* **2004**, *64*, 159–204. [CrossRef]
7. Kaatz, G.W.; McAleese, F.; Seo, S.M. Multidrug Resistance in *Staphylococcus aureus* Due to Overexpression of a Novel Multidrug and Toxin Extrusion (MATE) Transport Protein. *Antimicrob. Agents Chemother.* **2005**, *49*, 1857–1864. [CrossRef]
8. Blair, J.M.A.; Webber, M.A.; Baylay, A.J.; Ogbolu, D.O.; Piddock, L.J.V. Molecular Mechanisms of Antibiotic Resistance. *Nat. Rev. Microbiol.* **2015**, *13*, 42–51. [CrossRef]
9. Henderson, P.J.F.; Maher, C.; Elbourne, L.D.H.; Eijkelkamp, B.A.; Paulsen, I.T.; Hassan, K.A. Physiological Functions of Bacterial "Multidrug" Efflux Pumps. *Chem. Rev.* **2021**, *121*, 5417–5478. [CrossRef] [PubMed]
10. Domingues Almeida, G.; Pena Godoi, E.; Cardoso Santos, E.; Paes De Lima, L.R.; Emilia De Oliveira, M. Extrato Aquoso de *Allium sativum* Potencializa a Ação Dos Antibióticos Vancomicina, Gentamicina e Tetraciclina Frente *Staphylococcus aureus*. *Rev. Ciênc. Farm. Básica Apl.* **2013**, *34*, 487–492.
11. de Moura, D.F.; Rocha, T.A.; de Melo Barros, D.; da Silva, M.M.; dos Santos Santana, M.; Neta, B.M.; Cavalcanti, I.M.F.; Martins, R.D.; da Silva, M.V. Evaluation of the Antioxidant, Antibacterial, and Antibiofilm Activity of the Sesquiterpene Nerolidol. *Arch. Microbiol.* **2021**, *203*, 4303–4311. [CrossRef] [PubMed]

12. Oliveira-Tintino, C.D.d.M.; Santana, J.E.G.; Alencar, G.G.; Siqueira, G.M.; Gonçalves, S.A.; Tintino, S.R.; de Menezes, I.R.A.; Rodrigues, J.P.V.; Gonçalves, V.B.P.; Nicolete, R.; et al. Valencene, Nootkatone and Their Liposomal Nanoformulations as Potential Inhibitors of NorA, Tet(K), MsrA, and MepA Efflux Pumps in *Staphylococcus aureus* Strains. *Pharmaceutics* **2023**, *15*, 2400. [CrossRef]
13. Tran, S.; DeGiovanni, P.; Piel, B.; Rai, P. Cancer Nanomedicine: A Review of Recent Success in Drug Delivery. *Clin. Transl. Med.* **2017**, *6*, e44. [CrossRef]
14. Liu, G.; Hou, S.; Tong, P.; Li, J. Liposomes: Preparation, Characteristics, and Application Strategies in Analytical Chemistry. *Crit. Rev. Anal. Chem.* **2020**, *52*, 392–412. [CrossRef] [PubMed]
15. Azzi, J.; Auezova, L.; Danjou, P.E.; Fourmentin, S.; Greige-Gerges, H. First Evaluation of Drug-in-Cyclodextrin-in-Liposomes as an Encapsulating System for Nerolidol. *Food Chem.* **2018**, *255*, 399–404. [CrossRef]
16. Camilo, C.J.; Leite, D.O.D.; E Silva, A.R.A.; Menezes, I.R.A.; Coutinho, H.D.M.; da COSTA, J.G.M. Lipid Vesicles: Applications, Principal Components and Methods Used in Their Formulations: A Review. *Acta Biol. Colomb.* **2020**, *25*, 339–352. [CrossRef]
17. Choudhury, A.; Sonowal, K.; Laskar, R.E.; Deka, D.; Dey, B.K. Liposome: A Carrier for Effective Drug Delivery. *J. Appl. Pharm. Res.* **2020**, *8*, 22–28. [CrossRef]
18. Filipczak, N.; Pan, J.; Yalamarty, S.S.K.; Torchilin, V.P. Recent Advancements in Liposome Technology. *Adv. Drug Deliv. Rev.* **2020**, *156*, 4–22. [CrossRef]
19. Crommelin, D.J.A.; van Hoogevest, P.; Storm, G. The Role of Liposomes in Clinical Nanomedicine Development. What Now? Now What? *J. Control. Release* **2020**, *318*, 256–263. [CrossRef]
20. Ghosh, R.; De, M. Liposome-Based Antibacterial Delivery: An Emergent Approach to Combat Bacterial Infections. *ACS Omega* **2023**, *8*, 35442–35451. [CrossRef]
21. Huang, M.; Su, E.; Zheng, F.; Tan, C. Encapsulation of Flavonoids in Liposomal Delivery Systems: The Case of Quercetin, Kaempferol and Luteolin. *Food Funct.* **2017**, *8*, 3198–3208. [CrossRef]
22. Rackova, L.; Firakova, S.; Kostalova, D.; Stefek, M.; Sturdik, E.; Majekova, M. Oxidation of Liposomal Membrane Suppressed by Flavonoids: Quantitative Structure-Activity Relationship. *Bioorg. Med. Chem.* **2005**, *13*, 6477–6484. [CrossRef]
23. Tatipamula, V.B.; Kukavica, B. Phenolic Compounds as Antidiabetic, Anti-Inflammatory, and Anticancer Agents and Improvement of Their Bioavailability by Liposomes. *Cell Biochem. Funct.* **2021**, *39*, 926–944. [CrossRef]
24. Fang, J.Y.; Hung, C.F.; Hwang, T.L.; Huang, Y.L. Physicochemical Characteristics and In Vivo Deposition of Liposome-Encapsulated Tea Catechins by Topical and Intratumor Administrations. *J. Drug Target.* **2005**, *13*, 19–27. [CrossRef]
25. Cheng, X.; Yan, H.; Pang, S.; Ya, M.; Qiu, F.; Qin, P.; Zeng, C.; Lu, Y. Liposomes as Multifunctional Nano-Carriers for Medicinal Natural Products. *Front. Chem.* **2022**, *10*, 963004. [CrossRef] [PubMed]
26. Qin, J.; Chen, D.W.; Lu, W.G.; Xu, H.; Yan, C.Y.; Hu, H.Y.; Chen, B.Y.; Qiao, M.X.; Zhao, X.L. Preparation, Characterization, and Evaluation of Liposomal Ferulic Acid In Vitro and In Vivo. *Drug Dev. Ind. Pharm.* **2008**, *34*, 602–608. [CrossRef]
27. Di Sotto, A.; Paolicelli, P.; Nardoni, M.; Abete, L.; Garzoli, S.; Di Giacomo, S.; Mazzanti, G.; Casadei, M.A.; Petralito, S. SPC Liposomes as Possible Delivery Systems for Improving Bioavailability of the Natural Sesquiterpene β-Caryophyllene: Lamellarity and Drug-Loading as Key Features for a Rational Drug Delivery Design. *Pharmaceutics* **2018**, *10*, 274. [CrossRef]
28. Faridi Esfanjani, A.; Assadpour, E.; Jafari, S.M. Improving the Bioavailability of Phenolic Compounds by Loading Them within Lipid-Based Nanocarriers. *Trends Food Sci. Technol.* **2018**, *76*, 56–66. [CrossRef]
29. Mahmood, S.; Chatterjee, B.; Mandal, U.K. Pharmacokinetic Evaluation of the Synergistic Effect of Raloxifene Loaded Transfersomes for Transdermal Delivery. *J. Drug Deliv. Sci. Technol.* **2021**, *63*, 102545. [CrossRef]
30. Singh, S.; Verma, D.; Mirza, M.A.; Das, A.K.; Dudeja, M.; Anwer, M.K.; Sultana, Y.; Talegaonkar, S.; Iqbal, Z. Development and Optimization of Ketoconazole Loaded Nano-Transfersomal Gel for Vaginal Delivery Using Box-Behnken Design: In Vitro, Ex Vivo Characterization and Antimicrobial Evaluation. *J. Drug Deliv. Sci. Technol.* **2017**, *39*, 95–103. [CrossRef]
31. Caster, J.M.; Yu, S.K.; Patel, A.N.; Newman, N.J.; Lee, Z.J.; Warner, S.B.; Wagner, K.T.; Roche, K.C.; Tian, X.; Min, Y.; et al. Effect of Particle Size on the Biodistribution, Toxicity, and Efficacy of Drug-Loaded Polymeric Nanoparticles in Chemoradiotherapy. *Nanomed. Nanotechnol. Biol. Med.* **2017**, *13*, 1673–1683. [CrossRef] [PubMed]
32. Moazeni, M.; Kelidari, H.R.; Saeedi, M.; Morteza-Semnani, K.; Nabili, M.; Gohar, A.A.; Akbari, J.; Lotfali, E.; Nokhodchi, A. Time to Overcome Fluconazole Resistant Candida Isolates: Solid Lipid Nanoparticles as a Novel Antifungal Drug Delivery System. *Colloids Surf. B Biointerfaces* **2016**, *142*, 400–407. [CrossRef]
33. Li, C.; Li, X.; Li, S.; Weng, Y.; Wang, K.; Zhang, T.; Chen, S.; Lu, X.; Jiang, Y.; Liang, J. Development and Validation of a Method for Determination of Encapsulation Efficiency of CPT-11/DSPE-MPEG2000 Nanoparticles. *Med. Chem.* **2016**, *6*, 345–348. [CrossRef]
34. Honary, S.; Zahir, F. Effect of Zeta Potential on the Properties of Nano-Drug Delivery Systems—A Review (Part 2). *Trop. J. Pharm. Res.* **2013**, *12*, 265–273. [CrossRef]
35. Smith, M.C.; Crist, R.M.; Clogston, J.D.; McNeil, S.E. Zeta Potential: A Case Study of Cationic, Anionic, and Neutral Liposomes. *Anal. Bioanal. Chem.* **2017**, *409*, 5779–5787. [CrossRef]
36. Hufschmid, R.; Teeman, E.; Mehdi, B.L.; Krishnan, K.M.; Browning, N.D. Observing the Colloidal Stability of Iron Oxide Nanoparticles In Situ. *Nanoscale* **2019**, *11*, 13098–13107. [CrossRef] [PubMed]
37. Wilhelm, S.; Tavares, A.J.; Dai, Q.; Ohta, S.; Audet, J.; Dvorak, H.F.; Chan, W.C.W. Analysis of Nanoparticle Delivery to Tumours. *Nat. Rev. Mater.* **2016**, *1*, 16014. [CrossRef]

38. González-Vega, J.G.; García-Ramos, J.C.; Chavez-Santoscoy, R.A.; Castillo-Quiñones, J.E.; Arellano-Garcia, M.E.; Toledano-Magaña, Y. Lung Models to Evaluate Silver Nanoparticles' Toxicity and Their Impact on Human Health. *Nanomaterials* **2022**, *12*, 2316. [CrossRef]
39. Danaei, M.; Dehghankhold, M.; Ataei, S.; Hasanzadeh Davarani, F.; Javanmard, R.; Dokhani, A.; Khorasani, S.; Mozafari, M.R. Impact of Particle Size and Polydispersity Index on the Clinical Applications of Lipidic Nanocarrier Systems. *Pharmaceutics* **2018**, *10*, 57. [CrossRef]
40. Hoseini, B.; Jaafari, M.R.; Golabpour, A.; Momtazi-Borojeni, A.A.; Karimi, M.; Eslami, S. Application of Ensemble Machine Learning Approach to Assess the Factors Affecting Size and Polydispersity Index of Liposomal Nanoparticles. *Sci. Rep.* **2023**, *13*, 18012. [CrossRef]
41. Minelli, R.; Serpe, L.; Pettazzoni, P.; Minero, V.; Barrera, G.; Gigliotti, C.L.; Mesturini, R.; Rosa, A.C.; Gasco, P.; Vivenza, N.; et al. Cholesteryl Butyrate Solid Lipid Nanoparticles Inhibit the Adhesion and Migration of Colon Cancer Cells. *Br. J. Pharmacol.* **2012**, *166*, 587–601. [CrossRef] [PubMed]
42. Gonzalez Gomez, A.; Hosseinidoust, Z. Liposomes for Antibiotic Encapsulation and Delivery. *ACS Infect. Dis.* **2020**, *6*, 896–908. [CrossRef] [PubMed]
43. Gubernator, J. Active Methods of Drug Loading into Liposomes: Recent Strategies for Stable Drug Entrapment and Increased In Vivo Activity. *Expert Opin. Drug Deliv.* **2011**, *8*, 565–580. [CrossRef] [PubMed]
44. Grijalvo, S.; Mayr, J.; Eritja, R.; Díaz, D.D. Biodegradable Liposome-Encapsulated Hydrogels for Biomedical Applications: A Marriage of Convenience. *Biomater. Sci.* **2016**, *4*, 555–574. [CrossRef]
45. Kaatz, G.W.; Moudgal, V.V.; Seo, S.M.; Kristiansen, J.E. Phenothiazines and Thioxanthenes Inhibit Multidrug Efflux Pump Activity in *Staphylococcus aureus*. *Antimicrob. Agents Chemother.* **2003**, *47*, 719–726. [CrossRef] [PubMed]
46. Schindler, B.D.; Jacinto, P.; Kaatz, G.W. Inhibition of Drug Efflux Pumps in *Staphylococcus aureus*: Current Status of Potentiating Existing Antibiotics. *Future Microbiol.* **2013**, *8*, 491–507. [CrossRef]
47. DeMarco, C.E.; Cushing, L.A.; Frempong-Manso, E.; Seo, S.M.; Jaravaza, T.A.A.; Kaatz, G.W. Efflux-Related Resistance to Norfloxacin, Dyes, and Biocides in Bloodstream Isolates of *Staphylococcus aureus*. *Antimicrob. Agents Chemother.* **2007**, *51*, 3235–3239. [CrossRef] [PubMed]
48. Costa, S.; Falcão, C.; Viveiros, M.; MacHado, D.; Martins, M.; Melo-Cristino, J.; Amaral, L.; Couto, I. Exploring the Contribution of Efflux on the Resistance to Fluoroquinolones in Clinical Isolates of *Staphylococcus aureus*. *BMC Microbiol.* **2011**, *11*, 241. [CrossRef]
49. Kristiansen, J.E.; Thomsen, V.F.; Martins, A.; Viveiros, M.; Amaral, L. Non-Antibiotics Reverse Resistance of Bacteria to Antibiotics. *Vivo* **2010**, *24*, 751–754.
50. Patel, D.; Kosmidis, C.; Seo, S.M.; Kaatz, G.W. Ethidium Bromide MIC Screening for Enhanced Efflux Pump Gene Expression or Efflux Activity in *Staphylococcus aureus*. *Antimicrob. Agents Chemother.* **2010**, *54*, 5070–5073. [CrossRef]
51. Piddock, L.J.V. Clinically Relevant Chromosomally Encoded Multidrug Resistance Efflux Pumps in Bacteria. *Clin. Microbiol. Rev.* **2006**, *19*, 382–402. [CrossRef] [PubMed]
52. El-Baky, R.M.A.; Sandle, T.; John, J.; Abuo-Rahma, G.E.D.A.; Hetta, H.F. A Novel Mechanism of Action of Ketoconazole: Inhibition of the Nora Efflux Pump System and Biofilm Formation in Multidrug-Resistant *Staphylococcus aureus*. *Infect. Drug Resist.* **2019**, *12*, 1703–1718. [CrossRef] [PubMed]
53. Ramalhete, C.; Spengler, G.; Martins, A.; Martins, M.; Viveiros, M.; Mulhovo, S.; Ferreira, M.J.U.; Amaral, L. Inhibition of Efflux Pumps in Meticillin-Resistant *Staphylococcus aureus* and *Enterococcus faecalis* Resistant Strains by Triterpenoids from *Momordica balsamina*. *Int. J. Antimicrob. Agents* **2011**, *37*, 70–74. [CrossRef]
54. Prakash, B.; Kumar, A.; Singh, P.P.; Songachan, L.S. Antimicrobial and Antioxidant Properties of Phytochemicals. *Funct. Preserv. Prop. Phytochem.* **2020**, *1*, 1–45. [CrossRef]
55. Al Alsheikh, H.M.; Sultan, I.; Kumar, V.; Rather, I.A.; Al-sheikh, H.; Jan, A.T.; Haq, Q.M.R. Plant-Based Phytochemicals as Possible Alternative to Antibiotics in Combating Bacterial Drug Resistance. *Antibiotics* **2020**, *9*, 480. [CrossRef]
56. Kon, K.V.; Rai, M.K. Plant Essential Oils and Their Constituents in Coping with Multidrug-Resistant Bacteria. *Expert Rev. Anti-Infect. Ther.* **2012**, *10*, 775–790. [CrossRef]
57. Oliveira, D.; Borges, A.; Saavedra, M.J.; Borges, F.; Simões, M. Screening of Natural Molecules as Adjuvants to Topical Antibiotics to Treat *Staphylococcus aureus* from Diabetic Foot Ulcer Infections. *Antibiotics* **2022**, *11*, 620. [CrossRef] [PubMed]
58. da Cruz, R.P.; de Freitas, T.S.; Costa, M.D.S.; Dos Santos, A.T.L.; Campina, F.F.; Pereira, R.L.S.; Bezerra, J.W.A.; Quintans-Júnior, L.J.; Araújo, A.A.D.S.; De Siqueira Júnior, J.P.; et al. Effect of α-Bisabolol and Its β-Cyclodextrin Complex as TetK and NorA Efflux Pump Inhibitors in *Staphylococcus aureus* Strains. *Antibiotics* **2020**, *9*, 28. [CrossRef]
59. Barros, N.B.; Migliaccio, V.; Facundo, V.A.; Ciancaglini, P.; Stábeli, R.G.; Nicolete, R.; Silva-Jardim, I. Liposomal-Lupane System as Alternative Chemotherapy against Cutaneous Leishmaniasis: Macrophage as Target Cell. *Exp. Parasitol.* **2013**, *135*, 337–343. [CrossRef]
60. Dutta, D.; Paul, B.; Mukherjee, B.; Mondal, L.; Sen, S.; Chowdhury, C.; Debnath, M.C. Nanoencapsulated Betulinic Acid Analogue Distinctively Improves Colorectal Carcinoma In Vitro and In Vivo. *Sci. Rep.* **2019**, *9*, 11506. [CrossRef]
61. Alavi, S.E.; Koohi Moftakhari Esfahani, M.; Raza, A.; Adelnia, H.; Ebrahimi Shahmabadi, H. PEG-Grafted Liposomes for Enhanced Antibacterial and Antibiotic Activities: An In Vivo Study. *NanoImpact* **2022**, *25*, 100384. [CrossRef]

62. Webb, M.S.; Boman, N.L.; Wiseman, D.J.; Saxon, D.; Sutton, K.; Wong, K.F.; Logan, P.; Hope, M.J. Antibacterial Efficacy against an In Vivo *Salmonella typhimurium* Infection Model and Pharmacokinetics of a Liposomal Ciprofloxacin Formulation. *Antimicrob. Agents Chemother.* **1998**, *42*, 45–52. [CrossRef]
63. Shu, G.; Xu, D.; Zhang, W.; Zhao, X.; Li, H.; Xu, F.; Yin, L.; Peng, X.; Fu, H.; Chang, L.J.; et al. Preparation of Shikonin Liposome and Evaluation of Its In Vitro Antibacterial and In Vivo Infected Wound Healing Activity. *Phytomedicine* **2022**, *99*, 154035. [CrossRef] [PubMed]
64. Blair, J.M.A.; Piddock, L.J.V. How to Measure Export via Bacterial Multidrug Resistance Efflux Pumps. *mBio* **2016**, *7*, e00840-16. [CrossRef] [PubMed]
65. Gibbons, S.; Oluwatuyi, M.; Kaatz, G.W. A Novel Inhibitor of Multidrug Efflux Pumps in *Staphylococcus aureus*. *J. Antimicrob. Chemother.* **2003**, *51*, 13–17. [CrossRef] [PubMed]
66. Markham, P.N.; Westhaus, E.; Klyachko, K.; Johnson, M.E.; Neyfakh, A.A. Multiple Novel Inhibitors of the NorA Multidrug Transporter of *Staphylococcus aureus*. *Antimicrob. Agents Chemother.* **1999**, *43*, 2404–2408. [CrossRef] [PubMed]
67. Oliveira-Tintino, C.D.d.M.; Tintino, S.R.; Muniz, D.F.; Rodrigues dos Santos Barbosa, C.; Pereira, R.L.S.; Begnini, I.M.; Rebelo, R.A.; da Silva, L.E.; Mireski, S.L.; Nasato, M.C.; et al. Chemical Synthesis, Molecular Docking and MepA Efflux Pump Inhibitory Effect by 1,8-Naphthyridines Sulfonamides. *Eur. J. Pharm. Sci.* **2021**, *160*, 105753. [CrossRef]
68. *CLSI CLSI, M100ED29E*; Performance Standards for Antimicrobial Susceptibility Testing: 29th Informational Supplement 20. 29th ed. ANSI: Washington, DC, USA, 2019; ISBN 9781684400324.

molecules

MDPI

Article

Health Benefits, Antioxidant Activity, and Sensory Attributes of Selected Cold-Pressed Oils

Dobrochna Rabiej-Kozioł [1], Monika Momot-Ruppert [1,2], Barbara Stawicka [2] and Aleksandra Szydłowska-Czerniak [1,*]

[1] Department of Analytical Chemistry and Applied Spectroscopy, Faculty of Chemistry, Nicolaus Copernicus University in Toruń, Gagarina 7, 87-100 Toruń, Poland; d.rabiej@umk.pl (D.R.-K.); monika.momot@bunge.com (M.M.-R.)

[2] Bunge Polska Sp. z o.o., Niepodległości 42, 88-150 Kruszwica, Poland; barbara.stawicka@bunge.com

* Correspondence: olasz@umk.pl

Abstract: The consumption of cold-pressed oils (CPOs) has continuously increased due to their health-promoting compounds, such as polyunsaturated fatty acid (PUFA), tocopherols, sterols, and polyphenols. This study focused on the estimation and comparison of the physicochemical properties and sensory quality of six CPOs: linseed oil (CPLO), pumpkin oil (CPPO), milk thistle oil (CPMTO), rapeseed oil (CPRO), camelina oil (CPCO), and sunflower oil (CPSO), which are the most popular in the Polish market. These oils were analysed for their fatty acid composition (FAC), their tocopherol, sterol, polycyclic aromatic hydrocarbon (PAHs), water, and volatile matter (WVM) contents, as well as their antioxidant activity (AA) and oxidative stability parameters. Moreover, quantitative descriptive analysis (QDA) was performed to obtain detailed information on the sensory profiles and quantitative data on the CPOs' attributes that affected consumer acceptability and purchase intent. All of the analysed CPOs were rich in PUFA (27.94–68.42%). They were characterised by the different total amounts of health-beneficial compounds, such as tocopherols (TTC = 44.04–76.98 mg/100 g), sterols (TSC = 300–684 mg/100 g), and polyphenols (TPC = 2.93–8.32 mg GA/100 g). Additionally, their AA was determined using 2,2-diphenyl-1-picrylhydrazyl (DPPH), 2,2′-azino-bis(3-ethylbenzothiazoline-6-sulfonic acid) (ABTS), and ferric reducing antioxidant power (FRAP) methods, with results ranging between 185.36–396.63, 958.59–1638.58, and 61.93–119.21 µmol TE/100 g, respectively. However, the deterioration parameters of CPOs, such as peroxide values (PV = 0.24–4.61 meq O_2/kg), p-anisidine values (pAnV = 0.39–4.77), acid values (AV = 0.31–2.82 mg KOH/g), and impurity amounts (Σ4PAHs = 1.16–8.76 µg/kg and WVM = 0.020–0.090%), did not exceed the level recommended by the Codex Alimentarius Commission. The obtained results indicated that all of the investigated CPOs are valuable sources of health-promoting bioactive compounds.

Keywords: cold-pressed oils; antioxidants; antioxidant activity; nutritional indicates; oxidative stability; hazardous compounds; sensory analysis

1. Introduction

In recent years, cold-pressed oils (CPOs) have received increased attention due to their health-beneficial impact. The advantages of these oils over refined ones are related to higher amounts of bioactive substances, not being otherwise removed during refining processes. Another reason for the growing popularity of CPOs over refined oils is consumers choosing fewer processed products. According to the Codex Alimentarius [1], the CPOs have to be produced only by mechanical procedure, without heat treatment. These oils can be purified by washing them with water, then settling, filtering, and centrifuging them. Chemical and physical refining processes, such as degumming, neutralisation, bleaching, and deodorisation, as well as supplementation of CPOs with synthetic additives, are banned [2,3]. For this reason, CPOs are rich sources of minor constituents, such as the phytosterols, tocopherols, carotenoids, and polyphenols that are associated with health benefits. These oils are also distinguished by their high amounts of vitamins and minerals.

Additionally, CPOs have characteristic flavours and aromas, depending on the seeds or fruits from which they were pressed [4,5]. The characteristic aroma and taste has led to CPOs being mainly used as salad dressing, spread, or otherwise consumed directly. In addition, some CPOs containing high content of oleic acid can be applied for culinary operations at higher temperatures, such as for frying and cooking [6,7].

Furthermore, each CPO has a characteristic fatty acid composition (FAC) and profiles of accompanying compounds, such as tocopherols, sterols, and polyphenols. The unique minor compounds and FAC have been linked with the health-promoting attributes of these oils [4]. Cold-pressed camelina oil (CPCO) can be helpful for skin ailments, cardiovascular diseases, cancer, and chronic diseases [8]. Similarly, cold-pressed milk thistle oil (CPMTO) can be used for skin ailments, such as pruritic, psoriatic, or exsiccated skin, as well as facilitate the treatment of liver- and bile-related disorders due to its anti-inflammatory properties [9]. Additionally, CPMTO reduces paracetamol hepatotoxicity [4]. Cold-pressed linseed oil (CPLO) has preventive properties for coronary heart disease, some types of cancer, and neurological and hormonal disorders [10]. Cold-pressed pumpkin oil (CPPO) has become an effective means for the treatment of both cardiovascular diseases during menopause and benign prostatic hyperplasia in men, as well as androgenic alopecia [11]. Also, cold-pressed sunflower oil (CPSO) shows cardiovascular benefits associated with reduced total plasma cholesterol and low-density lipoprotein (LDL) cholesterol levels. In addition, this oil is helpful in lowering atherosclerosis, artery disease, and stroke, due to the presence of vitamin E at high levels [12]. However, the health-benefit properties of cold-pressed rapeseed oil (CPRO) are linked with regulating blood lipid profile, insulin sensitivity, and glycaemic control [13].

The health-promoting and sensory properties of CPOs have led to some oils being considered new functional and luxury products [14]. Especially niche CPOs should have an evident colour, clarity, and a fresh and clean aroma characteristic of the oil plants. However, it is well known that the origins of oilseeds and various cultivars, as well as their processing techniques, are associated with the composition of minor compounds and the organoleptic characteristics of oils. Therefore, previous studies have demonstrated that there have been differences in the overall quality, including the physicochemical and sensory properties, of commercially available CPOs derived from black cumin seeds, camelina seeds, rapeseed, and sunflower seeds [2,8,15,16]. Nevertheless, a sensory analysis allows the estimation of the proposed oil products' quality, freshness, and consumer acceptability. In recent studies, quantitative descriptive analysis (QDA) has been a well-known and essential tool for quantifying the sensory attributes of oils that have been cold-pressed from various oilseeds, as well as for facilitating the statistical description of the obtained results [2,8,15].

However, consumers are not willing to compromise on healthful properties or the sensory quality of functional products. Functional compounds present in CPOs, such as phenolics, are often associated with diminished acceptability to consumers, due to their bitter or pungent flavour [17]. Furthermore, one of the most critical factors influencing the sensory quality and nutritional value of CPOs is the oxidation process of unsaturated fatty acids (primarily linolenic acid), as well as the degradation and volatilisation of bioactive compounds. Primary (peroxides) and secondary (aldehydes, ketones, and hydrocarbons) oxidation products contribute to the unpleasant flavours and aromas associated with rancid oils and damage to health, lowering consumer acceptance. The secondary oxidation product, E-2-heptenal, was responsible for the unpleasant rancid or fried odour of sunflower oil, while cyclic octapeptide (consisting of eight amino acids: proline, leucine, phenylalanine, isoleucine, sulfoxidised methionine, leucine, valine, and phenylalanine) caused a bitter 'off' taste in oil cold-pressed from linseed [2,18]. Thus, there has been a negative relationship between the perceived healthiness of some oils and fats and their sensory attributes.

On the other hand, because of the lipophilic nature of polycyclic aromatic hydrocarbons (PAHs), unsafe levels of these carcinogenic contaminants can be found in oils and fats.

Oil contamination by PAHs may be attributed to the contact of the oilseeds with polluted air, the uptake by the oilseed plants through soil, and any contaminated element present in the production line or mineral oil residues from packaging [19].

Many research studies have demonstrated that the quality control plays an important role in the safety assessment of edible oils obtained by mechanical operations [2,20].

It is worth mentioning that information about the health properties of functional compounds (e.g., plant sterols, liposoluble vitamins, phenolics) in oils influence consumer sensory profiling and purchasing intentions. Hence, extrinsic properties and intrinsic sensory attributes of oils play a crucial role in purchasing intentions, and they can be a fundamental tool for the consumer's judgement of the oil quality.

To the best of our knowledge, the chemical composition, oxidative stability, and sensory properties of oils pressed from various oilseeds have been studied previously. However, relationships between the antioxidant, nutritional, and healthy features of CPOs and their sensory profiles have not yet been reported.

Therefore, following were the aims of this research: (I) To determine the profiles of fatty acids and liposoluble bioactive compounds (tocopherols and sterols), as well as total phenolic content (TPC) and antioxidant activity (AA) of six CPOs: CPLO, CPPO, CPMTO, CPRO, CPCO, and CPSO, which are the most common in the Polish retail market; (II) To investigate their oxidative status parameters and contaminants; (III) To calculate the nutritional quality indexes of oils based on their FACs for the prevention of cardiovascular diseases; (IV) To develop a vocabulary with adequate descriptors for oils and analysis of their sensory properties and consumers' acceptance. Moreover, multivariate analyses, such as principal component analysis (PCA) and hierarchical cluster analysis (HCA), were applied to check similarities and differences between the investigated CPOs and physicochemical, nutritional, and sensory properties.

For the first time, these studies have provided additional information to promote healthy oils containing bioactive compounds without undesirable compounds and have presented sensory attributes to create their unique position in the marketplace.

2. Results and Discussion

2.1. *Antioxidant Characteristics of Cold-Pressed Oils*

2.1.1. Fatty Acid Compositions and Nutrient Values

The fatty acid profiles have been frequently used to indicate the quality of CPOs, especially for the identification of their authenticity. Each year in the European Union, food frauds are detected where virgin oils have been adulterated with other vegetable oils. As can be seen in Table 1, all of the studied oils had typical fatty acid compositions recommended by Codex Alimentarius standard [1] and similar to those described by other authors [4,5,8,10,16,20].

On the other hand, the fatty acid profile of oil can affect its vulnerability to oxidation and deterioration reactions. Polyunsaturated fatty acids (PUFA) such as linoleic acid (C18:2) and linolenic acid (C18:3) are much more prone to oxidation than monounsaturated (MUFA) oleic acid (C18:1) and saturated (SFA) stearic acid (C18:0) [3,21–23].

The total SFA content in the investigated oil samples varied from 7.12 to 19.53%. Among SFA, palmitic (C16:0) and stearic (C18:0) acids were predominantly found in the analysed oils, having percentages of 4.63–12.41% and 1.61–5.90%, respectively. Similar moderate total SFA content in CPLO (11.63%) and CPSO (11.20%) were observed, while CPMTO (19.53%) and CPPO (18.92%) were the richest sources of SFA (Table 1). Moreover, arachidic acid (C20:0 = 2.90%) and behenic acid (C22:0 = 1.91%) reached the highest percentages in CPMTO from all of the evaluated oils.

Table 1. Fatty acid compositions and nutritional values of the studied cold-pressed oils.

Fatty Acid (%)	Oil Sample					
	CPLO	CPPO	CPMTO	CPRO	CPCO	CPSO
C16:0	5.93 ± 0.19 c	12.41 ± 0.13 f	8.37 ± 0.08 e	4.63 ± 0.03 a	5.41 ± 0.13 b	6.58 ± 0.06 d
C18:0	5.44 ± 0.19 d	5.90 ± 0.11 e	5.45 ± 0.21 d	1.61 ± 0.06 a	2.38 ± 0.07 b	3.52 ± 0.17 c
C20:0	0.22 ± 0.01 a	0.41 ± 0.01 b	2.90 ± 0.04 e	0.57 ± 0.01 c	1.62 ± 0.02 d	0.20 ± 0.00 a
C22:0	0.04 ± 0.00 a	0.10 ± 0.00 b,c	1.91 ± 0.33 e	0.31 ± 0.00 c	0.31 ± 0.00 c	0.70 ± 0.00 d
C24:0	<DL	0.10 ± 0.00 a	0.90 ± 0.00 c	<DL	<DL	0.20 ± 0.00 b
ΣSFA	11.63	18.92	19.53	7.12	9.72	11.20
C16:1	0.07 ± 0.00 a	0.10 ± 0.00 b	0.10 ± 0.00 b	0.18 ± 0.01 c	0.10 ± 0.00 b	0.10 ± 0.00 b
C18:1	19.85 ± 0.18 b	30.40 ± 0.26 d	24.69 ± 0.13 c	63.26 ± 0.06 e	18.11 ± 0.21 a	30.64 ± 0.54 d
C20:1	0.13 ± 0.00 a	0.10 ± 0.00 a	0.90 ± 0.00 b	1.28 ± 0.02 c	14.43 ± 0.19 d	0.10 ± 0.00 a
C22:1	<DL	0.10 ± 0.00 a	<DL	0.11 ± 0.00 a	2.69 ± 0.13 b	<DL
ΣMUFA	20.05	30.70	25.69	64.83	35.33	30.84
C18:2	16.30 ± 0.11 a	50.39 ± 0.36 d	56.46 ± 0.06 e	19.82 ± 0.09 c	19.38 ± 0.20 b	57.64 ± 0.08 f
C18:3	52.12 ± 0.28 d	0.20 ± 0.00 a	0.30 ± 0.00 a	8.12 ± 0.14 b	30.29 ± 0.21 c	0.10 ± 0.00 a
ΣPUFA	68.42	50.59	56.76	27.94	49.67	57.74
COX	13.14	5.54	6.13	4.43	8.72	6.26
AI	0.06	0.08	0.07	0.02	0.03	0.04
TI	0.06	0.45	0.33	0.09	0.07	0.23
HH	14.79	6.47	9.62	19.53	12.30	13.23

Results are expressed as mean percentage of total fatty acids ± standard deviation (SD) (n = 3); different letters (a–f) within the same rows indicate significant differences between the percentages of fatty acids of cold-pressed oils (one-way ANOVA and Duncan test, $p < 0.05$). Abbreviations: DL—detection limit; C16:0—palmitic acid; C18:0—stearic acid; C20:0—arachidic acid; C22:0—behenic acid; C16:1—palmitoleic acid; C18:1—oleic acid; C20:1—eicosenoic acid; C18:2—linoleic acid; C18:3—linolenic acid; SFA—saturated fatty acids; MUFA—monounsaturated fatty acids; PUFA—polyunsaturated fatty acids; COX—calculated oxidisability value; AI—atherogenicity index; TI—thrombogenicity index; HH—ratio of hypocholesterolemic to hypercholesterolemic fatty acids; CPLO—cold-pressed linseed oil; CPPO—cold-pressed pumpkin oil; CPMTO—cold-pressed milk thistle oil; CPRO—cold-pressed rapeseed oil; CPCO—cold-pressed camelina oil; CPSO—cold-pressed sunflower oil.

It is well known that high SFA intake increases the risk of developing metabolic syndrome, diabetes, cardiovascular diseases, heart failure, and mortality. Therefore, the consumption of oils with high SFA content should be limited in order to maintain a healthy life. Ulbricht and Southgate [24] classified short-chain SFA such as lauric (C12:0), myristic (C14:0), and C16:0 acids as atherogenic, while C14:0, C16:0 and C18:0 as thrombogenic SFA.

For comparison, a similar amounts of total SFA and C16:0 were determined in oils pressed from rapeseed (SFA = 6.73–7.32% and C16:0 = 4.13–4.33%) and pumpkin seeds (SFA = 18.27% and C16:0 = 12.23%) [23,25].

Moreover, CPRO revealed the highest total content of MUFA and the lowest total content of SFA, with values of 64.83% and 7.12%, respectively (Table 1). However, CPLO had the highest PUFA content (68.42%), among the investigated samples.

Choo et al. [26] reported similar values of C18:3 ranging between 51.8 and 60.4% for linseed oils.

It can be noted that the C18:1 acid (63.26%) was predominant in CPRO, while its percentage in CPSO and CPPO was approximately two times lower (30.64 and 30.40%, respectively) than in CPRO. Among the studied oils, the lowest concentration of C18:1 (18.11–24.69%) was found in CPCO, CPLO, and CPMTO. Interestingly, gadoleic acid (C20:1 = 14.43%) was at the highest level in CPCO. Similar results of C18:1 content (12.01–23.57%) in market linseed oils were reported by Symoniuk et al. [27].

Taking into consideration the health benefits of MUFA, especially C18:1, and their effect on the oxidative stability of the oils, there is a need to select CPOs with high amounts of MUFA. It has been documented that MUFA can reduce LDL cholesterol and increase high-density lipoprotein (HDL) cholesterol levels.

Generally, in all CPOs (except CPRO), PUFA (49.67–68.42%) dominated (Table 1). It is noteworthy that C18:2 (50.39–57.64%) was the major PUFA in CPPO, CPMTO, and CPSO.

However, C18:3 was the most abundant PUFA in CPLO and CPCO (52.12 and 30.29%, respectively).

Other researchers have presented similar amount of C18:2 in CPPO (44.95–53.59%) and CPMTO (51.78–53.87%) [20,23]. Furthermore, the content of C18:3 in 15 linseed oils ranged between 44.90 and 64.62% [27].

Although CPRO had a relatively low C18:3 level (8.12%), the nutritionally desirable ratio of omega-6 (ω-6) to omega-3 (ω-3) fatty acids (about 2:1) in this oil was observed [13]. The optimal ratio of ω-6:ω-3 fatty acids, close to balancing, is recommended for normal physiological functions in the body, and it is crucial in preventing many chronic diseases such as cancer, inflammatory and autoimmune, and cardiovascular diseases. This fact is of great importance for the regular intake of CPRO and its effect on human health. Nevertheless, the concentration of C18:3 in CPPO, CPMTO, and CPSO did not exceed 0.3% (Table 1).

Some recent studies have found that PUFA in vegetable oils can prevent coronary heart and cardiovascular diseases, cancer, inflammatory, thrombotic and autoimmune diseases, hypertension, type two diabetes, renal diseases, rheumatoid arthritis, ulcerative colitis, and Crohn's disease [12,13,17].

On the other hand, FACs provide information about the oxidative stability of oils and their nutritional values. In addition, based on the percentage of fatty acids in the studied CPOs, the oxidisability values (COX) associated with oxidative stability, as well as nutritional quality indicators related to cardiovascular and heart diseases, such as atherogenicity index (AI), thrombogenicity index (TI), and the ratio of hypocholesterolemic to hypercholesterolemic (HH) fatty acids, were calculated. These indicators were more helpful for the evaluation of the nutritional quality of oils and their autoxidation rates than the fatty acid profiles [6,28].

It can be noted that the COX values of the studied commercially available CPOs ranged from 4.43 for CPRO to 13.14 for CPLO (Table 1). These COX values increased in the following order: CPRO < CPPO < CPMTO < CPSO < CPCO < CPLO. The calculated COX results proved that CPLO had the highest susceptibility to oxidation, while CPRO was the most stable in autoxidation reactions among the investigated oils. The obtained COX values indicate that there were relationships between the amounts of SFA, MUFA, and PUFA in CPOs and their susceptibility to oxidation.

Generally, the COX values calculated for the six CPOs were similar to those observed by Symoniuk et al. [20,27] for CPLO (12.03–15.40), CPRO (4.27–4.41), CPMTO (5.64–5.86), CPCO (8.76–9.42), although CPPO revealed slightly higher COX indexes (6.07–6.32).

Additionally, levels of SFA, MUFA, and PUFA and the ω-6:ω-3 ratio significantly affect the dietary factors of oils. The AI and TI indices characterise the effects of individual fatty acids present in oils on human health, mainly the incidence of atherosclerosis, the development of blood clots, atheroma, and thrombus formation [29]. From the nutritional point of view, low values of AI and TI are preferred in the human diet. Furthermore, the HH ratio is an index related to cholesterol metabolism. Higher HH ratios are considered to be more beneficial to human health [30]. Hypocholesterolemic acids, such as unsaturated C18 and C20 fatty acids, are more effective in decreasing total cholesterol level, whereas C14:0 and C16:0 are classified as hypercholesterolemic acids increasing cholesterol level.

The AI and TI, amounting to between 0.02–0.08 and 0.06–0.45, respectively, were low for all investigated CPOs (Table 1). The lowest AI (0.02) and the highest HH ratio (19.53) had CPRO. In contrast, the highest values of AI (0.08) and TI (0.45), as well as the lowest HH ratio (6.47) were observed for CPPO. The calculated AI, TI, and HH results indicate that enriching the diet with studied oils can reduce coronary heart diseases.

Similar AI and TI values for rapeseed oil (0.05 and 0.09), sunflower oil (0.06 and 0.18), and linseed oil (0.06 and 0.05) were reported by Khalili Tilami et al. [29], while pumpkin oil revealed higher values of AI (0.21) and TI (0.47) than our results (AI = 0.08 and TI = 0.45).

For comparison, AI, TI, and HH values for refined rapeseed oil were 0.04, 0.09, and 21.91, respectively [6]. However, higher AI (0.23) and TI (0.53), as well as lower HH (4.15) for CPMTO were calculated by Ying et al. [31].

2.1.2. Tocopherol Profiles

Tocopherols play an important role in the antioxidant properties of oils rich in PUFA. These lipophilic antioxidants can chelate metal ions and scavenge free radicals. Thus, they are the most effective in protecting the oil from oxidation processes [4]. As can be seen in Table 2, total tocopherol content (TTC) in the tested CPOs ranged between 44.04 mg/100 g for CPLO and 76.98 mg/100 g for CPCO.

Table 2. Tocopherol compositions and contents in the studied cold-pressed oils.

Tocopherol Content (mg/100 g)	**Oil Sample**					
	CPLO	**CPPO**	**CPMTO**	**CPRO**	**CPCO**	**CPSO**
α-Tocopherol	1.78 ± 0.02 [a]	7.49 ± 0.34 [b]	38.91 ± 0.67 [d]	27.00 ± 1.18 [c]	1.20 ± 0.04 [a]	73.37 ± 2.61 [e]
β-Tocopherol	<DL	<DL	2.84 ± 0.08 [b]	<DL	<DL	2.56 ± 0.04 [a]
γ-Tocopherol	42.26 ± 1.02 [b]	56.96 ± 0.55 [c]	4.34 ± 0.20 [a]	42.11 ± 0.74 [b]	74.27 ± 1.85 [d]	<DL
δ-Tocopherol	<DL	<DL	<DL	1.10 ± 0.05 [a]	1.51 ± 0.03 [b]	<DL
TTC	44.04 ± 1.04 [a]	64.45 ± 0.50 [b]	46.09 ± 0.57 [a]	70.21 ± 1.04 [c]	76.98 ± 1.78 [d]	75.93 ± 2.57 [d]

Results are expressed as mean ± standard deviation (SD) (n = 3); different letters (a–e) within the same rows indicate significant differences between tocopherol contents in cold-pressed oils (one-way ANOVA and Duncan test, $p < 0.05$). Abbreviations: DL—detection limit; TTC—total tocopherol content; CPLO—cold-pressed linseed oil; CPPO—cold-pressed pumpkin oil; CPMTO—cold-pressed milk thistle oil; CPRO—cold-pressed rapeseed oil; CPCO—cold-pressed camelina oil; CPSO—cold-pressed sunflower oil.

It is noteworthy that CPCO, CPSO, and CPRO were the richest sources of TTC, containing approximately two times higher TTC (70.21–76.98 mg/100 g) than the CPLO (TTC = 44.04 mg/100 g) and CPMTO (TTC = 46.09 mg/100 g). However, TTC in CPPO was at the moderate level (64.45 mg/100 g). The results of TTC and individual tocopherols in the investigated oils demonstrated a wide variability. Evidence from large observational studies indicates that the climate during the growth and ripening of oilseeds, their genetic varieties, cold-pressing parameters, and oil storage conditions can affect amounts of tocopherols in CPOs [3,32,33].

Other researchers noticed similar TTC results in different CPOs. For instance, TTC values in CPLO, CPSO, CPRO, CPCO, CPPO, and CPMTO were 38.7–162.4 mg/100 g, 68.9–75.5 mg/100 g, 63.8–66.0 mg/100 g, 70.0–85.0 mg/100 g, 92.6 mg/100 g, and 42.9 mg/100 g, respectively [14,18,32,34].

Some recent studies have suggested that the antioxidant properties of tocopherol homologues were dependent on their concentrations in bulk oils. α-Tocopherol had maximum antioxidant activity at 100 mg/kg, whereas higher amounts of γ-tocopherol (250–500 mg/kg) revealed the best antioxidant activity in bulk oil without a prooxidant effect [4]. However, exceeding the optimum concentration of α-tocopherol in oil resulted in the generation of α-tocopherol peroxy, α-tocopherol oxy, α-tocopherolquinone oxy, and hydroxy radicals [1,35].

It is evident that the α- and γ-tocopherol homologues dominated in all of the studied CPOs (Table 2). Our results showed the highest amount of α-tocopherol in CPSO (73.37 mg/100 g), whereas CPCO (1.20 mg/100 g) and CPLO (1.78 mg/100 g) contained the lowest concentration of α-tocopherol. The α-tocopherol content in CPMTO and CPRO was almost two-fold lower compared to CPSO (Table 2). This homologue has the highest nutritional value because of the specificity of absorption and distribution within the human body [3,33]. Interestingly, the selective accumulation of α-tocopherol is mediated by α-tocopherol-transfer protein (α-TTP) (hepatic cytosolic protein), which preferentially binds to α-tocopherol over other homologues. For this reason, α-tocopherol is metabolized relatively slowly than other forms, and it is present in more quantities in human plasma. Moreover, α-tocopherol accumulated in areas with maximum free radical production, especially in the mitochondrial membranes and endoplasmic reticulum in the lungs and heart [36].

Nutrition standards for the Polish population recommend an all-day intake of α-tocopherol of 8 and 10 mg for women and men, respectively. However, some countries recommend higher amounts of α-tocopherol, ranging between 11 and 15 mg per adult [37]. The daily recommendation of E vitamin can be met by eating 7 spoons (55.5 g) of CPRO, almost 5 spoons (38.6 g) of CPMTO, and 2.5 spoons (20.4 g) of CPSO. From this point of view, some of the tested CPOs appear to be rich sources of α-tocopherol, which can protect against atherosclerosis, cardiovascular diseases, cataracts, neural tube defects, and cancer [33].

It is noteworthy that the γ-tocopherol predominated in CPCO (74.27 mg/100 g), whereas CPLO, CPPO, and CPRO contained an approximately two times lower amount of this homologue (42.11–56.96 mg/100 g). Among all of the studied oils, the γ-tocopherol was only undetected in CPSO (Table 2).

In contrast to α-tocopherol, γ-tocopherol can more effectively prevent lipid peroxidation and limit the prooxidant effect of α-tocopherol. The γ-tocopherol is consumed more slowly than α-tocopherol due to its higher stability [34]. Furthermore, γ-tocopherol has unique anti-inflammatory activity relevant to chronic disease prevention compared to α-tocopherol [3,33].

Based on the obtained results, close connections can be observed between profiles of tocopherols and fatty acids of the investigated oils. The high amount of α-tocopherol is linked with C18:2, while γ-tocopherol content is associated with the presence of C18:3. This can be explained by the fact that oil containing high PUFA content is highly prone to oxidation, and tocopherols can inhibit these undesirable reactions [3,4]. The obtained results confirmed that CPLO and CPCO contained the highest concentrations of C18:3 (52.12 and 30.29%) and γ–tocopherol (42.26 and 74.27 mg/100 g) (Tables 1 and 2). Moreover, CPSO and CPMTO were characterised by high amounts of C18:2 (57.64 and 56.46%) and α-tocopherol (73.37 and 38.91 mg/100 g). Unexpectedly, relationships between concentrations of fatty acids and tocopherol forms in CPPO were not observed. The C18:2 (50.39%) and γ-tocopherol (56.96 mg/100 g) were predominated in the studied CPPO. Other researchers noticed similar results in terms of the C18:2 (36.2–62.8%) and γ-tocopherol (52.3–644 μg/g) contents in pumpkin seed oils [38,39].

According to this study, β-tocopherol was detected only in CPMTO and CPSO in concentrations of 2.84 and 2.56 mg/100 g, respectively (Table 2). Moreover, β- and γ-tocopherols have similar reactivity with peroxyl radicals, while β-tocopherol does not inhibit smooth muscle proliferation [33,40].

Among all tocopherol homologues, δ-tocopherol was detected only in CPRO (1.10 mg/100 g) and CPCO (1.51 mg/100 g) (Table 2).

A recent studies have demonstrated that δ-tocopherol characterised more potent antiradical properties than γ-tocopherol [4]. For comparison, Gliszczyńska-Świgło et al. [33] reported that the concentration of δ-tocopherol in edible plant oils (except soybean oil) did not exceed 30 mg/kg.

The results of tocopherol profiles obtained in this study generally agree with the previously published data by other authors. Similar amounts of α-tocopherol (26.2–27.2 and 63.5–72.2 mg/100 g), γ-tocopherol (35.9–38.9 and 0.7–3.1 mg/100 g), β-tocopherol (not detected and 2.3–2.7 mg/100 g), and δ-tocopherol (0.8–0.9 mg/100 g and not detected) in CPRO and CPSO respectively were found by Franke et al. [32]. However, 17 different oils cold-pressed from rapeseed contained α-, γ-, and δ-tocopherols varying from 114.3 to 324.7 mg/kg, 155.1–508.2 mg/kg, and 5.3–18.0 mg/kg, respectively [16]. As expected, five extra virgin olive oil (EVOO) samples tested by the same authors had lower mean amounts of α- and γ-tocopherols (181.2 and 13.8 mg/kg), while δ-tocopherol was undetected. Moreover, α-, γ-, and δ-tocopherols in CPLO after 150 days of storage ranged between 0.8–3.2 mg/100 g, 28.1–41.0 mg/100 g, and 0.3–0.8 mg/100 g, respectively [18]. Generally, the tocopherol profile (α-tocopherol = 204.1 mg/kg, γ-tocopherol = 55.5 mg/kg, δ-tocopherol = 14.6 mg/kg) of CPMTO reported by Grajzer et al. [4] was similar, while levels of these homologues (α-tocopherol not detected, γ-tocopherol = 817.7 mg/kg, δ-tocopherol = 126 mg/kg) in CPCO were slightly higher than results presented in this study.

2.1.3. Sterol Profiles

The sterol compositions and the total sterol contents (TSC) in the six investigated oils are presented in Table 3.

Table 3. Sterol compositions and contents in the studied cold-pressed oils.

Sterol (mg/100 g)	Oil Sample					
	CPLO	CPPO	CPMTO	CPRO	CPCO	CPSO
Cholesterol	1 ± 0 [a,b]	1 ± 0 [a,b]	46 ± 1 [d]	2 ± 0 [b]	26 ± 1 [c]	<DL
Brassicasterol	3 ± 0 [b]	<DL	1 ± 0 [a,b]	73 ± 3 [d]	22 ± 1 [c]	<DL
Δ-5-Avenasterol	44 ± 2 [e]	8 ± 0 [a]	10 ± 0 [b]	10 ± 0 [b]	35 ± 1 [d]	15 ± 1 [c]
β-Sitosterol	166 ± 5 [a]	182 ± 10 [b]	192 ± 3 [b]	336 ± 19 [e]	265 ± 5 [d]	218 ± 4 [c]
Δ-7-Avenasterol	1 ± 0 [a]	57 ± 2 [d]	21 ± 1 [c]	<DL	<DL	16 ± 1 [b]
Δ-7-Stigmasterol	5 ± 0 [a]	21 ± 2 [b]	141 ± 4 [d]	2 ± 0 [a]	4 ± 0 [a]	61 ± 3 [c]
Stigmasterol	27 ± 1 [c]	2 ± 0 [a]	37 ± 1 [d]	3 ± 0 [a]	9 ± 0 [b]	26 ± 1 [c]
Campesterol	80 ± 2 [c]	7 ± 0 [a]	30 ± 1 [b]	249 ± 2 [e]	110 ± 3 [d]	32 ± 1 [b]
Unidentified steroles	8 ± 1 [a]	22 ± 1 [d]	31 ± 1 [e]	9 ± 0 [b]	8 ± 1 [a]	15 ± 1 [c]
TSC	335 ± 7 [b]	300 ± 10 [a]	509 ± 4 [e]	684 ± 13 [f]	479 ± 6 [d]	383 ± 6 [c]

Results are expressed as mean ± standard deviation (SD) (n = 3); different letters (a–f) within the same rows indicate significant differences between sterol contents in cold-pressed oils (one-way ANOVA and Duncan test, $p < 0.05$). Abbreviations: DL—detection limit; TSC—total sterol content; CPLO—cold-pressed linseed oil; CPPO—cold-pressed pumpkin oil; CPMTO—cold-pressed milk thistle oil; CPRO—cold-pressed rapeseed oil; CPCO—cold-pressed camelina oil; CPSO—cold-pressed sunflower oil.

Differences in TSC and the individual sterol compositions were evident for different CPO samples. The highest concentration of total sterols (684 mg/100 g) was revealed for CPRO, while CPPO had the lowest TSC (300 mg/100 g). Thus, CPRO was within the permissible TSC limits (450–1130 mg/100 g) quoted under the Codex Alimentarius [1]. However, Grajzer et al. [4] found approximately two times higher overall median quantity of total sterols (5459.9 mg/kg) in six commercially available CPPO samples.

From a nutritional and healthy point of view, sterols delay or inhibit the lipid oxidation process by scavenging free radicals. Moreover, they can reduce the LDL cholesterol fraction in human blood [41]. There is no doubt that the intake of sterols decreases LDL cholesterol and protects against cardiovascular diseases. On the other hand, sterol profile analyses can be used as a detection tool for adulteration of CPOs [42].

The sterol profiles of the tested CPOs can be affected by several factors, such as various cultivars, and growing, processing and storage conditions.

It can be noted that the β-sitosterol was the predominant compound in studied oils, with the highest amount in CPRO (336 mg/100 g), whereas CPLO (166 mg/100 g) comprised the lowest content of this sterol (Table 3). Moreover, a CPCO was characterised by a high level of β-sitosterol (265 mg/100 g). Szterk et al. [43] found a somewhat higher content of β-sitosterol in crude *Camelina sativa* oil (361.3 mg/100 g) and crude linseed oils (162.5 mg/100 g). However, refined rapeseed oil contained insignificantly lower β-sitosterol concentration (324.7 mg/100 g) [43]. Unexpectedly, technological processes did not significantly influence the highest level of β-sitosterol (approximately 94%) in EVOO samples obtained using super pressure (SP), continuous extraction techniques (2P and 3P), and traditional extraction, respectively [44].

On the other hand, brassicasterol was only found in oils cold-pressed from oilseeds belonging to the family of *Brassicaceae*, including CPRO (73 mg/100 g) and CPCO (22 mg/100 g), while campesterol (80–249 mg/100 g) dominated in both mentioned oil samples and CPLO (Table 3). Krygier et al. [45] reported similar amounts of brassicasterol (69.4–89.3 mg/100 g) and campesterol (216.1–231.3 mg/100 g) in three commercial CPROs. Additionally, campesterol concentrations in CPLO (80 mg/100 g) and CPCO (110 mg/100 g) were in close agreement with those for 30 commercial cold-pressed flaxseed oils from various manufacturers available on the Polish market (64.94–115.36 mg/100 g), crude linseed (97.5 mg/100 g), and *Camelina sativa* (109.6 mg/100 g) oils analysed by Szterk et al. [43] and Mikołajczak et al. [46].

However, Δ-7-stigmasterol was the principal sterol in CPMTO and CPSO, comprising up to 30% of the TSC (Table 3). The sterols identified in lower concentrations were Δ-5-avenasterol (44 and 35 mg/100 g in CPLO and CPCO, respectively) and Δ-7-avenasterol (57 and 21 mg/100 g in CPPO and CPMTO, respectively).

2.1.4. Total Phenolic Content

Phenolic compounds have strong antioxidant properties due to their ability to reduce other compounds, singlet oxygen quenching, hydrogen donating and metal chelating. On the other hand, polyphenols can behave as anti-inflammatory, anticancer and antibacterial agents. Therefore, phenolic compounds inhibit the oxidation reactions in oils, as well as affect their sensory and nutritional properties. It is known that CPOs are rich sources of phenolic compounds because they are not removed during a refining process [4,20,27].

The results of TPC in six CPOs analysed by the Folin–Ciocalteu (F–C) method are summarized in Table 4.

Table 4. Total phenolic content and antioxidant activity of the studied cold-pressed oils.

Oil Sample	Total Phenolic Content (mg GA/100 g)	Antioxidant Activity (μmol TE/100 g)		
	TPC	DPPH	ABTS	FRAP
CPLO	2.93 ± 0.20 [a]	185.36 ± 7.62 [a]	1040.86 ± 41.69 [b]	78.63 ± 1.64 [b]
CPPO	8.32 ± 0.12 [e]	396.63 ± 12.69 [d]	1638.58 ± 16.94 [d]	119.21 ± 3.49 [d]
CPMTO	5.42 ± 0.12 [d]	234.65 ± 9.85 [b]	958.59 ± 44.52 [a]	61.93 ± 2.56 [a]
CPRO	4.93 ± 0.11 [c]	293.10 ± 10.67 [c]	1328.00 ± 59.57 [c]	99.67 ± 1.48 [c]
CPCO	4.17 ± 0.23 [b]	396.04 ± 11.45 [d]	1367.50 ± 16.94 [c]	75.80 ± 1.95 [b]
CPSO	5.25 ± 0.15 [c,d]	241.06 ± 12.86 [b]	1085.10 ± 17.83 [b]	62.22 ± 2.59 [a]

Results are expressed as mean ± standard deviation (SD) (n = 3); different letters (a–e) within the same columns indicate significant differences between total phenolic content (TPC) and antioxidant activity of cold-pressed oils (one-way ANOVA and Duncan test, $p < 0.05$). Abbreviations: DPPH—2,2-diphenyl-1-picrylhydrazyl method; ABTS—2,2′-azino-bis(3-ethylbenzothiazoline-6-sulfonic acid); FRAP—ferric reducing antioxidant power; CPLO—cold-pressed linseed oil; CPPO—cold-pressed pumpkin oil; CPMTO—cold-pressed milk thistle oil; CPRO—cold-pressed rapeseed oil; CPCO—cold-pressed camelina oil; CPSO—cold-pressed sunflower oil; GA—gallic acid; TE—Trolox equivalent.

This analytical method is prevalent because it allows determining all phenolic compounds, regardless of their structure. The TPC values in the tested oils varied from 2.93 mg GA/100 g for CPLO to 8.32 mg GA/100 g for CPPO. The obtained results demonstrated that the type of raw material affected total phenolic concentrations in the studied CPOs. Insignificant differences in TPC results (4.93–5.42 mg GA/100 g) were observed between CPMTO, CPSO, and CPRO, while CPCO contained significantly lower amounts of phenolic compounds (TPC = 4.17 mg GA/100 g) (Table 4, Duncan test).

Interestingly, other researchers [4,20,27] noticed higher amounts of total polyphenols in CPLO (37.57–84.9 mg caffeic acid (CA)/kg, 60.25–115.12 mg/100 g, 60.3–89.6 mg ferulic acid (FA)/100 g), CPPO (53.67–184.6 mg CA/kg, 41.7–55.6 mg FA/100 g), CPMTO (71.7–124.7 mg CA/kg, 45.4–50.7 mg FA/kg), CPCO (34.12–138.9 mg CA/kg, 90.2–120.1 mg FA/100 g), CPRO (50.7–112.8 mg FA/100 g), and CPSO (43.8–52.3 mg FA/100 g) from Polish manufacturers or produced and purchased during their shelf life on the Polish market. However, the mean TPC (2.1 mmol GA/kg) in 17 samples of CPRO from England, Ireland, France, and other European Union countries was lower than the mean TPC (3.9 mmol GA/kg) in 5 samples of EVOO from Italy, Greece, and Spain [16].

2.1.5. Antioxidant Activity

The AA results obtained by 2,2-diphenyl-1-picrylhydrazyl (DPPH) and 2,2′-azino-bis(3-ethylbenzothiazoline-6-sulfonic acid) (ABTS) methods demonstrated differences in the antioxidant abilities of the studied CPOs to scavenge/deactivate DPPH and ABTS radicals. Moreover, the ferric reducing antioxidant power (FRAP) spectrophotometric single electron transfer (SET)-based assay measured the capacity of oil antioxidants to reduce an oxidant, which changed colour when reduced.

It is noteworthy that the AA results determined using three analytical methods, such as DPPH, ABTS, and FRAP, that differed significantly (by more than two orders of magnitude) for the same oil (Table 4). The differences between AA values for the studied CPOs can be explained by different mechanisms of the chosen analytical methods and their different affinities toward hydrophobic and hydrophilic antioxidants. The ABTS test was suitable for analysing both hydrophilic and lipophilic antioxidants, while DPPH measured antioxidants insoluble in water. In contrast, FRAP allows the quantification of the most hydrophilic antioxidants with redox potential not lower than that of the redox pair Fe^{3+}/Fe^{2+} [47]. For this reason, the FRAP values (61.93–119.21 µmol TE/100 g) of the studied CPOs were approximately 2–5 and 13–18 times lower than those of DPPH (185.36–396.63 µmol TE/100 g) and ABTS (958.59–1638.58 µmol TE/100 g), respectively (Table 4). Moreover, the ABTS cation radical scavenging ability of tested CPOs was 4 to 6 times higher than their DPPH radical scavenging activity. These differences indicated that CPOs contained many bioactive compounds that can be classified as lipophilic antioxidants (e.g., tocopherols). The same tendency was observed by Symoniuk et al. [23], although lower differences (1.5–2.8 times) between ABTS and DPPH results for CPLO, CPPO, CPRO, CPCO, and CPMTO were found.

As can be seen in Table 4, CPPO had the highest AA values determined using all analytical methods (DPPH = 396.63 µmol TE/100 g, ABTS = 1638.58 µmol TE/100 g, and FRAP = 119.21 µmol TE/100 g). However, CPLO revealed the lowest DPPH (185.36, µmol TE/100 g), while ABTS (958.59 µmol TE/100 g) and FRAP (61.93 µmol TE/100 g) were the lowest for CPMTO. Insignificant differences were observed between the FRAP results of CPMTO and CPSO, as well as CPLO and CPCO. In addition, the Duncan test indicated that ABTS values of CPLO and CPSO as well as CPRO and CPCO were similar (Table 4). Moreover, DPPH results of CPMTO and CPSO as well as CPPO and CPCO did not differ significantly. The results obtained by the DPPH method for CPLO (185.36 µmol TE/100 g), CPMTO (234.65 µmol TE/100 g) and CPSO (241.06 µmol TE/100 g) were in agreement with those (DPPH = 1.97–2.01, 2.14–2.56, and 1.76–2.34 mM TE/kg for CPLO, CPMTO, and CPSO, respectively) reported by Symoniuk et al. [20]. On the contrary, the same researchers reported lower DPPH values for CPPO (1.95–1.96 mM TE/kg), CPRO (1.81–1.96 mM TE/kg), and CPCO (1.88–2.44 mM TE/kg) than those demonstrated in our work. Unexpectedly, mean ABTS values for 17 oils cold-pressed from rapeseed (48.8 mmol TE/kg) and five samples of EVOO (50.9 mmol TE/kg) were lower than mean DPPH results (74.2 and 58.1 mmol TE/kg for CPROs and EVOOs, respectively) [16].

These AA differences can be explained by the fact that the antioxidant properties of CPOs were affected by many factors, such as agronomic, genetic, and environmental conditions, technological processing parameters, as well as differences in the mechanism of the applied assays for measuring their total antioxidant potential.

2.2. *Oxidative Stability and Quality of Cold-Pressed Oils*

2.2.1. Oxidative Stability

Oxidative stability is one of the most important parameters describing the susceptibility of oil to oxidation, and it is associated with its shelf life.

The oxidative stability of the studied CPOs was determined using the Rancimat method and expressed as the induction period (IP) at 100 °C (Table 5).

Table 5. Oxidative stability and quality parameters of the studied cold-pressed oils.

Parameter	Oil Sample					
	CPLO	CPPO	CPMTO	CPRO	CPCO	CPSO
IP (h)	4.87 ± 0.21 [a]	9.47 ± 0.25 [c]	9.03 ± 0.42 [c]	12.93 ± 0.15 [d]	5.37 ± 0.23 [b]	9.23 ± 0.25 [c]
PV (meq O_2/kg)	0.61 ± 0.02 [c]	2.44 ± 0.06 [d]	2.88 ± 0.14 [e]	0.42 ± 0.02 [b]	0.24 ± 0.01 [a]	4.61 ± 0.11 [f]
pAnV (-)	0.39 ± 0.02 [a]	4.77 ± 0.09 [f]	0.66 ± 0.07 [b]	0.88 ± 0.06 [d]	1.88 ± 0.16 [e]	0.79 ± 0.07 [c]
TOTOX	1.61	9.65	6.42	1.72	2.36	10.01
AV (mg KOH/g)	0.37 ± 0.01 [a]	1.55 ± 0.02 [d]	2.83 ± 0.03 [e]	0.42 ± 0.02 [b]	0.31 ± 0.08 [a]	1.16 ± 0.02 [c]
FFA (%)	0.18 ± 0.01 [a]	0.77 ± 0.01 [d]	1.42 ± 0.02 [e]	0.23 ± 0.01 [b]	0.15 ± 0.04 [a]	0.58 ± 0.01 [c]
WVC (%)	0.085 ± 0.002 [e]	0.030 ± 0.000 [b]	0.090 ± 0.000 [f]	0.045 ± 0.001 [c]	0.059 ± 0.002 [d]	0.020 ± 0.000 [a]
B(a)P (µg/kg)	0.41 ± 0.01 [c]	0.82 ± 0.02 [e]	0.32 ± 0.01 [b]	0.81 ± 0.02 [e]	0.20 ± 0.01 [a]	0.48 ± 0.02 [d]
Chry (µg/kg)	0.76 ± 0.02 [d]	0.54 ± 0.03 [c]	0.83 ± 0.03 [e]	0.28 ± 0.01 [b]	0.21 ± 0.01 [a]	0.54 ± 0.03 [c]
B(a)A (µg/kg)	7.16 ± 0.05 [f]	0.84 ± 0.01 [c]	0.75 ± 0.03 [b]	2.12 ± 0.03 [e]	0.51 ± 0.02 [a]	1.28 ± 0.06 [d]
B(b)F (µg/kg	0.43 ± 0.02 [d]	0.19 ± 0.01 [a]	0.22 ± 0.01 [b]	0.60 ± 0.02 [e]	0.24 ± 0.01 [b]	0.28 ± 0.01 [c]
Σ4PAHs (µg/kg)	8.76	2.39	2.12	3.81	1.16	2.58

Results are expressed as mean ± standard deviation (SD) (n = 3); different letters (a–f) within the same rows indicate significant differences between oxidative stability and quality parameters of cold-pressed oils (one-way ANOVA and Duncan test, $p < 0.05$). Abbreviations: IP—induction period; PV—peroxide value; pAnV—anisidine value; TOTOX—total oxidation value; AV—acid value; FFA—free fatty acids; WVC—water and volatile matter content; B(a)P—benzo(a)pyrene; Chry—chrysene; B(a)A—benzo(a)anthracene; B(b)F—benzo(b)fluoranthene; Σ4PAHs—sum of four specific polycyclic aromatic hydrocarbons; CPLO—cold-pressed linseed oil; CPPO—cold-pressed pumpkin oil; CPMTO—cold-pressed milk thistle oil; CPRO—cold-pressed rapeseed oil; CPCO—cold-pressed camelina oil; CPSO—cold-pressed sunflower oil.

It can be noted that the IP values varied from 4.87 h for CPLO to 12.93 h for CPRO. This suggests that CPRO had the highest resistance to oxidation reactions among the investigated oils. Other researchers [20,27] also reported similar IP results for CPRO (12.96–13.98 h) and CPLO (2.85–4.96 h). Evidently, the obtained IP results demonstrated that the oxidative stability of the studied oils was correlated with their degree of unsaturation. The high oxidative stability of CPRO was most likely due to its FAC with the lowest PUFA content (27.94%). In contrast, the shortest IP (4.87 h) was achieved in the case of CPLO containing the highest PUFA level (68.42%) and the lowest total amounts of tocopherols (44.04 mg/100 g) and phenolics (2.93 mg GA/100 g). Moreover, a high percentage of C18:3 (30.29%) caused the CPCO to oxidize quickly, exhibiting a low IP value (5.37 h). The oxidative stability of CPCO was similar to IP results (4.8–6.8 h) of different CPCOs reported in our previous work [8] and those (4.26–6.18 h) reported by Ratusz et al. [28].

Insignificant differences ($p > 0.05$, Duncan test) in IP values (9.47, 9.03, and 9.23 h) were observed between CPPO, CPMTO, and CPSO containing similar content of PUFA (50.59, 56.76, and 57.74%, respectively).

For comparison, CPPO, CPMTO, and CPSO analysed by other authors [20,23,43,48] had different IP results (13.63–34.39, 5.47–11.17, and 6.13–19.87 h, respectively). These differences can be explained by the fact that the IP values are highly dependent on the Rancimat test parameters (temperature and airflow rate) and profiles of fatty acids, antioxidants, oxidation products, and impurities in oil samples.

Additionally, the IP results of the investigated CPOs determined using the Rancimat method increased in the opposite order of their calculated COX values (Tables 1 and 5).

2.2.2. Amounts of Primary and Secondary Oxidation Products and Free Fatty Acids

Oxidative parameters such as peroxide value (PV), p-anisidine value (pAnV), total oxidation value (TOTOX), acid value (AV), and free fatty acids (FFA) were used to determine the oxidation state of the analysed CPOs.

Primary oxidation processes in oils mainly form hydroperoxides, (measured via PV), while secondary oxidation occurs when the peroxides decompose to form aldehydes and ketones (measured via pAnV). The TOTOX calculated from the PV and pAnV data is a measure of the overall oxidation profile of each oil. However, AV determines the amount of FFA, released from the triacylglycerol molecules during processing and storage conditions. These important oxidative indicators affecting the CPO quality are presented in Table 5.

As can be seen, amounts of the primary oxidation products in all of the studied CPOs (PV = 0.24–4.61 meq O_2/kg) were below the legal limit (15 meq O_2/kg) permitted for cold-pressed and virgin oils, according to Codex Alimentarius [1]. Interestingly, CPLO, CPRO, and CPCO reached one-order lower PV results (0.24–0.61 meq O_2/kg) when compared to other studied oils (PV = 2.44–4.61 meq O_2/kg) (Table 5). In addition, PV for CPSO (4.61 meq O_2/kg) was approximately two times higher than those obtained for CPPO and CPMTO (PV = 2.44 and 2.88 meq O_2/kg, respectively).

The PV results below 1.0 meq O_2/kg for CPRO (PV = 0.85 meq O_2/kg) and CPLO (PV = 0.95 meq O_2/kg) were also observed by Symoniuk et al. [23]. On the contrary, the content of primary oxidation products (PV = 4.91 meq O_2/kg) in camelina oil analysed by the same authors was higher than PV (0.24 meq O_2/kg) for the CPCO sample tested in this work. However, PV (0.42 meq O_2/kg) for CPRO was significantly lower than those reported by McDowell et al. [16] for 17 various oils cold-pressed from rapeseed (mean PV = 5.5 meq O_2/kg) and five samples of EVOO (mean PV = 10.4 meq O_2/kg). Moreover, PV results for EVOOs produced by different technologies, namely, super pressure (SP), two-phase (2P), and three-phase (3P) systems, and traditional extraction system ranged between 16.10–19.40 meq O_2/kg [44]. This can be explained by the fact that lipid oxidation is a complex process, depending on many factors, such as oilseed variety and quality, the FAC of the pressed oil and the processing technology applied. In addition, the content of peroxides in oils can depend on their shelf life. The PV generally increases with the storage time [16]. Therefore, oils with low PV values prolong their shelf life before consumption [49]. The PV of oil is also more closely associated with its IP determined using the Rancimat test; thus, oil with initially low PV is more prone to oxidation processes than oil containing high levels of peroxides [23,50].

It is interesting to note that amounts of secondary oxidation products in CPRO, CPCO, and CPPO were higher (pAnV = 0.88–4.77) than contents of primary oxidation products in these oils (PV = 0.24–2.44 meq O_2/kg) (Table 5). This suggests that the primary oxidation products were decomposed to secondary oxidation products in the mentioned oils. Unexpectedly, CPLO contained the lowest amounts of secondary oxidation products (pAnV = 0.39), while pAnV (4.77) was the highest for CPPO. For comparison, other authors [23] reported significantly higher pAnV (7.14) for CPPO. Moreover, CPOs analysed by various researchers had pAnV ranging between 0.07–1.43, 0.58–8.60, 0.13–2.56, 0.53–3.55, 0.22–1.93, and 0.45–0.89 for CPLO, CPPO, CPMTO, CPRO, CPCO, and CPSO, respectively [4,5,20,23,25,27].

The calculated TOTOX indexes showed different degrees of oxidation, varying from 1.61 for CPLO to 10.01 for CPSO (Table 5). Based on obtained TOTOX results, all of the studied oils revealed acceptable quality, although the overall oxidation profiles of CPSO (TOTOX = 10.01) and CPPO (TOTOX = 9.65) were significantly higher than the oxidative status of other oils (TOTOX < 6.5).

Similar TOTOX results (2.38–11.74) for oils cold-pressed from linseed, milk thistle, rapeseed, camelina, and sunflower seeds were reported by other authors [20,23,27].

Furthermore, AV and FFA levels in the tested CPOs can be good indicators of their status in terms of oxidative stability. A higher concentration of triacylglycerol hydrolysis products indicates a lower oil quality level. The AV (0.31 and 0.37 mg KOH/g) and FFA (0.15 and 0.18%) results were the lowest for CPCO and CPLO and did not differ significantly among each other (Duncan test, Table 5). On the contrary, CPMTO had the highest results of AV (2.83 mg KOH/g) and FFA (1.42%). Nevertheless, the amounts of hydrolysis products in CPRO (AV = 0.42 mg KOH/g and FFA = 0.23%) were approximately three times lower than those in CPSO (AV = 1.16 mg KOH/g and FFA = 0.58%) and CPPO (AV = 1.55 mg KOH/g and FFA = 0.77%).

Krygier et al. [45] and Symoniuk et al. [20] reported similar AV values for CPRO (0.29–2.46 mg KOH/g) and CPSO (0.40–1.40 mg KOH/g), whereas Meddeb et al. [51] observed much higher AV results (5.48–8.34 mg KOH/g) for oils cold-pressed from some varieties of milk thistle seeds growing in different areas in Tunisia. However, CPRO samples from various geographic locations, such as England, Ireland, France, and other European Union countries, had comparable mean AV value (2.2 mg KOH/g) with that of EVOOs (mean AV = 2.0 mg KOH/g) [16]. Moreover, the type of technological process affected the acidity of EVOOs (0.54–0.72%) [44]. These results were in agreement with FFA levels analysed in CPSO (0.58%) and CPPO (0.77%).

Importantly, all of the tested oils had AV values below the maximum permissible limit (4.0 mg KOH/g) prescribed by the Codex Alimentarius Commission [1].

It is well known that oil acidity is related to the presence of FFAs, initially formed during enzymatic hydrolysis caused by lipases, naturally occurring in seeds. For this reason, AV depends on the general quality, varieties, and humidity of seeds, the shelf life of oils pressed from them, and their light exposure during storage in bottles [8,52,53]. The FFAs are much more susceptible to oxidation than those connected to a triacylglyceride moiety, causing oil deterioration and influencing sensory characteristics; thus, high AV values are not desired for total oil quality [51].

The consumption of lipid oxidation products may cause serious health risks to consumers due to their reactions with proteins, DNA, and phospholipids, which promote chronic diseases such as inflammation, cancer, and cardiovascular diseases [54].

2.2.3. Water and Volatile Matter Contents

It can be noted that levels of water and volatile compounds (WVC) in the investigated CPOs ranged between 0.020% and 0.090% (Table 5), which are below the legal requirement recommended by Codex Alimentarius (0.20%) [1].

Similar WVC content (779 ppm) in oil pressed from milk thistle seeds was observed by Rokosik et al. [30]. Somewhat higher WVC values (0.05–0.17%) were determined for eight CPCO samples produced by the most famous Polish manufacturers, as previously described in our report [8]. In addition, traditionally extracted EVOO revealed significantly lower humidity (0.04%) than super-pressure extracted EVOOs (0.07–0.15%) [44]. It is known that water excess in vegetable oils favours hydrolysis of triacylglycerols, causing oil deterioration. A high amount of water in oil can be observed when raw material is too wet, filtration after cold-pressing is inefficient, or the oil shelf life is prolonged.

2.2.4. Polycyclic Aromatic Hydrocarbon Content

The concentrations of four polycyclic aromatic hydrocarbons (PAHs) (benzo(a)pyrene (B(a)P), chrysene (Chr), benzo(a)anthracene (B(a)A), and benzo(b)fluoranthene (B(b)F) in six CPOs were determined, and the obtained results are listed in Table 5. It should be emphasized that these compounds may be toxic to healthy tissues causing carcinogenic and mutagenic effects. Therefore, limits of 10 μg/kg for the sum of four PAHs (Σ4PAHs) and 2 μg/kg for B(a)P in vegetable oils were established by European Commission [55]. It is noteworthy that B(a)P (0.20–0.82 μg/kg) and Σ4PAHs (1.16–8.76 μg/kg) contents in the analysed oil samples were significantly lower than that recommended by the European Commission [55]. The CPLO revealed the highest amounts of Σ4PAHs (8.76 μg/kg) and B(a)A (7.16 μg/kg), while CPCO had the lowest levels of Σ4PAHs (1.16 μg/kg) and B(a)A 0.51 μg/kg). In addition, Chry concentration was high in CPLO (0.76 μg/kg) and CPMTO (0.83 μg/kg). However, the same Chry content (0.54 μg/kg) was observed for CPPO and CPSO (Table 5).

For comparison, amounts of Σ4PAHs in rapeseed oils pressed from conventional and organic farming seeds varied from 3.13 to 6.15 μg/kg [56]. Significantly higher concentrations of total PAHs (17.85–37.16 μg/kg) in different cold-pressed vegetable oils were reported by Roszko et al. [57]. However, contamination of these oils with B[a]P (0.02–1.25 μg/kg) was similar to those found in our study.

All of the analysed CPOs had good quality due to low levels of dangerous compounds, such as PAHs affecting human health.

2.3. Relationships between Descriptive Attributes and Acceptance Test

The PCA was applied to find the relationships between descriptive attributes of the investigated CPOs, overall flavour intensity (OFI), and hedonic responses: overall liking (OL) and purchase intent (PI). The first two principal components accounted for 63.71% of the experimental data variance (Figure 1).

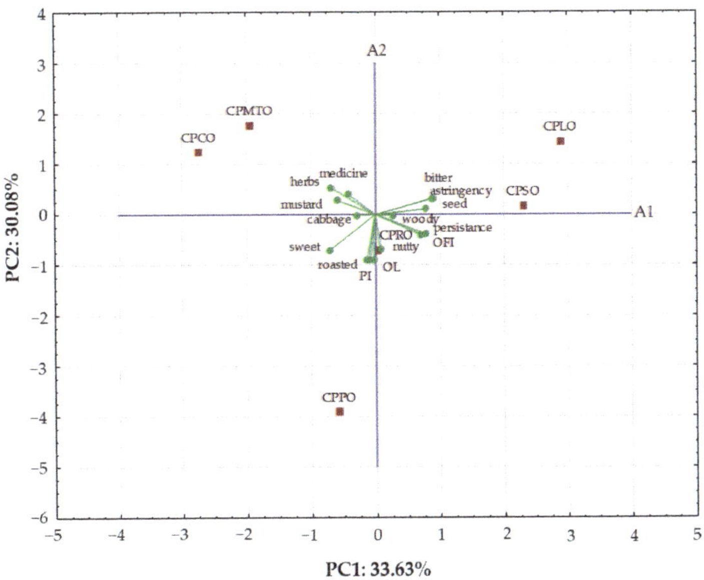

Figure 1. Principal component analysis of the sensory descriptors, hedonic overall liking (OL), purchase intent (PI), and overall flavour intensity (OFI) for six cold-pressed oils. Abbreviations: CPLO—cold-pressed linseed oil; CPPO—cold-pressed pumpkin oil; CPMTO—cold-pressed milk thistle oil; CPRO—cold-pressed rapeseed oil; CPCO—cold-pressed camelina oil; CPSO—cold-pressed sunflower oil.

As can be seen in Figure 1, the CPPO characterised by very high scores of sweet taste and roasted and nutty flavour intensity created an evidently distinct cluster. Furthermore, the CPSO and CPLO having the highest bitterness perception were located close to each other on the right side of the map. Both CPMTO and CPCO were situated on the left side of the PCA map due to the presence of herb-like flavour in the sensory profile. Additionally, the medicine-like flavour was identified only in CPMTO, while differences in the mustard-like flavour were found for CPMTO and CPCO. However, CPRO with a dominance of cabbage-like and woody-like sensory attributes was located in the middle of the PCA map.

The current study found a high positive correlation (r = 0.9831, $p < 0.001$) between OL and PI. The attributes of sweet taste (r = 0.6387) and roasted flavour (r = 0.7187) contributed insignificantly ($p > 0.05$) positively to consumer acceptance (OL). These attributes appeared in CPPO and received the highest consumer scores. On the contrary, the attributes of bitter taste, astringency, and herbs-like flavour were responsible for the lower OL (r = −0.3601, −0.3404, and −0.4222, respectively, $p > 0.05$) and PI (r = −0.4470, −0.4307, and −0.4425, respectively, $p > 0.05$) of the tested CPOs.

2.4. Chemometrics Analysis

Multivariate mathematical approaches, such as PCA and HCA, were powerful tools that permitted a relatively simple representation of similarities and discrepancies between the investigated oil samples based on complex analytical and sensory data.

2.4.1. Principal Component Analysis

PCA was applied to the matrix formed by all physicochemical parameters and three sensory characteristics corresponding to the studied CPOs. As can be seen in Figure 2, a biplot depicted the relationships between observed data (six CPOs: CPLO, CPPO, CPMTO, CPRO, CPCO, and CPSO) and dependent variables (SFA, MUFA, PUFA, COX, AI, TI, HH, TTC, TSC, TPC, DPPH, ABTS, FRAP, IP, PV, pAnV, TOTOX, AV, FFA, WVC, Σ4PAHs, OFI, OL, and PI) in terms of principal components. The first two principal components with the highest eigenvalue contributions (9.07 and 7.18, respectively) for the observed variations were taken from the 24 principal components. A combination of PC1 and PC2 described 67.73% (PC1 = 37.79%, PC2 = 29.94%) of the total variance in the dataset. It was found that PC1 highly negatively correlated with TPC (−0.9918), OL (−0.9049), TI (−0.8813), and PI (−0.8591), while PC2 showed the highest positive correlation with MUFA (0.9175) and inversely contributed by PUFA (−0.8679) and AI (−0.8283).

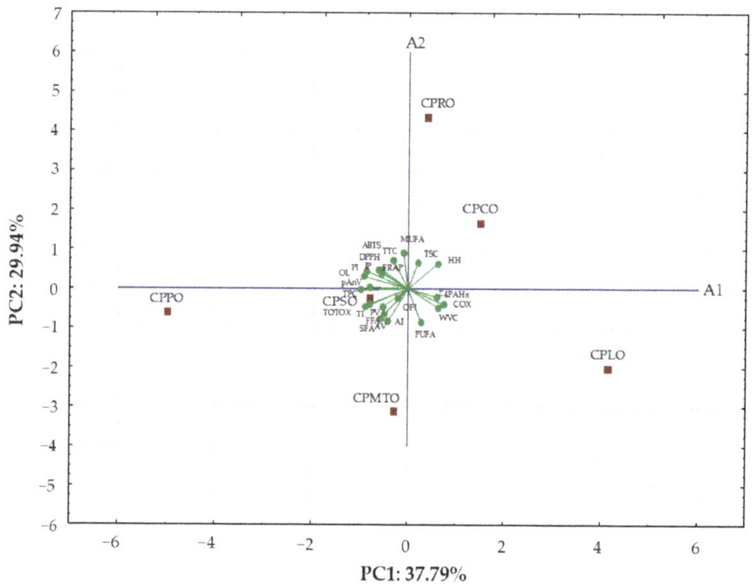

Figure 2. Biplot of scores and loadings of data obtained for physicochemical and overall sensory characteristics of six cold-pressed oils. Abbreviations: SFA—saturated fatty acids; MUFA—monounsaturated fatty acids; PUFA—polyunsaturated fatty acids; COX—calculated oxidisability value; AI—atherogenicity index; TI—thrombogenicity index; HH—ratio of hypocholesterolemic to hypercholesterolemic fatty acids; TTC—total tocopherol content; TSC—total sterol content; TPC—total phenolic content; DPPH—2,2-diphenyl-1-picrylhydrazyl method; ABTS—2,2′-azino-bis(3-ethylbenzothiazoline-6-sulfonic acid); FRAP—ferric reducing antioxidant power; IP—induction period; PV—peroxide value; pAnV—anisidine value; TOTOX—total oxidation value; AV—acid value; FFA—free fatty acids; WVC—water and volatile matter content; Σ4PAHs—sum of four specific polycyclic aromatic hydrocarbons; OL—overall liking; PI—purchase intent; OFI—overall flavour intensity; CPLO—cold-pressed linseed oil; CPPO—cold-pressed pumpkin oil; CPMTO—cold-pressed milk thistle oil; CPRO—cold-pressed rapeseed oil; CPCO—cold-pressed camelina oil; CPSO—cold-pressed sunflower oil.

The biplot depicted that CPRO and CPCO with the highest MUFA contents (64.83 and 35.33%) and the lowest amounts of SFA (7.12 and 9.72%), AI (0.02 and 0.03), and PV (0.42 and 0.24 meq O_2/kg) were clustered in the right upper quarter. CPLO containing the highest levels of PUFA (68.42%), COX (13.14), Σ4PAHs (8.76 μg/kg), and OFI (6.83) and the lowest values of TTC (44.04 mg/100 g), TPC (2.93 mg GA/100 g), DPPH

(185.36 μmol TE/100 g), IP (4.87 h), pAnV (0.39), TOTOX (1.61), OL (4.18), and PI (3.73) was also situated on the same side under the A1 axis. Interestingly, the lowest oxidative stability and antioxidant contents in CPLO as well as the highest OFI affected low scores of its acceptance (via low scores of OL and PI). However, CPPO, CPSO, and CPMTO containing similar high concentrations of SFA (11.20–19.53%), PUFA (50.59–57.74%), TI (0.23–0.45), TPC (5.25–8.32 mg GA/100 g), IP (9.03–9.47 h), PV (2.44–4.61 meq O_2/kg), AV (1.16–2.83 mg KOH/g), FFA (0.58–1.42%), and Σ4PAHs (2.12–2.58 μg/kg) were located in the bottom left quarter of the biplot (Figure 2). The PCA results revealed that the investigated oil samples were clearly categorised into three distinct groups according to different raw materials and their botanical origin, which indicated differences in determined physicochemical parameters and sensory characteristics.

2.4.2. Hierarchical Cluster Analysis

The HCA was used to group the studied oils per their similarities and discrepancies based on 24 physicochemical and sensory variables. The dendrograms in Figure 3 showed the clustering patterns of six oils and variable sets. It can be noted that CPO samples were segregated into two main clusters (Figure 3a). The first cluster included an inter-cluster of CPLO, CPSO, and CPMTO. It is noteworthy that CPLO and CPSO had similar content of SFA (11.63 and 11.20%) and ABTS values (1040.86 and 1085.10 μmol TE/100 g), while CPMTO revealed somewhat lower ABTS (958.59 μmol TE/100 g). In addition, insignificant differences between TPC (5.42 and 5.25 mg GA/100 g), DPPH (234.65 and 241.06 μmol TE/100 g), FRAP (61.93 and 62.22 μmol TE/100 g), and IP (9.03 and 9.23 h) were observed for CPMTO and CPSO. Evidently, CPCO, CPRO, and CPPO were arranged in the second cluster. This group was associated with the highest AC (DPPH = 293.10–396.63 μmol TE/100 g, ABTS = 1328.00–1638.58 μmol TE/100 g, and FRAP = 75.80–119.21 μmol TE/100 g) together with high TTC (64.45–76.98 mg/100 g) in CPCO, CPRO, and CPPO. CPPO was distanced from this cluster because this oil sample had approximately two times higher amounts of TPC (8.32 mg GA/100 g) and SFA (18.92%) as well as sensory acceptance score (OL = 8.21) than those of CPRO and CPCO (TPC = 4.17–4.93 mg GA/100 g, SFA = 7.12–9.72%, OL = 5.00–6.72).

On the other hand, cluster analysis results indicate that the determined 24 physicochemical and sensory parameters as variables comprised three main groups (Figure 3b). The dendrogram clearly separated the ABTS method allowing the determination of both hydrophilic and lipophilic antioxidants present in oil samples from the other variables. However, DPPH method and TSC formed one cluster. This suggests that hydrophobic compounds such as sterols influenced the DPPH results of the investigated CPOs. However, the third cluster was composed of two subgroups consisting of (I) FRAP, TTC, PUFA, and MUFA and (II) all studied oxidative stability parameters (five characteristic quality values, oxidisability value, three nutritional quality indexes, IP), sensory characteristics (OFI, OL, PI), TPC, SFA, and amounts of undesirable compounds (Σ4PAHs and WVC). Unexpectedly, the results of HCA depicted that lipophilic antioxidants, such as tocopherols (TTC), PUFA, and MUFA levels, affected the FRAP values of the studied CPOs (Figure 3b). In addition, phenolic compounds (TPC) contributed more to sensory characteristics (OFI, OL, and PI) of CPOs than impurities (Σ4PAHs, WVC), oxidation and hydrolysis products (PV, pAnV, TOTOX, IP, COX, AV, FFA), nutritional quality indexes (AI, TI, HH), and SFA content. Considering the Euclidean distances, the HCA may capture physicochemical and sensory similarities between groups of the studied CPOs and variables more efficiently visually than the PCA.

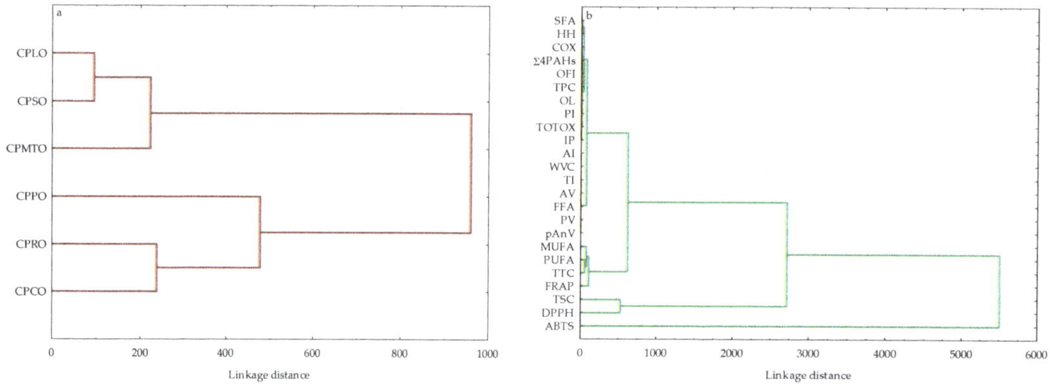

Figure 3. Dendrograms of hierarchical cluster analysis for (**a**) six cold-pressed oils and (**b**) the studied variables. Abbreviations: SFA—saturated fatty acids; MUFA—monounsaturated fatty acids; PUFA—polyunsaturated fatty acids; COX—calculated oxidisability value; AI—atherogenicity index; TI—thrombogenicity index; HH—ratio of hypocholesterolemic to hypercholesterolemic fatty acids; TTC—total tocopherol content; TSC—total sterol content; TPC—total phenolic content; DPPH—2,2-diphenyl-1-picrylhydrazyl method; ABTS—2,2′-azino-bis(3-ethylbenzothiazoline-6-sulfonic acid); FRAP—ferric reducing antioxidant power; IP—induction period; PV—peroxide value; pAnV—anisidine value; TOTOX—total oxidation value; AV—acid value; FFA—free fatty acids; WVC—water and volatile matter content; Σ4PAHs—sum of four specific polycyclic aromatic hydrocarbons; OL—overall liking; PI—purchase intent; OFI—overall flavour intensity; CPLO—cold-pressed linseed oil; CPPO—cold-pressed pumpkin oil; CPMTO—cold-pressed milk thistle oil; CPRO—cold-pressed rapeseed oil; CPCO—cold-pressed camelina oil; CPSO—cold-pressed sunflower oil.

2.4.3. Correlation Analysis

The positive and negative correlations between physicochemical, nutritional, and sensory characteristics of six tested CPOs are presented as a correlation matrix in Figure 4.

It is noteworthy that positive significant relationships were found between SFA–AI (r = 0.9257, p = 0.008), SFA–TI (r = 0.8714, p = 0.024), SFA–AV (r = 0.8630, p = 0.027), and SFA–FFA (r = 0.8553, p = 0.030), whereas SFA negatively associated with HH ratio (r = −0.8815, p = 0.020). This suggests that nutritional quality indexes (AI and TI) and acidity indexes (AV and FFA) increased with increasing SFA content in CPOs. Thus, the intake of oils rich in SFA increases the risk of cardiovascular diseases.

Expectedly, COX values providing information about the oxidative stability of six CPOs inversely correlated with their IP results determined by the Rancimat method (r = −0.8810, p = 0.020). Moreover, a high negative correlation was found between MUFA and PUFA (r = −0.9676, p = 0.002), while TSC positively affected the MUFA amounts in the studied oils (r = 0.8200, p = 0.046). Additionally, the calculated r values indicated significant associations among AA results of all oils determined using ABTS and DPPH (r = 0.8899, p = 0.018) as well as ABTS and FRAP (0.8699, p = 0.024) methods. This can be explained by the fact that at the same time, antioxidants in the tested oils were capable of reducing ferric-tripyridyltriazine (Fe^{3+}-TPTZ) to an intense blue colour ferrous-tripyridyltriazine complex (Fe^{2+}-TPTZ) and scavenging ABTS and DPPH radicals. However, the effect of TPC on AA of six CPOs determined using three analytical methods was insignificant (r = 0.5629–06428, p > 0.05). This suggests that phenolic compounds in the studied oils contributed insignificantly to their AA. In contrast, high correlation coefficients were observed between TPC–OL (r = 0.8797, p = 0.021), TPC–PI (r = 0.8414, p = 0.036), and TPC–TI (r = 0.8986, p = 0.015). Unexpectedly, TPC significantly influenced the prothrombotic index and enhanced consumer acceptance, purchase intent, and positive emotions. However, these hedonic sensory scores significantly decreased with increasing water content in the tested CPOs (r = −0.8400,

$p = 0.036$ and $r = -0.8132$, $p = 0.049$ for relationships OL—WVC, PI—WVC, respectively). It can be seen that the high sensory acceptability (OL) was essential for the consumer to intend to buy (PI) the CPOs ($r = 0.9831$, $p < 0.0001$).

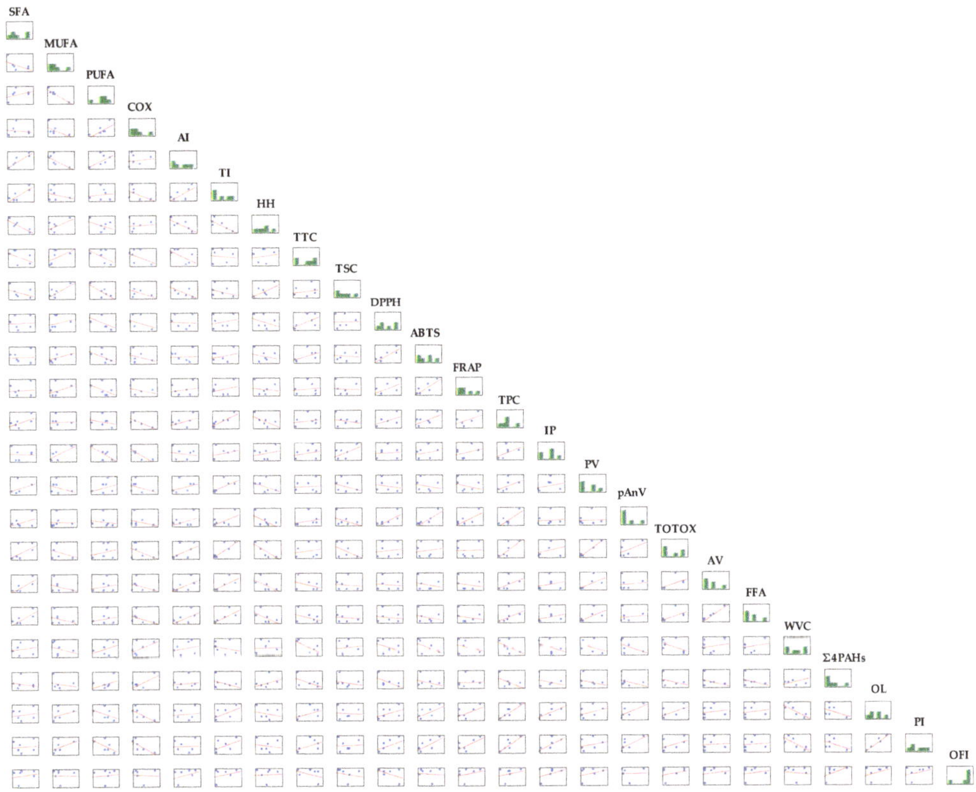

Figure 4. Correlation matrix between physicochemical parameters and sensory quality of six cold-pressed oils. Abbreviations: SFA—saturated fatty acids; MUFA—monounsaturated fatty acids; PUFA—polyunsaturated fatty acids; COX—calculated oxidisability value; AI—atherogenicity index; TI—thrombogenicity index; HH—ratio of hypocholesterolemic to hypercholesterolemic fatty acids; TTC—total tocopherol content; TSC—total sterol content; TPC—total phenolic content; DPPH—2,2-diphenyl-1-picrylhydrazyl method; ABTS—2,2′-azino-bis(3-ethylbenzothiazoline-6-sulfonic acid); FRAP—ferric reducing antioxidant power; IP—induction period; PV—peroxide value; pAnV—anisidine value; TOTOX—total oxidation value; AV—acid value; FFA—free fatty acids; WVC—water and volatile matter content; Σ4PAHs—sum of four specific polycyclic aromatic hydrocarbons; OL—overall liking; PI—purchase intent; OFI—overall flavour intensity.

Further studies are required to understand better interactions and relationships between bioactive compounds, other physicochemical, and sensory parameters of studied CPOs to predict their nutritional effectiveness and health benefits.

3. Materials and Methods

3.1. Reagents and Samples

All of the reagents and chemicals utilized in this research were purchased from Merck Sp. z o. o. (Warszawa, Poland). Six different CPOs in the original sealed dark glass bottles (250 mL) were donated by domestic manufacturers, which are leaders in the Polish market. These oil samples were obtained by mechanically pressing unheated seeds without additional extraction with solvents, followed by washing with water, precipitation, filtration, and centrifugation. The oils with a valid shelf life of at least 6 months were coded (CPLO—cold-pressed linseed oil, CPPO—cold-pressed pumpkin oil, CPMTO—cold-pressed milk thistle oil, CPRO—cold-pressed rapeseed oil, CPCO—cold-pressed camelina oil, and CPSO—cold-pressed sunflower oil) and kept refrigerated (4 °C) until analysis.

3.2. Determination of Fatty Acid Compositions

The FAC of each CPO was determined according to the official ISO 5508:1996 method [58]. Fatty acid methyl esters were prepared in accordance with the standard ISO 5509:2000 procedure [59] and analysed using a gas chromatography (HP 5890 GC) fitted with capillary column BPX 70 (60 m × 0.25 mm, 0.25 μm) and equipped with a flame-ionization detector (FID) (Hewlett-Packard, Avondale, PA, USA). The furnace temperature at 1.3 °C/min was 210 °C, initially starting from 150 °C. The temperatures of the injector and the detector were set at 250 °C. Helium was used as the carrier gas at a flow rate of 0.6 mL/min. The individual peaks were identified by comparison of their retention times with those of fatty acid methyl ester standards.

3.3. Calculated Oxidisability (COX) Value and Nutritional Quality Indexes

The *COX* values of the studied CPOs were calculated based on the Formula (1) proposed by Fatemi and Hammond [60].

$$COX = \frac{C18:1 + 10.3 \times C18:2 + 21.6 \times C18:3}{100} \quad (1)$$

Nutritional quality indexes, including *AI* and *TI*, were calculated by Formulas (2) and (3), respectively, given by Ulbricht and Southgate [24].

$$AI = \frac{C12:0 + 4 \times C14:0 + C16:0}{\Sigma MUFA + \Sigma(\omega - 3) + \Sigma(\omega - 6)} \quad (2)$$

$$TI = \frac{C14:0 + C16:0 + C18:0}{0.5 \times MUFA + 0.5 \times \Sigma(\omega - 6) + 3 \times \Sigma(\omega - 3) + \left(\frac{\omega-3}{\omega-6}\right)} \quad (3)$$

However, the *HH* ratio was calculated according to the Formula (4) proposed by Santos-Silva et al. [61] to evaluate the impact of fatty acids on human cholesterol levels.

$$HH = \frac{C18:1 + C18:2 + C18:3 + C18:4 + C20:4}{C14:0 + C16:0} \quad (4)$$

3.4. Determination of Tocopherol Compositions

The tocopherol compositions of the studied CPOs were determined chromatographically according to the official ISO 9936:2016 method [62] with some modifications. The oil sample (0.5 g) was dissolved in hexane (95 mL), injected (5–20 μL) into LiChrospher 100 Diol (125 × 4 mm, 5 μm particle size) column (Merc, Darmstadt, Germany), and analysed using Agilent 1100 HPLC system (Agilent Technologies, Palo Alto, CA, USA) with an autosampler and fluorescence detector. The mobile phase was a mixture of hexane with tetrahydrofuran (96:4, *v*/*v*%) at a flow rate of 0.8 mL/min. The fluorescence detector was set at 280 nm excitation wavelength and 340 nm emission wavelength. The tocopherol homologues were identified by comparison of the retention times with standards of the α-, β-, γ-, and δ-tocopherols.

3.5. Determination of Sterol Compositions

The sterol compositions were analysed using a gas chromatography (GC) according to the official ISO 12228-1:2014 method [63]. Each studied oil sample was saponified by adding methanolic potassium hydroxide solution (c = 1 mol/L). The sterol fraction was extracted with a hexane/methyl tert-butyl ether (1:1) mixture. A capillary gas chromatography

(Agilent Technologies 6890, Wilmington, DE, USA) system equipped with a capillary column (25 m, 0.20 mm i.d. and 0.33 μm film thickness) and a flame ionization detector (FID) was applied for separation and quantification of the silylated sterol fraction. The temperatures of the injector and detector were both set at 300 °C. The carrier gas was hydrogen at a 1.5 mL/min flow rate. The sterols were identified by comparing the retention times with those of standard samples, and 5α-cholestane was used as an internal standard for quantification.

3.6. Determination of Total Phenolic Content and Antioxidant Activity

The methanolic extracts of oils were prepared for TPC and AA analysis based on the previously reported procedure [15]. Briefly, 2.00 g of oils were weighed in test tubes and extracted with 5 mL of methanol for 30 min using an orbital shaker (SHKA25081 CE, Labo Plus, Warszawa, Poland). Then, the samples were stored in a refrigerator for 30 min to separate methanolic extracts from the oil layers. The extraction was repeated three times, whereas obtained extracts were combined and collected into glass bottles.

The TPC and AA of oil samples were determined using the F–C, DPPH, ABTS, and FRAP methods, respectively, according to procedures described in our previous work [64]. The UV–Vis spectra were recorded using a Hitachi U-2900 spectrophotometer (Hitachi, Tokyo, Japan) in a 1 cm quartz cell. The TPC results were expressed as mg of gallic acid equivalents per 100 g of oil sample (mg GA/100 g), while the AA was expressed as micromoles of Trolox equivalents per 100 g of oil sample (μmol TE/100 g).

3.7. Determination of Oxidative Stability

The oxidative stability of each studied CPO was analysed with 743 Rancimat apparatus (Metrohm, Herisau, Switzerland) according to the AOCS official method Cd12b-92 [65]. The oil sample (3.00 ± 0.01 g) was oxidised by air (flow 20 L/h) at 100 ± 0.3 °C in a measuring cell supplied with distilled water. The conductivity of the oxidation products dissolved in the water was measured, and the induction period (IP) (expressed in hours) was determined.

The PV was determined potentiometrically by the official ISO 27107:2010 method [66] and expressed as milliequivalents of active oxygen per kilogram of oil (meq O_2/kg).

The pAnV was measured spectrophotometrically according to the official ISO 6885:2016 method [67].

The TOTOX as a characteristic value of the total oxidation of triacylglycerols was calculated by the expression: TOTOX = (2PV + pAnV).

The amounts of FFA produced during the hydrolysis processes of oils and the AV were measured according to the official ISO 660:2020 method [68].

3.8. Determination of Water and Volatile Matter Contents

The WVC content in the analysed CPOs was determined by drying 10 g of oil in an oven at 103 °C according to the official ISO 662:2016 procedure [69].

3.9. Determination of Polycyclic Aromatic Hydrocarbons

A high-performance liquid chromatography with a fluorescence detector (HPLC-FLD, Shimadzu, Kyoto, Japan) was applied to determine four PAHs, including B(a)P, Chry, B(a)A, and B(b)F, in the studied CPOs. Each oil was dissolved in cyclohexane and extracted to dimethyl formaldehyde. Data acquisition and calculations were conducted in Open-LAB CDS ChemStation Edition Software version 10.1 (Agilent Technologies, Waldbronn, Germany). The reversed-phase Zorbax Eclipse PAH column (particle size 3.5 μm, length 150 mm, diameter 4.6 mm, Agilent, Santa Clara, CA, USA) with a precolumn Eclipse XDB-C18 (3.5 μm, 4.6 × 150 mm, Agilent) at an oven temperature of 30 °C were utilized for PAHs analysis in six CPOs. Benzo(b)chrysene diluted in acetonitrile was used as an internal standard. Calibration curves were prepared by using the peak areas as a function of the PAH concentration standards in the range between 0.25 and 8.50 μg/kg.

3.10. Hedonic Consumer Test

The consumer home-use test was conducted with 100 untrained consumers of both sexes aged between 21 and 65 years. The studied CPOs in 250 mL bottles were tested by consumers for six days (1 bottle per day). Overall acceptability (overall liking, OL) was evaluated by using the 9-point hedonic scale anchored by: 1 = "dislike extremely" and 9 = "like extremely". Additionally, purchase intent (PI) was measured by using the 9-point scale (1 = "I would definitely not buy" and 9 = "I would definitely buy"). At the end of the tasting, consumers answered questions about their social demographic characteristics and CPOs' consumption frequency. Before the tasting session, the sensory sensitivity of each consumer was screened and trained.

3.11. Sensory Profiling

The sensory profiling of each CPO was determined by quantitative descriptive analysis (QDA). Ten well-trained assessors (seven women and three men) evaluated thirteen sensory attributes (overall flavour intensity (OFI), two basic tastes, eight flavours, and two mouths feel sensory terms) of six investigated CPOs (Table 6). Each attribute was assessed using a 10 cm unstructured intensity scale anchored at the ends by "no intensity" on the left and "high intensity" on the right.

Table 6. Sensory attributes for the studied cold-pressed oils.

Sensory Attributes	Description
OFI	The intensity of all flavour and taste attributes taken together
Sweet taste	The basic taste simulated by sugar
Bitter taste	The basic taste elicited by quinine and caffeine
Herbs-like flavour	The flavour reminiscent of herbs
Cabbage-like flavour	The flavour associated with asparagus, cabbage, or fresh green vegetables
Seed-like flavour	The flavour associated with fresh seeds
Mustard-like flavour	The flavour associated with mustard, onion, and spiciness
Nutty flavour	The flavour associated with fresh nuts
Roasted flavour	The flavour associated with roasted oils
Wood-like flavour	The flavour associated with fresh, dry, cut wood
Medicine-like flavour	The flavour reminiscent of medicine, hospital, and pharmacy
Persistence	How long do flavour sensations remain as aftertaste
Astringency	The shrinking or drying effect on the tongue surface elicited by tannins

All of the studied oils were evaluated in duplicate by a trained sensory panel in a sensory laboratory under conditions described by ISO 8589:2007 [70]. Oil samples were labelled with random three-digit codes and served in a transparent glass jars (250 mL) at room temperature (about 20 °C). During each evaluation, warm dark tea and fresh apple were provided to panellists to clear their palates and avoid the carry-over effect. Before the tasting session, the sensory sensitivity of each panellist was screened and trained on sensory attributes typical for CPOs.

3.12. Statistical Analysis

All chemical analyses were conducted in triplicate on the same day. The obtained results were presented as mean (c) ± standard deviation (SD). The experimental data were statistically evaluated using analysis of variance (ANOVA) test. A post hoc Duncan test was applied to calculate the significant differences among the mean values of physicochemical and sensory parameters of oils at a probability level of $p < 0.05$.

Two chemometric techniques, such as PCA and HCA, were utilized to identify the possible links between the physicochemical, nutritional, and sensory characteristics of the studied CPOs. The scores and loadings of the data analysed using PCA were displayed as biplots. However, the Ward technique and the squared Euclidean distance matrix were performed to define each cluster, resulting in hierarchical dendrograms from HCA.

The statistical analysis was carried out using Statistica 8.0 software (StatSoft Inc., Tulsa, OK, USA), while Fizz software (Biosystemes, Courtenon, France) was used to collect all sensory data.

4. Conclusions

The similarities and discrepancies between amounts of the health-beneficial compounds (tocopherols, sterols, and polyphenols), total antioxidant potential, oxidative stability, compounds hazardous to health, and sensory attributes of the most popular CPOs in the Polish market were observed. The studied CPLO and CPCO were rich sources of PUFA with high content of health-promoted linolenic acid belonging to the ω-3 acid family (PUFA = 68.42 and 49.67%, C18:3 = 52.12% and 30.29%, respectively). However, CPPO, CPMTO, and CPSO had a large amount of ω-6 linoleic acid (C18:2 > 50%). Additionally, the fatty acid profiles of CPOs affected their nutritional quality and human health. The calculated AI, TI, and HH indexes can help in the evaluation of the risk of chronic diseases and metabolic syndromes. The CPRO, CPCO, CPSO, and CPLO revealed low AI and TI but a high HH ratio. In fact, the consumption of these oils can be associated with a reduced risk of several chronic diseases. On the other hand, the highest PUFA level in CPLO decreased its oxidative stability, which resulted in the lowest IP value (4.87 h). However, high amounts of antioxidants in the studied CPOs increased their AA and oxidation stability. Furthermore, the highest TSC (684 mg/100 g) in CPRO enhanced its health-promoting properties. Most importantly, concentrations of hazardous oxidation products and toxic PAHs in the investigated CPOs were below the legal limits recommended by EU food legislation and Codex Alimentarius standards.

In addition, the chemometric analyses clarified the interactions between the physicochemical parameters of the studied CPOs, their sensory attributes, and consumer acceptance. The bitter taste, astringency, and herbs-like flavour were negative descriptors, while the consumers favoured the sweet taste and roasted flavour. Based on sensory analysis, the CPPO showed the highest level of consumer preference, whereas the sensory attributes of CPLO had a negative impact on consumer attitude and purchase behaviour.

Author Contributions: Conceptualization, A.S.-C. and M.M.-R.; methodology, D.R.-K., B.S. and M.M.-R.; software, A.S.-C.; validation, A.S.-C.; formal analysis, D.R.-K., B.S. and M.M.-R.; investigation, A.S.-C., D.R.-K., B.S. and M.M.-R.; resources, A.S.-C., M.M.-R. and B.S.; data curation, A.S.-C., D.R.-K., M.M.-R. and B.S.; writing—original draft preparation, D.R.-K., B.S. and M.M.-R.; writing—review and editing, A.S.-C.; visualization, A.S.-C. and D.R.-K. supervision, A.S.-C.; project administration, A.S.-C.; funding acquisition, A.S.-C. All authors have read and agreed to the published version of the manuscript.

Funding: This research received no external funding.

Institutional Review Board Statement: Not applicable.

Informed Consent Statement: Not applicable.

Data Availability Statement: The data presented in this study are available on request from the corresponding author.

Conflicts of Interest: The authors declare no conflict of interest.

Sample Availability: Samples of the compounds are available from the authors.

References

1. Codex Alimentarius. Codex Standard for Named Vegetable Oils. Codex-Stan 210-1999. Available online: https://tinyurl.com/46s68uzu (accessed on 25 April 2023).
2. Bendini, A.; Barbieri, S.; Valli, E.; Buchecker, K.; Canavari, M.; Toschi, T.G. Quality Evaluation of Cold Pressed Sunflower Oils by Sensory and Chemical Analysis. *Eur. J. Lipid Sci. Technol.* **2011**, *113*, 1375–1384. [CrossRef]
3. Redondo-Cuevas, L.; Castellano, G.; Torrens, F.; Raikos, V. Revealing the Relationship between Vegetable Oil Composition and Oxidative Stability: A Multifactorial Approach. *J. Food Compos. Anal.* **2018**, *66*, 221–229. [CrossRef]

4. Grajzer, M.; Szmalcel, K.; Kuźmiński, Ł.; Witkowski, M.; Kulma, A.; Prescha, A. Characteristics and Antioxidant Potential of Cold-Pressed Oils—Possible Strategies to Improve Oil Stability. *Foods* **2020**, *9*, 1630. [CrossRef] [PubMed]
5. Prescha, A.; Grajzer, M.; Dedyk, M.; Grajeta, H. The Antioxidant Activity and Oxidative Stability of Cold-Pressed Oils. *J. Am. Oil Chem. Soc.* **2014**, *91*, 1291–1301. [CrossRef] [PubMed]
6. Wroniak, M.; Raczyk, M.; Kruszewski, B.; Symoniuk, E.; Dach, D. Effect of Deep Frying of Potatoes and Tofu on Thermo-Oxidative Changes of Cold Pressed Rapeseed Oil, Cold Pressed High Oleic Rapeseed Oil and Palm Olein. *Antioxidants* **2021**, *10*, 1637. [CrossRef] [PubMed]
7. Tauferova, A.; Dordevic, D.; Jancikova, S.; Tremlova, B.; Kulawik, P. Fortified Cold-Pressed Oils: The Effect on Sensory Quality and Functional Properties. *Separations* **2021**, *8*, 55. [CrossRef]
8. Momot, M.; Stawicka, B.; Szydłowska-Czerniak, A. Physicochemical Properties and Sensory Attributes of Cold-Pressed Camelina Oils from the Polish Retail Market. *Appl. Sci.* **2023**, *13*, 1924. [CrossRef]
9. Kachel, M.; Krajewska, M.; Stryjecka, M.; Ślusarczyk, L.; Matwijczuk, A.; Rudy, S.; Domin, M. Comparative Analysis of Phytochemicals and Antioxidant Properties of Borage Oil (*Borago officinalis* L.) and Milk Thistle (*Silybum marianum* Gaertn). *Appl. Sci.* **2023**, *13*, 2560. [CrossRef]
10. Kasote, D.M.; Badhe, Y.S.; Hegde, M.V. Effect of Mechanical Press Oil Extraction Processing on Quality of Linseed Oil. *Ind. Crops Prod.* **2013**, *42*, 10–13. [CrossRef]
11. Šamec, D.; Loizzo, M.R.; Gortzi, O.; Çankaya, İ.T.; Tundis, R.; Suntar, İ.; Shirooie, S.; Zengin, G.; Devkota, H.P.; Reboredo-Rodríguez, P.; et al. The Potential of Pumpkin Seed Oil as a Functional Food—A Comprehensive Review of Chemical Composition, Health Benefits, and Safety. *Compr. Rev. Food Sci. Food Saf.* **2022**, *21*, 4422–4446. [CrossRef]
12. Adeleke, B.S.; Babalola, O.O. Oilseed Crop Sunflower (*Helianthus annuus*) as a Source of Food: Nutritional and Health Benefits. *Food Sci. Nutr.* **2020**, *8*, 4666–4684. [CrossRef]
13. Chew, S.C. Cold-Pressed Rapeseed (*Brassica napus*) Oil: Chemistry and Functionality. *Food Res. Int.* **2020**, *131*, 108997. [CrossRef]
14. Czwartkowski, K.; Wierzbic, A.; Golimowski, W. Quality, Key Production Factors, and Consumption Volume of Niche Edible Oils Marketed in the European Union. *Sustainability* **2022**, *14*, 1846. [CrossRef]
15. Szydłowska-Czerniak, A.; Momot, M.; Stawicka, B.; Rabiej-Kozioł, D. Effects of the Chemical Composition on the Antioxidant and Sensory Characteristics and Oxidative Stability of Cold-Pressed Black Cumin Oils. *Antioxidants* **2022**, *11*, 1556. [CrossRef] [PubMed]
16. McDowell, D.; Elliott, C.T.; Koidis, A. Characterization and Comparison of UK, Irish, and French Cold Pressed Rapeseed Oils with Refined Rapeseed Oils and Extra Virgin Olive Oils. *Eur. J. Lipid Sci. Technol.* **2017**, *119*, 1600327. [CrossRef]
17. Bou Fakhreddine, L.; Sánchez, M. The Interplay between Health Claims and Sensory Attributes in Determining Consumers' Purchase Intentions for Extra Virgin Olive Oil. *Food Qual. Prefer.* **2023**, *106*, 104819. [CrossRef]
18. Brühl, L.; Matthäus, B.; Scheipers, A.; Hofmann, T. Bitter Off-Taste in Stored Cold-Pressed Linseed Oil Obtained from Different Varieties. *Eur. J. Lipid Sci. Technol.* **2008**, *110*, 625–631. [CrossRef]
19. Sánchez-Arévalo, C.M.; Olmo-García, L.; Fernández-Sánchez, J.F.; Carrasco-Pancorbo, A. Polycyclic Aromatic Hydrocarbons in Edible Oils: An Overview on Sample Preparation, Determination Strategies, and Relative Abundance of Prevalent Compounds. *Compr. Rev. Food Sci. Food Saf.* **2020**, *19*, 3528–3573. [CrossRef]
20. Symoniuk, E.; Ratusz, K.; Ostrowska-Ligęza, E.; Krygier, K. Impact of Selected Chemical Characteristics of Cold-Pressed Oils on Their Oxidative Stability Determined Using the Rancimat and Pressure Differential Scanning Calorimetry Method. *Food Anal. Methods* **2018**, *11*, 1095–1104. [CrossRef]
21. Gharby, S.; Oubannin, S.; Ait Bouzid, H.; Bijla, L.; Ibourki, M.; Gagour, J.; Koubachi, J.; Sakar, E.H.; Majourhat, K.; Lee, L.-H.; et al. An Overview on the Use of Extracts from Medicinal and Aromatic Plants to Improve Nutritional Value and Oxidative Stability of Vegetable Oils. *Foods* **2022**, *11*, 3258. [CrossRef]
22. Fadda, A.; Sanna, D.; Sakar, E.H.; Gharby, S.; Mulas, M.; Medda, S.; Yesilcubuk, N.S.; Karaca, A.C.; Gozukirmizi, C.K.; Lucarini, M.; et al. Innovative and Sustainable Technologies to Enhance the Oxidative Stability of Vegetable Oils. *Sustainability* **2022**, *14*, 849. [CrossRef]
23. Symoniuk, E.; Wroniak, M.; Napiórkowska, K.; Brzezińska, R.; Ratusz, K. Oxidative Stability and Antioxidant Activity of Selected Cold-Pressed Oils and Oils Mixtures. *Foods* **2022**, *11*, 1597. [CrossRef]
24. Ulbricht, T.L.V.; Southgate, D.A.T. Coronary Heart Disease: Seven Dietary Factors. *Lancet* **1991**, *338*, 985–992. [CrossRef]
25. Symoniuk, E.; Ratusz, K.; Krygier, K. Comparison of the Oxidative Stability of Cold-Pressed Rapeseed Oil Using Pressure Differential Scanning Calorimetry and Rancimat Methods. *Eur. J. Lipid Sci. Technol.* **2017**, *119*, 1600182. [CrossRef]
26. Choo, W.-S.; Birch, J.; Dufour, J.-P. Physicochemical and Quality Characteristics of Cold-Pressed Flaxseed Oils. *J. Food Compos. Anal.* **2007**, *20*, 202–211. [CrossRef]
27. Symoniuk, E.; Ratusz, K.; Krygier, K. Oxidative Stability and the Chemical Composition of Market Cold-pressed Linseed Oil. *Eur. J. Lipid Sci. Technol.* **2017**, *119*, 1700055. [CrossRef]
28. Ratusz, K.; Symoniuk, E.; Wroniak, M.; Rudzińska, M. Bioactive Compounds, Nutritional Quality and Oxidative Stability of Cold-Pressed Camelina (*Camelina sativa* L.) Oils. *Appl. Sci.* **2018**, *8*, 2606. [CrossRef]
29. Khalili Tilami, S.; Kouřimská, L. Assessment of the Nutritional Quality of Plant Lipids Using Atherogenicity and Thrombogenicity Indices. *Nutrients* **2022**, *14*, 3795. [CrossRef] [PubMed]

30. Rokosik, E.; Dwiecki, K.; Siger, A. Nutritional Quality and Phytochemical Contents of Cold Pressed Oil Obtained from Chia, Milk Thistle, Nigella, and White and Black Poppy Seeds. *Grasas Aceites* **2020**, *71*, 368. [CrossRef]
31. Ying, Q.; Wojciechowska, P.; Siger, A.; Kaczmarek, A.; Rudzińska, M. Phytochemical Content, Oxidative Stability, and Nutritional Properties of Unconventional Cold-Pressed Edible Oils. *J. Food Nutr. Res.* **2018**, *6*, 476–485. [CrossRef]
32. Franke, S.; Fröhlich, K.; Werner, S.; Böhm, V.; Schöne, F. Analysis of Carotenoids and Vitamin E in Selected Oilseeds, Press Cakes and Oils. *Eur. J. Lipid Sci. Technol.* **2010**, *112*, 1122–1129. [CrossRef]
33. Gliszczyńska-Świgło, A.; Sikorska, E.; Khmelinskii, I.; Sikorski, M. Tocopherol Content in Edible Plant Oils. *Pol. J. Food Nutr. Sci.* **2007**, *57*, 157–161.
34. Abramovič, H.; Butinar, B.; Nikolič, V. Changes Occurring in Phenolic Content, Tocopherol Composition and Oxidative Stability of *Camelina Sativa* Oil during Storage. *Food Chem.* **2007**, *104*, 903–909. [CrossRef]
35. Kim, H.J.; Lee, H.O.; Min, D.B. Effects and Prooxidant Mechanisms of Oxidized α-Tocopherol on the Oxidative Stability of Soybean Oil. *J. Food Sci.* **2007**, *72*, C223–C230. [CrossRef] [PubMed]
36. Aksoz, E.; Korkut, O.; Aksit, D.; Gokbulut, C. Vitamin E (α-, β + γ- and δ-tocopherol) Levels in Plant Oils. *Flavour Fragr. J.* **2020**, *35*, 504–510. [CrossRef]
37. Office of Dietary Supplements—Vitamin, E. Available online: https://ods.od.nih.gov/factsheets/VitaminE-HealthProfessional/ (accessed on 30 April 2023).
38. Stevenson, D.G.; Eller, F.J.; Wang, L.; Jane, J.-L.; Wang, T.; Inglett, G.E. Oil and Tocopherol Content and Composition of Pumpkin Seed Oil in 12 Cultivars. *J. Agric. Food Chem.* **2007**, *55*, 4005–4013. [CrossRef] [PubMed]
39. Murkovic, M.; Pfannhauser, W. Stability of Pumpkin Seed Oil. *Eur. J. Lipid Sci. Technol.* **2000**, *102*, 607–611. [CrossRef]
40. Evans, J.C.; Kodali, D.R.; Addis, P.B. Optimal Tocopherol Concentrations to Inhibit Soybean Oil Oxidation. *J. Am. Oil Chem. Soc.* **2002**, *79*, 47–51. [CrossRef]
41. Choe, E.; Min, D.B. Mechanisms of Antioxidants in the Oxidation of Foods. *Compr. Rev. Food Sci. Food Saf.* **2009**, *8*, 345–358. [CrossRef]
42. Aksoylu Özbek, Z.; Günç Ergönül, P. Cold Pressed Pumpkin Seed Oil. In *Cold Pressed Oils*; Elsevier: Amsterdam, The Netherlands, 2020; pp. 219–229.
43. Szterk, A.; Roszko, M.; Sosińska, E.; Derewiaka, D.; Lewicki, P.P. Chemical Composition and Oxidative Stability of Selected Plant Oils. *J. Am. Oil Chem. Soc.* **2010**, *87*, 637–645. [CrossRef]
44. Sakar, E.H.; Khtira, A.; Aalam, Z.; Zeroual, A.; Gagour, J.; Gharby, S. Variations in Physicochemical Characteristics of Olive Oil (cv 'Moroccan Picholine') According to Extraction Technology as Revealed by Multivariate Analysis. *AgriEngineering* **2022**, *4*, 922–938. [CrossRef]
45. Krygier, K.; Wroniak, M.; Dobczyński, K.; Kiełt, I.; Grześkiewicz, S.; Obiedziński, M. Characteristic of Commercial Cold Pressed Vegetable Oils. *Rośliny Oleiste* **1998**, *19*, 573–582.
46. Mikołajczak, N.; Tańska, M. Effect of Initial Quality and Bioactive Compounds Content in Cold-Pressed Flaxseed Oils on Oxidative Stability and Oxidation Products Formation during One-Month Storage with Light Exposure. *NFS J.* **2022**, *26*, 10–21. [CrossRef]
47. Çelik, S.E.; Özyürek, M.; Güçlü, K.; Apak, R. Solvent Effects on the Antioxidant Capacity of Lipophilic and Hydrophilic Antioxidants Measured by CUPRAC, ABTS/Persulphate and FRAP Methods. *Talanta* **2010**, *81*, 1300–1309. [CrossRef] [PubMed]
48. Ghosh, M.; Upadhyay, R.; Mahato, D.K.; Mishra, H.N. Kinetics of Lipid Oxidation in Omega Fatty Acids Rich Blends of Sunflower and Sesame Oils Using Rancimat. *Food Chem.* **2019**, *272*, 471–477. [CrossRef]
49. Matthäus, B.; Brühl, L. Quality of Cold-Pressed Edible Rapeseed Oil in Germany. *Nahrung/Food* **2003**, *47*, 413–419. [CrossRef]
50. Kamal-Eldin, A. Effect of Fatty Acids and Tocopherols on the Oxidative Stability of Vegetable Oils. *Eur. J. Lipid Sci. Technol.* **2006**, *108*, 1051–1061. [CrossRef]
51. Meddeb, W.; Rezig, L.; Abderrabbą, M.; Lizard, G.; Mejri, M. Tunisian Milk Thistle: An Investigation of the Chemical Composition and the Characterization of Its Cold-Pressed Seed Oils. *Int. J. Mol. Sci.* **2017**, *18*, 2582. [CrossRef]
52. Nyam, K.L.; Tan, C.P.; Lai, O.M.; Long, K.; Che Man, Y.B. Physicochemical Properties and Bioactive Compounds of Selected Seed Oils. *LWT—Food Sci. Technol.* **2009**, *42*, 1396–1403. [CrossRef]
53. Choe, E.; Min, D.B. Mechanisms and Factors for Edible Oil Oxidation. *Compr. Rev. Food Sci. Food Saf.* **2006**, *5*, 169–186. [CrossRef]
54. Vieira, S.A.; Zhang, G.; Decker, E.A. Biological Implications of Lipid Oxidation Products. *J. Am. Oil Chem. Soc.* **2017**, *94*, 339–351. [CrossRef]
55. Commission Regulation (EU) No 835/2011 of 19 August 2011 Amending Regulation (EC) No 1881/2006 as Regards Maximum Levels for Polycyclic Aromatic Hydrocarbons in FoodstuffsText with EEA Relevance. Available online: https://eur-lex.europa.eu/legal-content/EN/TXT/?uri=CELEX%3A32011R0835 (accessed on 30 April 2023).
56. Wroniak, M.; Rękas, A. A Preliminary Study of PCBs, PAHs, Pesticides and Trace Metals Contamination in Cold-Pressed Rapeseed Oils from Conventional and Ecological Cultivations. *J. Food Sci. Technol.* **2017**, *54*, 1350–1356. [CrossRef] [PubMed]
57. Roszko, M.; Szterk, A.; Szymczyk, K.; Waszkiewicz-Robak, B. PAHs, PCBs, PBDEs and Pesticides in Cold-Pressed Vegetable Oils. *J. Am. Oil Chem. Soc.* **2012**, *89*, 389–400. [CrossRef]
58. *ISO 5508:1996*; Animal and Vegetable Fats and Oils-Analysis by Gas Chromatography of Methyl Esters of Fatty Acids. ISO: Geneva, Switzerland,, 1996.
59. *ISO 5509:2000*; Animal and Vegetable Fats and Oils-Preparation of Methyl Esters of Fatty Acids. ISO: Geneva, Switzerland, 2000.

60. Fatemi, S.H.; Hammond, E.G. Analysis of Oleate, Linoleate and Linolenate Hydroperoxides in Oxidized Ester Mixtures. *Lipids* **1980**, *15*, 379–385. [CrossRef]
61. Santos-Silva, J.; Mendes, I.A.; Portugal, P.V.; Bessa, R.J.B. Effect of Particle Size and Soybean Oil Supplementation on Growth Performance, Carcass and Meat Quality and Fatty Acid Composition of Intramuscular Lipids of Lambs. *Livest. Prod. Sci.* **2004**, *90*, 79–88. [CrossRef]
62. *ISO 9936:2016*; Animal and Vegetable Fats and Oils-Determination of Tocopherol and Tocotrienol Contents by High Performance Liquid Chromatography. ISO: Geneva, Switzerland, 2016.
63. *ISO 12228-1:2014*; Animal and Vegetable Fats and Oils-Determination of Individual and Total Sterols Contents-Gas Chromatographic Method-Part 1. ISO: Geneva, Switzerland, 2014.
64. Szydłowska-Czerniak, A.; Łaszewska, A. Effect of Refining Process on Antioxidant Capacity, Total Phenolics and Prooxidants Contents in Rapeseed Oils. *LWT—Food Sci. Technol.* **2015**, *64*, 853–859. [CrossRef]
65. AOCS. Official Method Cd 12b-92: Oil stability index. In *Official Method and Recommended Practices of the American Oil Chemist's Society*; AOCS Publishing: Champaing, IL, USA, 2017.
66. *ISO 27107:2010*; Animal and Vegetable Fats and Oils-Determination of Peroxide Value-Potentiometric End-Point Determination. ISO: Geneva, Switzerland, 2010.
67. *ISO 6885:2016*; Animal and Vegetable Fats and Oils—Determination of Anisidine Value. International Organization for Standardization: Geneva, Switzerland, 2016.
68. *ISO 660:2020*; Animal and Vegetable Fats and Oils—Determination of Acid Value and Acidity. International Organization for Standardization: Geneva, Switzerland, 2020.
69. *ISO: 662:2016*; Animal and Vegetable Fats and Oils—Determination of Moisture and Volatile Matter Content. International Organization for Standardization: Geneva, Switzerland, 2016.
70. *ISO 8589:2007*; Sensory analysis—General guidance for the design of test rooms. International Organization for Standardization: Geneva, Switzerland, 2007.

molecules

MDPI

Article

Micronized Powder of Raspberry Pomace as a Source of Bioactive Compounds

Renata Różyło [1,*], Ryszard Amarowicz [2], Michał Adam Janiak [2], Marek Domin [3], Sławomir Gawłowski [1], Ryszard Kulig [1], Grzegorz Łysiak [1], Klaudia Rząd [4] and Arkadiusz Matwijczuk [4,5]

1 Department of Food Engineering and Machines, University of Life Sciences in Lublin, Głęboka 28, 20-612 Lublin, Poland; slawomir.gawlowski@up.lublin.pl (S.G.); ryszard.kulig@up.lublin.pl (R.K.); grzegorz.lysiak@up.lublin.pl (G.Ł.)
2 Department of Chemical and Physical Properties of Food, Institute of Animal Reproduction and Food Research, Polish Academy of Sciences, Tuwima 10, 10-748 Olsztyn, Poland; r.amarowicz@pan.olsztyn.pl (R.A.); m.janiak@pan.olsztyn.pl (M.A.J.)
3 Department of Biological Bases of Food and Feed Technologies, University of Life Sciences in Lublin, 28 Głęboka Str., 20-612 Lublin, Poland; marek.domin@up.lublin.pl
4 Department of Biophysics, Institute of Molecular Biophysics, Faculty of Environmental Biology, University of Life Sciences in Lublin, Akademicka 13, 20-950 Lublin, Poland; k.terlecka.98@wp.pl (K.R.); arkadiusz.matwijczuk@up.lublin.pl (A.M.)
5 ECOTECH-COMPLEX—Analytical and Programme Centre for Advanced Environmentally-Friendly Tech-Nologies, Maria Curie-Sklodowska University, Głęboka 39, 20-033 Lublin, Poland
* Correspondence: renata.rozylo@up.lublin.pl

Abstract: Red raspberries, which contain a variety of nutrients and phytochemicals that are beneficial for human health, can be utilized as a raw material in the creation of several supplements. This research suggests micronized powder of raspberry pomace production. The molecular characteristics (FTIR), sugar, and biological potential (phenolic compounds and antioxidant activity) of micronized raspberry powders were investigated. FTIR spectroscopy results revealed spectral changes in the ranges with maxima at ~1720, 1635, and 1326, as well as intensity changes in practically the entire spectral range analyzed. The discrepancies clearly indicate that the micronization of the raspberry byproduct samples cleaved the intramolecular hydrogen bonds in the polysaccharides present in the samples, thus increasing the respective content of simple saccharides. In comparison to the control powders, more glucose and fructose were recovered from the micronized samples of the raspberry powders. The study's micronized powders were found to contain nine different types of phenolic compounds, including rutin, elagic acid derivatives, cyanidin-3-sophoroside, cyanidin-3-(2-glucosylrutinoside), cyanidin-3-rutinoside, pelargonidin-3-rutinoside, and elagic acid derivatives. Significantly higher concentrations of ellagic acid and ellagic acid derivatives and rutin were found in the micronized samples than in the control sample. The antioxidant potential assessed by ABTS and FRAP significantly increased following the micronization procedure.

Keywords: micronization; raspberry pomace; FTIR spectra; phenolic identification; antioxidant activity

1. Introduction

Red raspberries (*Rubus idaeus* L.) contain a variety of nutrients and phytochemicals important to human health [1–3]. Raspberry consumption is effective in reducing the levels of oxidative and inflammatory stress that promote heart morphological changes in old age, thereby preventing or delaying heart disease [4]. These fruits are unique berries with a rich nutritional and bioactive composition. They are a source of several essential micronutrients and dietary fiber [5]. Raspberry fruit is rich in antioxidant compounds, especially polyphenols [6–8]. The polyphenols present in the fruit are mainly ellagitannins and anthocyanins. Due to the high content of anthocyanins, these fruits have a red color [5].

The antioxidant activity of raspberries is directly related to the total amount of phenolic compounds found in raspberries [9].

Many health benefits of raspberry fruit have been established. An increasing number of studies suggest that red raspberries may play a role in lowering the risk of metabolically related chronic diseases [5]. In a study on obese individuals with type 2 diabetes, the effects of earing raspberries daily were discovered to potentially lower postprandial hyperglycemia and inflammation in diabetic adults, as well as to have anti-inflammatory properties [10]. A diet high in raspberries has been demonstrated to improve immunological function and phospholipid metabolism in obese patients in trials [11]. Other clinical trials have demonstrated that including fresh raspberry extract in the diet of old rats lowers indicators of aging, improves psychomotor coordination and balance, and boosts muscle tone and endurance [12].

Red raspberries are commonly used for producing dietary supplements because of their health benefits. They can be purchased in the form of dried or liquid products [13]. Raspberry ketone (4-(4-hydroxyphenyl)-2-butanone) supplements have also gained popularity among customers due to their ability to burn fat and aid in weight loss. Raspberry ketone is found naturally in raspberry fruit (up to 4.3 mg/kg) [14] and is used as a flavoring substance [15]. Processing raspberries extends their shelf life and culinary uses, although the nutritional information among processed forms is limited [16,17]. Raspberries are often processed into juices. After producing the juices, the pomace is obtained, which is a recyclable by-product [18]. Fresh pomace is perishable; thus, drying it can significantly extend its shelf life and increase reusability. Raspberry powders obtained after the dehydration of whole fruit or pomace can be added as functional additives (e.g., natural dyes) to food preparations. In most studies, powders were obtained from drying whole fruits [19,20]; only a few studies concern the drying of raspberry pomace [21]. Previous findings suggest that powders obtained by spray-drying raspberry extracts could be used as natural colorants or antioxidants [22]. Drying studies of whole raspberries have shown that these techniques affect the physical properties, bioactive compounds, and antioxidant activity of the resulting powders differently [19]. Previous studies have shown that freeze-drying causes changes in the physical properties of raspberry pomace [21]. Only a few data describe the antioxidant properties of processed raspberry byproducts [22,23].

There is no information on micronized forms of raspberry byproducts. Whole raspberry fruit micronization tests were carried out only by fluidized bed jet milling with drying [24]. Micronization is a grinding process that reduces the particle size of material from microns down to the nanometer range. This technology has developed rapidly in food production in recent decades [25–27]. Previous research has proven that the micronization process can be applied to plant materials, resulting in increased functionality [28,29].

The process of micronization or fine grinding has been applied to other plant materials, and it has been demonstrated that techniques leading to particle size reduction also cause changes in powder properties such as viscosity and porosity, and such powders can be used for a variety of purposes [30].

Additionally, the use of infrared spectroscopy has provided insight into the structure/interactions or packing of molecules during this and other similar processes. This information can facilitate the identification of fast molecular markers indicative of such changes. Notably, there are still rather few publications on the topic available in the literature, although some authors are beginning to publish findings in this area.

Micronization in our studies was carried out with a ball mill; such studies on raspberry pomace powders are missing in the available literature. The research hypothesis assumes that raspberry byproduct micronization with the use of a ball mill results in a significant reduction in particle size and may have a positive effect on the biochemical properties of the obtained powders. The obtained powders were analyzed using the technique of ATR/FTIR spectroscopy (attenuated total reflectance Fourier transform infrared spectroscopy) to identify any changes that could potentially occur at the molecular level during the process of micronization. The presented results will contribute to the identification of marker changes occurring at the molecular level during the analyzed process of powder micronization. Furthermore, the presented molecular analysis will facilitate the easy identification of such changes in subsequent studies pertaining to this and similar types of samples, even if the micronization process itself is modified.

2. Results

2.1. Particle Size and Color Results of Micronized Byproduct Raspberry Powders

The investigation demonstrated that, as expected, the micronization process with the utilized ball mill induced considerable changes in the particle size distribution of the studied samples (Figure 1). The mean particle size (D[4;3]) was equal to 277 μm for the control sample (CRP). After 10 min (10 MRP) of micronization, this value dropped to 29.8 μm, and after 20 min of micronization (20 MRP) to 11 μm. More than 90% of the particles (d90) in the control sample (CRP) had dimensions below 578 μm. In the sample with 10 min of micronization, this parameter was 59.4 μm, and with 20 min of micronization, only 19.2 μm. The particle sizes for the 50% share (d50) were less than 225 μm for CRP, 25 μm for 10 MRP, and 10.5 μm for 20 MRP.

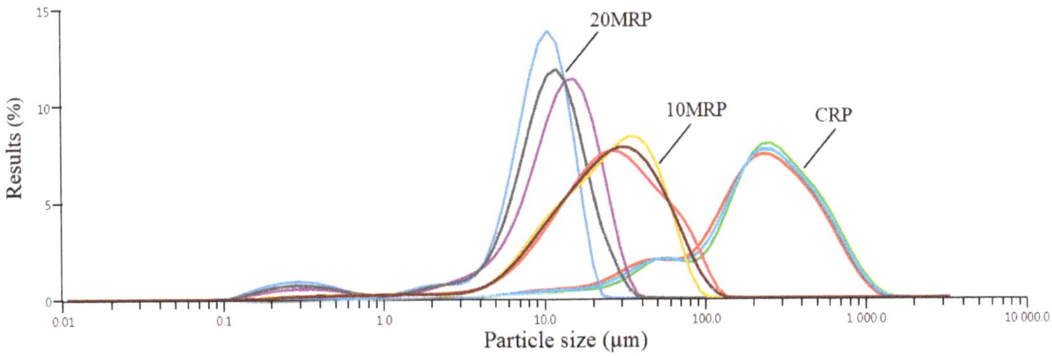

Figure 1. Particle size distribution for raspberry byproduct powders. CRP—control (without micronization) raspberry powder; 10 MRP—raspberry byproduct micronized for 10 min; 20 MRP—raspberry byproduct micronized for 20 min. Different line colors grouped with the same name mean single measurements. Different line colors grouped with the same name mean single measurements.

In other studies in which superfine grinding of apple pomace [31] or herbal plant [32] was used, similarly to our tests, a significant reduction in particle size was observed. In these tests, the control apple pomace had a d50 particle size below 326 μm, and fine grinding reduced the d50 particle size parameter below 51.5 μm. In our research, at 10 and 20 min of micronization (10 MRP; 20 MRP), we obtained much lower dimensions of d50, i.e., 25 and 10.5 μm. This was due to different characteristics of the raw material. Here, we micronized raspberry pomace after removing the seeds. For cryogrinding buckwheat hulls [33], the d50 was below 15.1 μm. Micronization of oat husks [34] helped to obtain d50 dimensions below 15.5 μm.

The color results revealed that the color of the raspberry powders after micronization differed marginally, but this difference was significant, particularly in the case of the L* and b* parameters. The control (CRP) sample had the lightest color (L* = 49.2), and the longest micronized sample (20 MRP) had the darkest color (L* = 48.3) (Figure 2).

There were no significant differences in this parameter between 10 and 20 min of micronization. For the longest time (20 MRP), the share of red color a* was highest in the micronized samples, and the share of yellow color b* was highest in the micronized samples (20 MRP). Previous research [28] has shown that the micronization method affects the color of spinach stems and leaves. In comparison to the control sample and dry micronization, wet micronization produced a darker color.

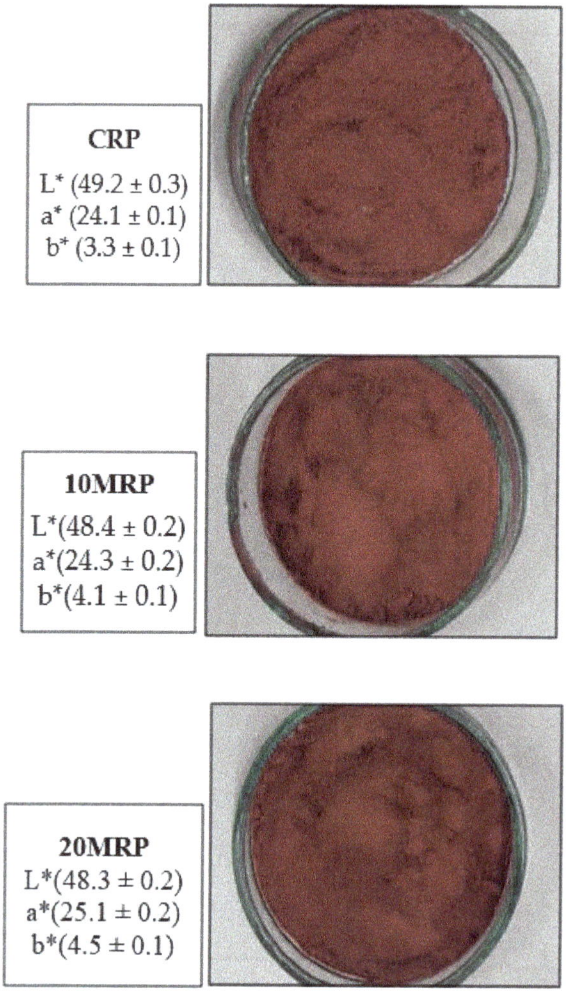

Figure 2. Appearance and color parameters of raspberry byproduct powders. CRP—control (without micronization) raspberry powder; 10 MRP—raspberry byproduct micronized for 10 min; 20 MRP—raspberry byproduct micronized for 20 min. L*—brightness color parameter; a*—redness color parameter; b*—blueness color parameter.

2.2. FTIR Results of Micronized Byproduct Raspberry Powders

At the next stage of the study, spectroscopic ATR-FTIR measurements were also performed (Figure 3), which assessed the impact of the sample treatment employed to facilitate more effective micronization. The results obtained from the spectral analysis in the infrared range might suggest that changes occur at the molecular level in the chemical structure of the raspberry samples in question.

Table 1 presents all the characteristic bands present in the obtained spectra and correlates them with specific vibrations of their respective functional groups, based on a detailed literature review [31–33,35–41] as well as on the careful structural analysis of the molecules present in our samples.

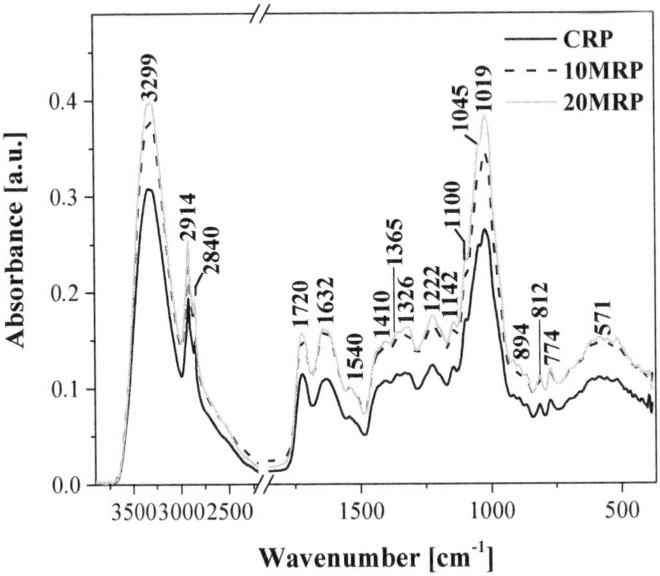

Figure 3. FTIR spectra for the analyzed raspberry samples: CRP—control (without micronization) raspberry powder; 10 MRP—raspberry byproduct micronized for 10 min; 20 MRP—raspberry byproduct micronized for 20 min. The measurements were performed within the spectral range from 450 to 3700 cm^{-1}. The spectra were registered at room temperature (See Section 3).

Table 1. The location of the maxima of the FTIR absorption bands, with assignments of particular vibrations to the respective raspberry samples corresponding to the data in Figure 4, registered within the spectral range of 450–3700 cm^{-1}.

FTIR Positioning of Band (cm^{-1})	Type and Origin of Vibrations
3312	ν(O-H) in H_2O and *intra*-molecular hydrogen bonding
2914	ν(C-H) in CH_2 and CH_3 asymmetrical and symmetrical
2845	
1723	ν(C=O)
1635	ν(C=C) or/and δ (O-H) adsorbed H_2O
1541	ν(C=C)
1407	δ(-OH in plane), δ(CH_2), δ(C-H)
1363	
1326	δ(C-H) and δ (O-H)
1222	δ(C-H) and asymmetrical bridge oxygen stretching -OH in-plane bending
1146	asymmetrical in phase ring stretching and ν(C-O-C) and ν(C-O) and ring stretching modes
1017	
912/893/866	*β*-linkage of cellulose ring breathing and asymmetrical out of phase stretching -OH *out-of-plane* bending and CH_2 rocking
813/772	
613/580/550/511	

Note: ν—stretching vibrations; δ—deformation vibrations

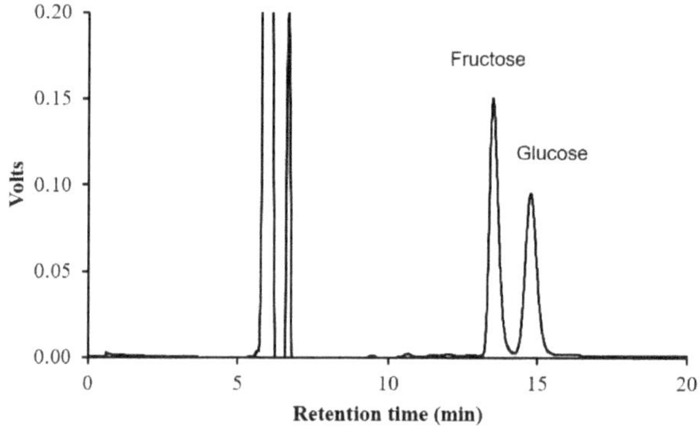

Sample	Fructose	Glucose
CRP	118 ± 1[b]	103 ± 1[b]
10MRP	132 ± 6[a]	115 ± 5[a]
20MRP	131 ± 4[a]	114 ± 4[a]

Figure 4. HPLC chromatogram and sugar content in micronized byproduct raspberry powders (mg/g). Values in the same column marked with different letters differ significantly ($p < 0.05$).

The infrared spectra recorded for the analyzed raspberry byproduct contain clear bands that can be fairly accurately associated with identifiable vibrations of particular functional groups that are characteristic of ingredients present in products rich in polysaccharides and similar nutrients [31–33,38].

Starting from the highest wave numbers, we first observed characteristic, wide bands that are present in the range from approx. 2500 to 3600 cm^{-1}. The maximum of these bands, located at ~3300 cm^{-1}, is typical of the vibrations of asymmetric bonds induced by the stretching vibrations of the hydroxy -OH groups, which are characteristic of polysaccharide molecules that are predominant in the analyzed samples [38,41]. Moreover, the groups are involved in the formation of hydrogen bonds between smaller units of polysaccharide molecules. Notably, the band corresponding to the stretching vibrations of the -OH groups enhances the stretching vibrations of the C-H groups [36,41]. There are two very sharp bands present in this region, with wave numbers of 2914 and 2848 cm^{-1}, that are particularly characteristic of asymmetric and symmetric stretching vibrations in the CH_2 groups present in this type of food sample. The highest intensity of this particular band, similar to the band corresponding to the -OH vibrations, could be observed in the sample subjected to micronization for 20 min, with a lower value registered for the sample micronized for 10 min and the lowest for the control.

Next, we proceed to the analysis of the essential group of vibrations known as the fingerprint region. The first two key bands corresponded to vibrations with the maxima at 1720 and 1632 cm^{-1}, respectively. The former was associated with the characteristic stretching vibrations of the carbonyl group ν(C=O) [38] found in the molecules of simple sugars present in the samples. The latter band with the maximum at approx. 1630 cm^{-1} [37,38] corresponded to the deformation vibrations characteristic of water molecules, δ(-OH). During the micronization process, the relation between the two bands was noticeably altered. In the control, the respective ratio of the 1632/1720 bands was: 0.94, whereas in the sample subjected to 20 min., micronization increased to as much as 1.04. This pair of bands can therefore be treated as an excellent spectroscopic marker for the processing treatment in question.

Furthermore, the bands observed in the region from 1550 to 900 cm^{-1} corresponded to the strong vibrations of the C-O, C-C, C-O-H, and C-O-C groups, various oligomolecules or polysaccharides [32,33]. Even though the spectra recorded in this region seemed similar in terms of shape, two significant changes ought to be noted. Firstly, there was a change in the shape and increase in the intensity of the band with the maximum at 1540 cm^{-1}, characteristic of ν(C=C) vibrations. Secondly, we observed changes in terms of the intensity and, most importantly, the shape of the bands with the maximum at 1326 cm^{-1}, characteristic of δ(C-H) vibrations and potentially enhanced by δ(O-H) vibrations in the molecules of poly- and oligosaccharides that were the primary ingredients of the analyzed samples. While other vibrations in this region retained their original shapes, their intensity was significantly increased with the growing duration of the micronization process. On addition, the intensity of the bands with the maximum at 1412 cm^{-1} was characteristic of δ(C-H) vibrations. Next, we observed an increase in intensity of the vibrations with the maxima at 1360, 1222, and 1144 cm^{-1}. These are characteristic of the vibrations of, respectively, δ(-OH in plane), δ(CH2) groups, and δ(C-H) groups, as well as of the stretching vibrations in the C-O-C system found in oligo- and polysaccharides present in raspberries. We also observed increased intensity of the bands, with the main maximum at 1019 cm^{-1}. The bands in this spectral region are primarily associated with C-O and C-O-C stretching vibrations in polysaccharide molecules. The highest intensity of these vibrations was observed in the samples subjected to 20 min processing.

The last spectral "fingerprint" region below 930 cm^{-1} corresponds primarily to the crystalline regions and indicates conformational changes occurring in the analyzed material through possible changes to the β-1,4-glycoside bonds in polysaccharide molecules [38]. As we know, a fingerprint region is a spectral infrared range where each organic compound produces its unique absorption band. Such bands provide information as to the presence of various functional groups found in the given analyzed sample. As can be seen in our results, in the discussed case, the region below 930 cm^{-1} was characterized by a relatively low intensity of the bands with the maxima at ~891, 808, 774, or 580 cm^{-1}. Apart from intensity variations, no particularly significant changes were observed here. Nonetheless, the changes in vibration intensity observed in this region, particularly in samples subjected to micronization, clearly indicate effects on the bonds between individual units in polysaccharide molecules [31–33].

To briefly recapitulate the already discussed results obtained from the spectral FTIR measurements, the observed discrepancies in terms of band intensity and, in some cases, slight shifts thereof indicate that the mechanical strength of finely ground material had a significant impact in the molecular properties of the analyzed samples. Firstly, the method of micronization employed for the raspberry byproduct samples resulted in cleaving the intramolecular hydrogen bonds in cellulose, hemicellulose, and polysaccharides predominantly present in the samples in question, likely leading to an increase in their content of amorphic cellulose and simple saccharides [31–33]. This was evidenced mainly by the changes in the intensity of bands characteristic of the stretching vibrations of the -OH group and the altered ratio of the 1720/1632 cm^{-1} bands, but also by the increased intensity of bands such as those with maxima at 1326, 1222, or 1019 cm^{-1}. As follows from the literature data, mechanical strength usually cleaves only the amorphic region on the ordered surface in a crystalline substance [31–33]. Due to the same, the stiff and ordered structure of cellulose was slightly deteriorated by very fine grinding. All the mentioned shifts were related to the cleavage of hydrogen bonds present in the polymer chain during the grinding process. Additional cleavage of bonds and structural changes occurring in the polysaccharides were also facilitated by the observed increase in sample temperature during the micronization process. However, the observed spectral changes also clearly indicated that very fine grinding had no effect on the primary functional groups in cellulose, and the observed discrepancies were associated mainly with modifications to polysaccharide chains. This observation was further corroborated by the vibrations recorded below 930 cm^{-1}, where the only effects noted related to band intensity. The bands in this region are characteristic of the vibrations on β-1,4-glycoside bonds, as already discussed above. As for the registered changes in the intensity of said vibrations, they evidence the same susceptibility to factors related to the micronization of the tested raspberry samples.

As evidenced by the above, the use of FTIR spectroscopy allowed us to identify the bands of marker changes characterizing the micronization process employed. The presented results may allow for better optimization of the process in the course of future research with a view of obtaining even higher-quality products.

2.3. Identification of Sugars in Micronized Byproduct Raspberry Powders

A study of the identification of sugars in raspberry powders (Figure 4) showed that, compared to control samples, micronization, regardless of the time of 10 or 20 min, resulted in significantly higher amounts of fructose and glucose. The disintegration of cell membranes during physical destruction, but also to some extent heat treatment, which facilitates the extraction of the solute, can be used to explain changes in the levels of sugars [42].

Additionally, our analysis of the FTIR spectra indicated that it is likely that the polysaccharides' intramolecular hydrogen bonds had broken, boosting the proportion of simple saccharides in the mixture. According to a different study [43], micronizing soybean fibers resulted in a considerable reduction in the amount of polysaccharides, including cellulose. Research on the effects of ball mill micronization on the characteristics of polygonatum powder revealed that the process' extension led to a much higher concentration of soluble sugar in these powders [44].

2.4. Identification of Phenolic Compounds in Micronized Byproduct Raspberry Powders

The list of compounds identified in this study using UHPLC-Q-TOFMS/MS is presented in Table 2, and the results of the content of individual phenolic compounds of micronized byproduct raspberry powders are shown in Table 3.

Table 2. Identification of the main phenolic compounds of raspberry powders by UHPLC-Q-TOFMS/MS.

Compound Number	Ionization	Compound Name	MS	MS/MS
1	$[M-H]^-$	Ellagic acid derivative	571	301, 229
2	$[M-H]^+$	Cyanidin-3-sophoroside	611	449, 287, 269
3	$[M-H]^+$	Cyanidin-3-(2-glucosylrutinoside)	757	611, 287
4	$[M-H]^+$	Cyanidin-3-rutinoside	449	287
5	$[M-H]^+$	Pelargonidin-3-rutinoside	579	271
6	$[M-H]^-$	Ellagic acid derivative	934	1235, 934, 633, 315, 301
7	$[M-H]^-$	Ellagic acid derivative	934	1235, 934, 633, 315, 301
8	$[M-H]^-$	Ellagic acid	301	229
9	$[M-H]^-$	Rutin	609	301

Table 3. Content of individual phenolic compounds, total phenolics and antioxidant potential of micronized byproduct raspberry powders (mg/g).

Compound Number	Compound Name	CRP	10 MRP	20 MRP
1	Ellagic acid derivative	0.132 ± 0.003 ᵇ	0.161 ± 0.010 ᵃ	0.166 ± 0.008 ᵃ
2	Cyanidin-3-sophoroside	0.354 ± 0.001 ᵃ	0.372 ± 0.014 ᵃ	0.347 ± 0.013 ᵃ
3	Cyanidin-3-(2-glucosylrutinoside)	0.134 ± 0.001 ᵃ	0.142 ± 0.005 ᵃ	0.132 ± 0.005 ᵃ
4	Cyanidin-3-rutinoside	0.402 ± 0.003 ᵃ	0.418 ± 0.015 ᵃ	0.385 ± 0.010 ᵇ
5	Pelargonidin-3-rutinoside	0.162 ± 0.001 ᵃ	0.169 ± 0.006 ᵃ	0.157 ± 0.002 ᵇ
6	Ellagic acid derivative	0.697 ± 0.024 ᵇ	1.034 ± 0.068 ᵃ	0.878 ± 0.082 ᵃ
7	Ellagic acid derivative	2.164 ± 0.041 ᵇ	2.631 ± 0.120 ᵃ	2.627 ± 0.128 ᵃ
8	Ellagic acid	0.079 ± 0.001 ᵇ	0.093 ± 0.004 ᵃ	0.098 ± 0.003 ᵃ
9	Rutin	0.012 ± 0.001 ᵇ	0.013 ± 0.001 ᵃ	0.014 ± 0.001 ᵃ

Values in the same row marked with different letters differ significantly ($p < 0.05$). Content of compounds **1**, **6**, and **7** is expressed as ellagic acid equivalents, and compounds **2**, **3**, **4**, and **5** as cyaniding-3-glucoside.

The two main peaks showcased in Figure 5 can be attributed to ellagitannins such as lambertianin C, sanguiin H6, H10, casuarictin or their isomers. Similar results were obtained for red and black raspberry extracts [45,46]. Other researchers also confirmed that dimeric sanguiin H6 and trimeric lamberatianin C are the main ellagotannins in raspberries [47]. In the black raspberry seed extracts, lamberatianin C is not present [48]; however, in Siberian raspberries, they can be detected within the leave extracts [49].

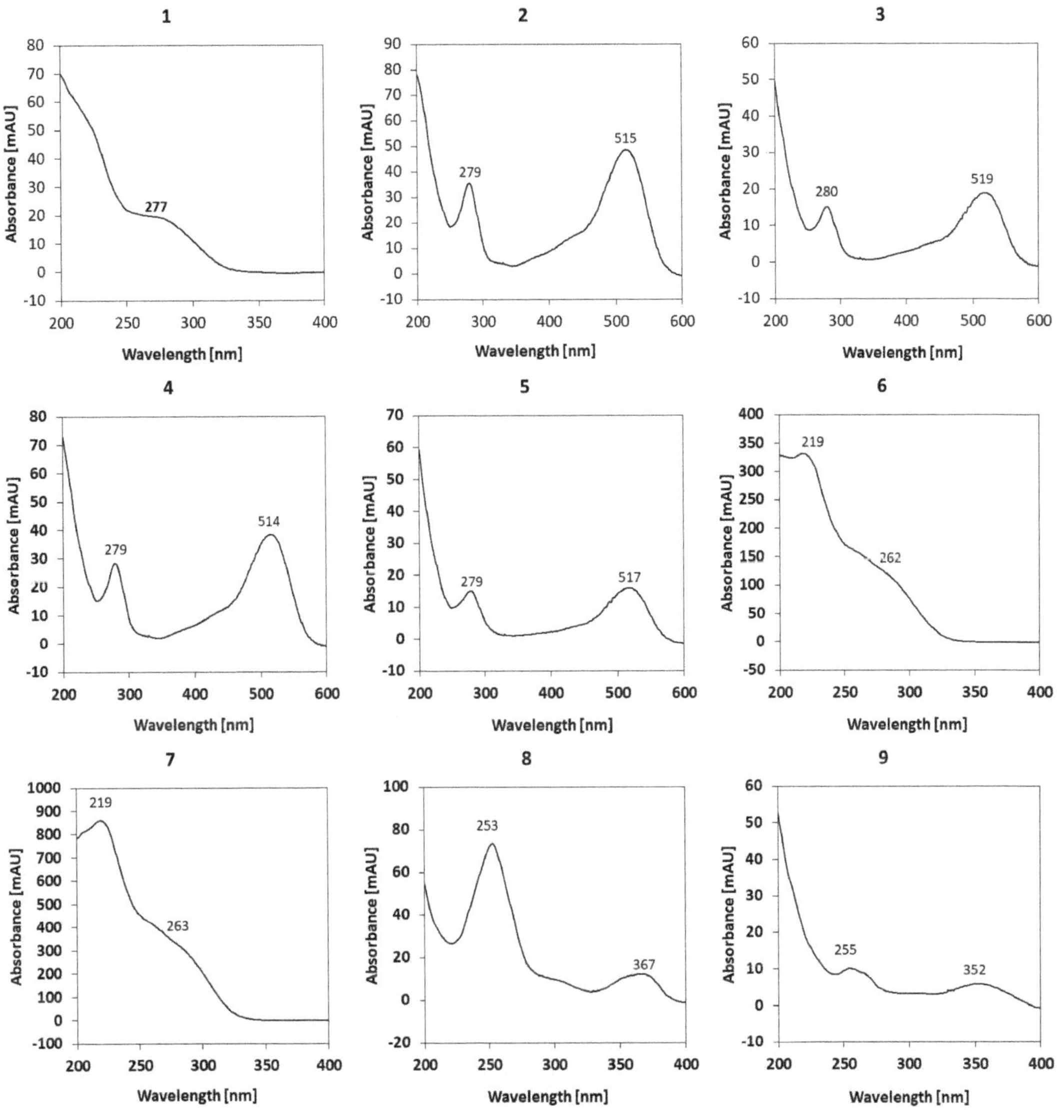

Figure 5. UV spectra of individual phenolic compounds of raspberry powders: **1**—ellagic acid derivative; **2**—cyanidin-3-sophoroside; **3**—cyanidin-3-(2-glucosylrutinoside); **4**—cyanidin-3-rutinoside; **5**—pelargonidin-3-rutinoside; **6**—ellagic acid derivative; **7**—ellagic acid derivative; **8**—ellagic acid; **9**—rutin.

Recorded UV spectrums for these compounds are depicted in Figure 5. We did not observe parent ions nor daughter ions above m/z 1235; however, according to the abovementioned team of Ross et al. [45], the peaks can be attributed to ellagotannins. It can be assumed that ion 934 is a daughter ion derived from the cleavage of complex ellagotannins [16].

The identification of phenolic compounds carried out showed that in the studied micronized powders from raspberry byproduct, there are nine different types of such compounds, including: ellagic acid derivative (**1**), cyanidin-3-sophoroside, cyanidin-3-(2-glucosylrutinoside), cyanidin-3-rutinoside, pelargonidin-3-rutinoside, ellagic acid derivative (**6**), ellagic acid derivative (**7**), ellagic acid and rutin. Significantly higher contents of ellagic acid and ellagic acid derivatives (**1**, **6**, **7**) and rutin were detected in micronized samples (10 and 20 MRP) than in the control (CRP) sample, regardless of micronization time in the tested range of 10 to 20 min. Longer micronization carried out for 20 min compared to the sample micronized for 10 min and the control sample reduced the content of cyanidin-3-rutinoside and pelargonidin-3-rutinoside. The content of the other phenolic compounds was not significantly different for all samples tested. In our previous studies [28] on wet and dry micronization of spinach leaves and stems, it was observed that both dry and wet micronization affected the contents of o-coumaric acid and gallic acid. In addition, dry micronization of spinach leaves increased the content of 3-hydroxyphenylacetic acid, 4-hydroxyphenylacetic acid, and p-coumaric acid. Other studies on micronization of grape pomace and fiber concentrates have explained that micronization increases the extractability of phenolic compounds, especially catechin and epicatechin [50].

We also found several anthocyanins that were tentatively identified by analyzing their fragmentation patterns. The presence of those compounds was also detected by RP-HPLC-DAD (Figure 6).

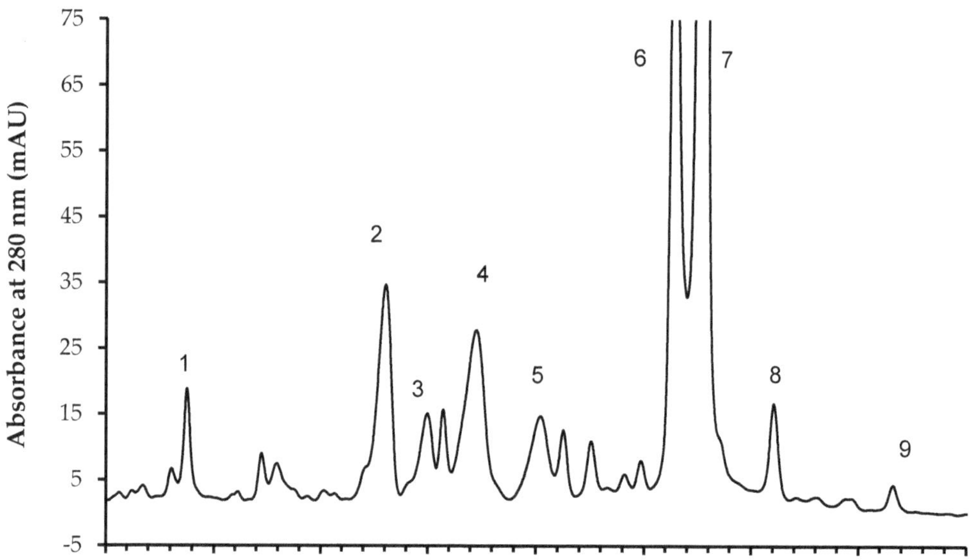

Figure 6. HPLC-DAD chromatogram of phenolic compounds of raspberry powder. **1**—ellagic acid derivative; **2**—cyanidin-3-sophoroside; **3**—cyanidin-3-(2-glucosylrutinoside); **4**—cyanidin-3-rutinoside; **5**—pelargonidin-3-rutinoside; **6**—ellagic acid derivative; **7**—ellagic acid derivative; **8**—ellagic acid; **9**—rutin.

Other teams that have analyzed raspberries reported similar anthocyanin profiles [16,45]. The profile of the compounds that we observed in our study demonstrates that other compounds can be detected in raspberry pomace. We observed derivatives of phenolic acids and flavonoids, e.g., glucuronides, hexosides, and pentosides. The general profile of the compounds that are present in the raspberry pomace is comparable with other studies [16,45,46,49]. However, more flavonoid aglycones were recently identified in raspberry puree [51], probably due to the fact that most of them are removed when puree/juice is produced and when pomace is obtained.

2.5. Antioxidant Potential of Micronized Byproduct Raspberry Powders

The results of the total phenolic and antioxidant potential of micronized byproduct raspberry powders are shown in Table 4.

Table 4. Total phenolics and antioxidant potential of micronized byproduct raspberry powders (mg/g).

Antioxidant Potential	CRP	10 MRP	20 MRP
Total phenolics (mg GAE/g)	19.74 ± 0.055 [b]	22.79 ± 0.78 [a]	23.94 ± 0.95 [a]
ABTS (mmol TE/g)	0.180 ± 0.005 [b]	0.221 ± 0.004 [a]	0.216 ± 0.005 [a]
DPPH (mmol TE/g)	0.241 ± 0.004 [a]	0.201 ± 0.004 [c]	0.218 ± 0.007 [b]
FRAP (mmol Fe^{2+}/g)	0.151 ± 0.003 [b]	0.181 ± 0.004 [a]	0.179 ± 0.004 [a]
ACL (mmol TE/g)	0.183 ± 0.006 [a]	0.178 ± 0.003 [a]	0.180 ± 0.007 [a]

Values in the same row marked with different letters differ significantly ($p < 0.05$).

After the micronization process, the antioxidant potential measured in our study by ABTS and FRAP greatly enhanced; there were no significant differences between the micronization times of 10 and 20 min. The scavenging activity index (ACL) did not significantly vary for any of the raspberry powder test samples. However, after the application of micronization, a considerable drop in the value of the DPPH indicator was seen.

Whole raspberry micronized fruits obtained by fluidized bed jet milling with drying were characterized by a higher content of anthocyanins and polyphenols as well as by higher antioxidant properties compared to the powders obtained by convection and spray-drying methods [24].

Sheng et al. [31] found that grape pomace treated by superfine grinding treatment had lower total phenolic content and proanthocyanidins values than control samples. The heat degradation of phenolic compounds was indicated by the authors as the explanation for the decrease in the concentration of these compounds. In our investigation, micronizing raspberry powder for 10 min did not cause it to reach a temperature of 41 °C, and micronizing it for 20 min did not cause it to reach a temperature of 55 °C.

3. Materials and Methods

3.1. Materials

The raw material for the research was raspberries of the Polesie variety from a plantation in the Lublin region. Raspberry pomace as a byproduct during juice production was obtained with the use of a slow-running press. The research material was obtained by freeze-drying raspberry pomace. The resulting pomace was molded into dies and frozen at −30 °C under free convection conditions. In this way, frozen cuboidal solids with dimensions of 2 × 2 × 4 cm were obtained, which were subjected to a freeze-drying process at a pressure of 20 Pa for 72 h without heating the shelves in the Christ Alpha 2–4 LD plus device. The obtained lyophilisate was crushed and sieved to separate the seeds. In further analyses, only pomace without seeds was used.

3.2. *Micronization of Raspberry Byproduct Powders*

The micronization of freeze-dried raspberry byproduct (without seeds) was carried out on a ball mill (Pulverisette 6, Fritsh, Idar-Oberstein, Germany) (Figure 7). The Planetary Mill with one working station uses one 500 ml grinding bowl that rotates with a transmission ratio of 1:1.82 relative to the main disk. The primary disk's rotational speed can be adjusted from 100 to 650 revolutions per minute. Grinding balls with a diameter of 10 mm made of hardened, stainless steel, FE-CR, were utilized in the process. Freeze-dried raspberry pomace powder without seeds (50 g) was placed into a bowl filled with 15 balls and micronized for 10 or 20 min with a speed of 600 rpm. The process was carried out with simultaneous monitoring of the particle size and temperature after micronization. It was observed that 10 min of micronization already significantly reduced the particle size. Further extension of micronization to 10 min increased the temperature of the raw material to 41 °C, and after 20 min to 55 °C. We did not want to cause a large degradation of the compounds; thus, we finished the process after 20 min. As a result, three samples were chosen for further study: a control sample of freeze-dried non-micronized raspberry pomace (CRP), a sample of freeze-dried raspberry pomace micronized for 10 min (10 MRP), and a sample of freeze-dried raspberry pomace micronized for 20 min (20 MRP).

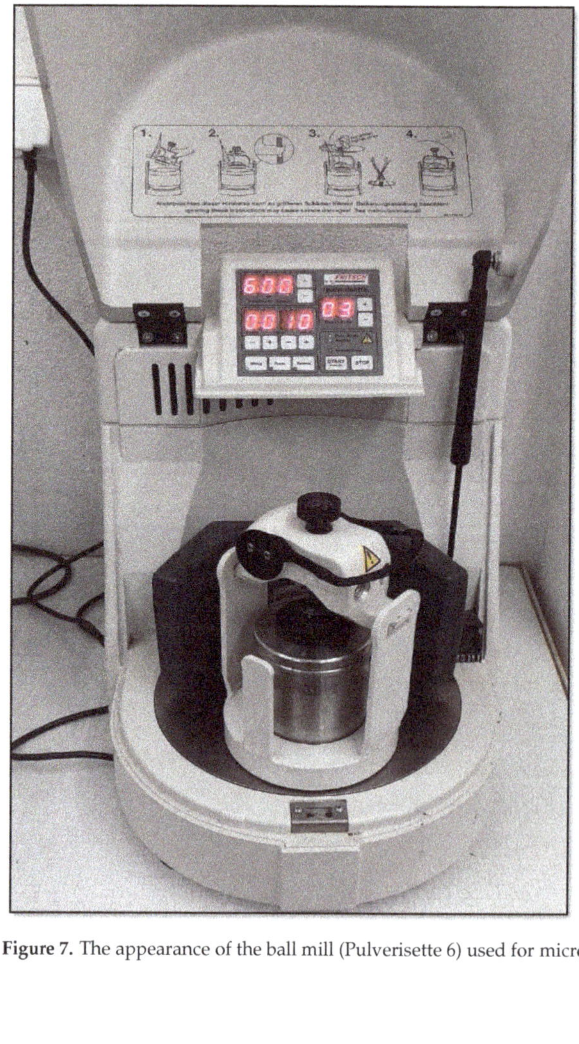

Figure 7. The appearance of the ball mill (Pulverisette 6) used for micronization.

3.3. Particle Size Analysis

Analyses of the particle size of control raspberry pomace powder (CRP) and micronized raspberry powders (10 and 20 MRP) were performed on a Mastersizer 3000 (Malvern Instruments Ltd., Malvern, UK) [28,52]. Measurements were made using a dry dispersion adapter (Aero S). As a result of the measurements, the mean particle dimensions weighed by volume (D[4;3] (µm)) or surface area D[3;2] (µm) were obtained, and the dimensions of the specific surface area (SSA) ($m^2 \cdot kg^{-1}$) and the distributions were obtained. D50 is the particle size in microns at which 50% of the sample is smaller or larger. d10 is the particle size, with 10% of particles being smaller than this dimension. D90 is the particle size at which 90% of the particles in the sample are smaller than this value. Particle size analyses of raspberry powders were performed in three replications.

3.4. Color Measurements

The color measurements of the analyzed samples were determined on the CIE L*a*b*, scale using the Precise Color Reader (4 Wave CR30-16, Planeta, Tychy, Poland) colorimeter [48]. The L* parameter, meaning the brightness of the material, was in the range of 0–100. The a* color index ranged from −150 to +100, and negative values indicated green and positive values red. The b* index, determined in the range from −100 to +150, determined the share of blue color when it was negative and yellow when it was positive.

3.5. Infrared Spectra Measurements

Measurements of the ATR-FTIR spectra registered for the analyzed samples were performed using an IRSprit spectrometer from Shimatzu (Tokyo, Japan). An ATR (Attenuated Total Reflection) attachment in the form of a Zn Se crystal with adequate geometry (45°) was used to facilitate multiple internal reflections of the laser beam. The micronized powder samples were placed on the crystal. The spectrometer attachment allowed for a very exact measurement owing to the very precisely controlled contact between the sample and the crystal, with the possibility of regulating the amount of pressure. The attachment facilitates considerably more precise measurements in samples of this particular type. During the measurements, 24 scans were registered for each of the samples. Subsequently, the software automatically averaged the obtained spectra. Before and after each measurement, the crystal was spotless using ultrapure solvents. All the solvents were purchased from Sigma-Aldrich (Poznań, Poland). The spectra were registered within the range from 450 to 3800 cm^{-1} at a resolution of 2 cm^{-1}. The spectral measurements were conducted at the Laboratory of the Department of Biophysics, Molecular Biophysics Institute, University of Life Sciences in Lublin. All the clearly discernible bands were associated with corresponding vibrations based on a detailed review of available literature as well as on information regarding the structure of the molecules present in the analyzed samples. All the spectra were processed and prepared for publication using Grams AI software (Version 9.1) from ThermoGalactic Industries (San Jose, CA, USA).

3.6. Sugar Identification

Sugars were extracted from raspberry seeds with hot 85% (v/v) methanol [53,54]. Individual sugars were determined using the HPLC method. Individual sugars were separated using an HPLC Shimadzu system (Shimadzu, Kyoto, Japan), which consisted of an SCL-10A controller, an LC-10AD pump, and a RID-10A detector. A portion of 20 µL of the extract was injected into a Luna Omega 3 µm SUGAR column (4.6 × 250 mm) (Phenomenex, Torrance, CA, USA). The flow rate of the mobile phase (acetonitrile–water, 25:75, v/v) was 1 mL/min. For calibration, the external standard method was used.

3.7. Phenolic Compounds Extraction

Phenolic compounds were extracted from raspberry byproduct powder, according to Tomas [51]. Raspberry powder (2 g) was extracted in 10 mL of methanol–water solution (75:25, v/v) containing 1% of formic acid in an ultrasonic bath for 15 min (Ultron U-509,

Dywity, Poland). Then, the sample was centrifuged at 2700× g at 4 °C for 10 min, the supernatant was collected, and the sample was adjusted to 25 mL with the solvent used for extraction.

The content of total phenolic compounds in the extract was determined using Folin–Ciocalteou's phenol reagent [55]. The results were expressed as gallic acid equivalents per gram of raspberry powder.

3.8. HPLC-DAD Analysis of Phenolic Compounds

Polyphenolic compounds were analyzed using RP-HPLC-DAD. Extracts were injected (1 µL) into the Shimadzu Nexera system that consisted of a degassing unit (DGV-20A 5R), two pumps (LC-30AD), an autosampler (SIL-30AC), column oven, PDA detector (SPD-M30A) and a controlling unit (CBM-20A). The flow rate was set to 1 mL/mL. Separation (Kinetex, SHIM-POL, Warsaw, Poland C18 2.6 µm, 100 A, 75 × 3 mm) was monitored at 280 and 520 nm and was conducted under binary gradient conditions. The eluents used were (A) water:acetonitrile:trifluoroacetic acid (95:5:0.1, $v/v/v$) and (B) acetonitrile:trifluoroacetic acid (100:0.1, v/v). The gradient was set up for eluent B as follows: 0–10 min: 0–18.8%; 10.5 min: 0%; 12 min: 0%. Peak areas were recorded and compared with those of ellagic acid and cyaniding-3-glucoside obtained from prepared calibration curves. Results were expressed as milligrams of the standard per gram of extract per gram of D.W.

3.9. Identification of Phenolic Compounds

To identify more of the compounds from raspberry pomace, samples were analyzed using an Exigent microLC 200 system coupled with a TripleTOF 5600+ mass spectrometer (AB Sciex, Framingham, MA, USA). Electrospray ionization was performed in positive and negative. The operating MS conditions were as follows: ion spray voltage: 4.5 kV; turbo spray temperature: 350 °C; nebulizer gas (GS1) and curtain gas flow rate: 30 L/min; heater gas (GS2) flow rate: 35 L/min; declustering potential (DP) and collision energy (CE) for the full-scan MS: 90 or −90 V and 10 or −10 eV, respectively; and for MS2 mode: 80 or −80 V and 30 or −30 eV, respectively. The TOF MS scan was scanned in the mass range of 100–1250 m/z. Compounds were separated using an Exigent Halo C18 column (0.5 × 50 mm, 2.7 µm; AB Sciex). The binary gradient that was employed consisted of 0.1% (v/v) formic acid in water (eluent A) and 0.1% (v/v) formic acid in acetonitrile (eluent B), and it was set up from 5 to 90% B within 3 min, maintained to 3.8 min and 5% within 4 min to finally be maintained to 5 min.

3.10. Antiradical Activity Evaluation

Antiradical activity against ABTS•+ and DPPH• was determined using the methods described by Re et al. [56] and Amarowicz et al. [57]. The results were expressed as millimoles of Trolox equivalents (TE) per gram of powder. The method of Benzie and Strain [58] was used for the determination of ferric-reducing antioxidant power (FRAP). The results were expressed as mmol Fe^{2+} per gram of raspberry powder.

3.11. Photochemiluminescence Assay

The scavenging activity of raspberry byproduct powder samples was evaluated by a photochemiluminescence (PCL-ACL) method [59] in which superoxide radical anions (O2•−) are generated from luminol. The reactions were carried out using kits from Analytic Jena, (Jena, Germany). Measurement was performed on a Photochem device with PCLsoft 5.1 software (Analytic Jena). The results were expressed as mmol of Trolox equivalents per g of raspberry powders.

3.12. Statistical Analysis

Measurements were made in triplicate, means and deviations were calculated, and other statistical analyses were performed in Statistica 12.0 (StatSoft, Kraków, Poland).

One-way analysis of variance (ANOVA) was performed, and Tukey's test was performed to determine the significance of differences ($p < 0.05$) between the means. Significantly different means were marked with different letters (a, b, c, etc.) in the figures and tables.

4. Conclusions

The conducted research proved that the micronization process of freeze-dried raspberry pomace byproducts with the applied ball mill caused significant changes in the particle size distribution of the tested samples. The use of micronization for 10 and 20 min allowed for a significant reduction of the d50 particle size from 225 μm (CRP) to 25 μm (10 MRP) and 10.5 μm (20 MRP).

The micronization procedure significantly increased the amount of recognized sugars, such as glucose and fructose, in the raspberry powders. Despite a large reduction in particle size, the amount of these two sugars increased by approximately 12 percent after 10 and 20 min of micronization. There were no significant differences in glucose and fructose content between these samples.

Significantly higher contents of ellagic acid and ellagic acid derivatives and rutin were detected in the samples micronized compared to the control sample. Longer micronization reduced the content of cyanidin-3-rutinoside and pelargonidin-3-rutinoside.

The antioxidant potential of ABTS and FRAP significantly increased after 10 min of micronization, and extending this process to 20 min did not cause significant changes. According to these investigations, controlling the raspberry byproduct micronization process can result in powders with higher antioxidant potential. As a result, these powders can be utilized to create innovative functional foods and dietary supplements. Because of their low humidity, such powders may be stored for an extended period and can be simply utilized in processing. Furthermore, because of their finely separated particles and intense color, they can be used as functional food colorants with high antioxidant activity. In turn, the most significant discrepancies in the registered FTIR spectra were observed in the bands with the maxima at ~1720, 1635, and 1326. The process of micronization of raspberry byproduct samples resulted in the cleavage of intramolecular hydrogen bonds in the polysaccharide molecules present in the samples.

The research findings have significant industrial implications because they can be used to create innovative micronized raspberry powder that can be used as nutritionally valuable dietary supplements. The reuse of this pomace as juice-processing byproducts is a significant environmental issue.

For future research, it will be beneficial to run broader analyses on a larger number of samples, such as different raspberry varieties, so that correlation relationships or PCA tests can be conducted. Furthermore, additional investigation of the physical properties of the powders during storage, as well as in vivo testing, should be carried out.

Author Contributions: Conceptualization, R.R.; methodology, R.R. and R.A.; methodology of ATR-FTIR, A.M.; software, R.A., A.M., R.K. and G.Ł.; validation, R.A., G.Ł. and R.K.; formal analysis, R.A., M.A.J., A.M. and K.R.; investigation, M.A.J., M.D., S.G., R.K., A.M. and K.R.; resources, R.R., R.A., K.R. and A.M.; data curation, R.R., R.A. and A.M.; writing—original draft preparation, R.R., R.A. and A.M.; writing—review and editing, R.R., R.A., K.R. and A.M.; visualization, R.R., R.A., S.G. and A.M.; supervision, R.R. and R.A.; project administration, R.R.; funding acquisition, R.R. and R.A. All authors have read and agreed to the published version of the manuscript.

Funding: This research received no external funding.

Institutional Review Board Statement: Not applicable.

Informed Consent Statement: Not applicable.

Data Availability Statement: Not applicable.

Conflicts of Interest: The authors declare no conflict of interest.

Sample Availability: Not available.

References

1. Wu, Q.; Naeem, A.; Zou, J.; Yu, C.; Wang, Y.; Chen, J.; Ping, Y. Isolation of Phenolic Compounds from Raspberry Based on Molecular Imprinting Techniques and Investigation of Their Anti-Alzheimer's Disease Properties. *Molecules* **2022**, *27*, 6893. [CrossRef] [PubMed]
2. Yang, Y.; Yin, X.; Zhang, D.; Zhang, B.; Lu, J.; Wang, X. Structural Characteristics, Antioxidant, and Immunostimulatory Activities of an Acidic Polysaccharide from Raspberry Pulp. *Molecules* **2022**, *27*, 4385. [CrossRef] [PubMed]
3. Rao, A.V.; Snyder, D.M. Raspberries and human health: A review. *J. Agric. Food Chem.* **2010**, *58*, 3871–3883. [CrossRef] [PubMed]
4. Noratto, G.; Chew, B.P.; Ivanov, I. Red raspberry decreases heart biomarkers of cardiac remodeling associated with oxidative and inflammatory stress in obese diabetic db/db mice. *Food Funct.* **2016**, *7*, 4944–4955. [CrossRef] [PubMed]
5. Burton-Freeman, B.M.; Sandhu, A.K.; Edirisinghe, I. Red raspberries and their bioactive polyphenols: Cardiometabolic and neuronal health links. *Adv. Nutr.* **2016**, *7*, 44–65. [CrossRef]
6. Beekwilder, J.; Jonker, H.; Meesters, P.; Hall, R.D.; Van Der Meer, I.M.; De Vos, C.H.R. Antioxidants in raspberry: On-line analysis links antioxidant activity to a diversity of individual metabolites. *J. Agric. Food Chem.* **2005**, *53*, 3313–3320. [CrossRef]
7. Frías-Moreno, M.N.; Parra-Quezada, R.A.; González-Aguilar, G.; Ruíz-Canizales, J.; Molina-Corral, F.J.; Sepulveda, D.R.; Salas-Salazar, N.; Olivas, G.I. Quality, bioactive compounds, antioxidant capacity, and enzymes of raspberries at different maturity stages, effects of organic vs. Conventional fertilization. *Foods* **2021**, *10*, 953. [CrossRef]
8. Shoukat, S.; Mahmudiono, T.; Al-Shawi, S.G.; Abdelbasset, W.K.; Yasin, G.; Shichiyakh, R.A.; Iswanto, A.H.; Kadhim, A.J.; Kadhim, M.M.; Al–Rekaby, H.Q. Determination of the antioxidant and mineral contents of raspberry varieties. *Food Sci. Technol.* **2022**, *42*, e118521. [CrossRef]
9. Sariburun, E.; Şahin, S.; Demir, C.; Türkben, C.; Uylaşer, V. Phenolic content and antioxidant activity of raspberry and blackberry cultivars. *J. Food Sci.* **2010**, *75*, 328–335. [CrossRef]
10. Schell, J.; Betts, N.M.; Lyons, T.J.; Basu, A. Raspberries improve postprandial glucose and acute and chronic inflammation in adults with type 2 diabetes. *Ann. Nutr. Metab.* **2019**, *74*, 165–174. [CrossRef]
11. Franck, M.; de Toro-Martín, J.; Garneau, V.; Guay, V.; Kearney, M.; Pilon, G.; Roy, D.; Couture, P.; Couillard, C.; Marette, A.; et al. Effects of daily raspberry consumption on immune-metabolic health in subjects at risk of metabolic syndrome: A randomized controlled trial. *Nutrients* **2020**, *12*, 3858. [CrossRef] [PubMed]
12. Galli, R.L.; Carey, A.N.; Luskin, K.A.; Bielinski, D.F.; Shukitt-Hale, B. Red raspberries can improve motor function in aged rats. *J. Berry Res.* **2016**, *6*, 97–103. [CrossRef]
13. Lee, J. Rosaceae products: Anthocyanin quality and comparisons between dietary supplements and foods. *NFS J.* **2016**, *4*, 1–8. [CrossRef]
14. Bredsdorff, L.; Wedebye, E.B.; Nikolov, N.G.; Hallas-Møller, T.; Pilegaard, K. Raspberry ketone in food supplements—High intake, few toxicity data—A cause for safety concern? *Regul. Toxicol. Pharmacol.* **2015**, *73*, 196–200. [CrossRef]
15. Abdelaal, S.H.; El Azab, N.F.; Hassan, S.A.; El-Kosasy, A.M. Quality control of dietary supplements: An economic green spectrofluorimetric assay of Raspberry ketone and its application to weight variation testing. *Spectrochim. Acta Part A Mol. Biomol. Spectrosc.* **2021**, *261*, 120032. [CrossRef]
16. Piechowiak, T.; Grzelak-Błaszczyk, K.; Sójka, M.; Balawejder, M. Changes in phenolic compounds profile and glutathione status in raspberry fruit during storage in ozone-enriched atmosphere. *Postharvest Biol. Technol.* **2020**, *168*, e111277. [CrossRef]
17. Żyżelewicz, D.; Oracz, J.; Bilicka, M.; Kulbat-Warycha, K.; Klewicka, E. Influence of Freeze-Dried Phenolic-Rich Plant Powders on the Bioactive Compounds Profile, Antioxidant Activity and Aroma of Different Types of Chocolates. *Molecules* **2021**, *26*, 7058. [CrossRef]
18. Bélafi-Bakó, K.; Cserjési, P.; Beszédes, S.; Csanádi, Z.; Hodúr, C. Berry Pectins: Microwave-Assisted Extraction and Rheological Properties. *Food Bioprocess Technol.* **2012**, *5*, 1100–1105. [CrossRef]
19. Si, X.; Chen, Q.; Bi, J.; Wu, X.; Yi, J.; Zhou, L.; Li, Z. Comparison of different drying methods on the physical properties, bioactive compounds and antioxidant activity of raspberry powders. *J. Sci. Food Agric.* **2016**, *96*, 2055–2062. [CrossRef]
20. Szadzińska, J.; Łechtańska, J.; Pashminehazar, R.; Kharaghani, A.; Tsotsas, E. Microwave- and ultrasound-assisted convective drying of raspberries: Drying kinetics and microstructural changes. *Dry. Technol.* **2019**, *37*, 1–12. [CrossRef]
21. Domin, M.; Ćwiklińska, M.; Góral-Kowalczyk, M. Reeze-Drying Impact on Hardness of Selected Soft Fruit Liophilisates. *Agric. Eng.* **2021**, *25*, 135–146. [CrossRef]
22. Gagneten, M.; Corfield, R.; Mattson, M.G.; Sozzi, A.; Leiva, G.; Salvatori, D.; Schebor, C. Spray-dried powders from berries extracts obtained upon several processing steps to improve the bioactive components content. *Powder Technol.* **2019**, *342*, 1008–1015. [CrossRef]
23. Szymanowska, U.; Baraniak, B. Antioxidant and potentially anti-inflammatory activity of anthocyanin fractions from pomace obtained from enzymatically treated raspberries. *Antioxidants* **2019**, *8*, 299. [CrossRef]
24. Sadowska, A.; Świderski, F.; Hallmann, E. Properties of raspberry powder obtained by a new method of fluidized-bed jet milling and drying compared to other drying methods. *J. Sci. Food Agric.* **2020**, *100*, 4303–4309. [CrossRef] [PubMed]
25. Alam, S.A.; Järvinen, J.; Kirjoranta, S.; Jouppila, K.; Poutanen, K.; Sozer, N. Influence of particle size reduction on structural and mechanical properties of extruded rye bran. *Food Bioprocess Technol.* **2014**, *7*, 2121–2133. [CrossRef]
26. Chen, T.; Zhang, M.; Bhandari, B.; Yang, Z. Micronization and nanosizing of particles for an enhanced quality of food: A review. *Crit. Rev. Food Sci. Nutr.* **2018**, *58*, 993–1001. [CrossRef]

27. Protonotariou, S.; Mandala, I.; Rosell, C.M. Jet Milling Effect on Functionality, Quality and In Vitro Digestibility of Whole Wheat Flour and Bread. *Food Bioprocess Technol.* **2015**, *8*, 1319–1329. [CrossRef]
28. Różyło, R.; Piekut, J.; Dziki, D.; Smolewska, M.; Gawłowski, S.; Wójtowicz, A.; Gawlik-Dziki, U. Effects of Wet and Dry Micronization on the GC-MS Identification of the Phenolic Compounds and Antioxidant Properties of Freeze-Dried Spinach Leaves and Stems. *Molecules* **2022**, *27*, 8174. [CrossRef]
29. Zhang, J.; Dong, Y.; Nisar, T.; Fang, Z.; Wang, Z.C.; Guo, Y. Effect of superfine-grinding on the physicochemical and antioxidant properties of *Lycium ruthenicum* Murray powders. *Powder Technol.* **2020**, *372*, 68–75. [CrossRef]
30. Silventoinen, P.; Rommi, K.; Holopainen-Mantila, U.; Poutanen, K.; Nordlund, E. Biochemical and Techno-Functional Properties of Protein- and Fibre-Rich Hybrid Ingredients Produced by Dry Fractionation from Rice Bran. *Food Bioprocess Technol.* **2019**, *12*, 1487–1499. [CrossRef]
31. Sheng, K.; Qu, H.; Liu, C.; Yan, L.; You, J.; Shui, S.; Zheng, L. A comparative assess of high hydrostatic pressure and superfine grinding on physicochemical and antioxidant properties of grape pomace. *Int. J. Food Sci. Technol.* **2017**, *52*, 2106–2114. [CrossRef]
32. Meng, Q.; Fan, H.; Chen, F.; Xiao, T.; Zhang, L. Preparation and characterization of *Dendrobium officinale* powders through superfine grinding. *J. Sci. Food Agric.* **2018**, *98*, 1906–1913. [CrossRef] [PubMed]
33. Radoš, K.; Čukelj Mustač, N.; Drakula, S.; Novotni, D.; Benković, M.; Kraljić, K.; Štifter, S.; Voučko, B.; Ćurić, D. The effect of cryo-grinding and size separation on bioactive profile of buckwheat hulls. *Int. J. Food Sci. Technol.* **2022**, *57*, 1911–1919. [CrossRef]
34. Dziki, D.; Tarasiuk, W.; Gawlik-Dziki, U. Micronized oat husk: Particle size distribution, phenolic acid profile and antioxidant properties. *Materials* **2021**, *14*, 5443. [PubMed]
35. Andrianjaka-Camps, Z.N.; Baumgartner, D.; Camps, C.; Guyer, E.; Arrigoni, E.; Carlen, C. Prediction of raspberries puree quality traits by Fourier transform infrared spectroscopy. *LWT-Food Sci. Technol.* **2015**, *63*, 1056–1062. [CrossRef]
36. Chen, F.; Zhang, M.; Mujumdar, A.S.; Guo, C.; Yu, D. Comparative analysis of composition and hygroscopic properties of infrared freeze-dried blueberries, cranberries and raspberries. *Dry. Technol.* **2021**, *39*, 1261–1270.
37. Gales, O.; Rodemann, T.; Jones, J.; Swarts, N. Application of near infra-red spectroscopy as an instantaneous and simultaneous prediction tool for anthocyanins and sugar in whole fresh raspberry. *J. Sci. Food Agric.* **2021**, *101*, 2449–2454. [CrossRef]
38. Kupryaniuk, K.; Wójtowicz, A.; Mazurkiewicz, J.; Słowik, T.; Matwijczuk, A. The Influence of the Pressure-Thermal Agglomeration Methods of Corn Bran on Their Selected Physicochemical Properties and Biogas Efficiency. *Energies* **2021**, *14*, 6997. [CrossRef]
39. Przybył, K.; Koszela, K.; Adamski, F.; Samborska, K.; Walkowiak, K.; Polarczyk, M. Deep and machine learning using SEM, FTIR, and texture analysis to detect polysaccharide in raspberry powders. *Sensors* **2021**, *21*, 5823. [CrossRef]
40. Tian, Q.; Giusti, M.M.; Stoner, G.D.; Schwartz, S.J. Characterization of a new anthocyanin in black raspberries (*Rubus occidentalis*) by liquid chromatography electrospray ionization tandem mass spectrometry. *Food Chem.* **2006**, *94*, 465–468. [CrossRef]
41. Xu, Y.; Liu, N.; Fu, X.; Wang, L.; Yang, Y.; Ren, Y.; Liu, J.; Wang, L. Structural characteristics, biological, rheological and thermal properties of the polysaccharide and the degraded polysaccharide from raspberry fruits. *Int. J. Biol. Macromol.* **2019**, *132*, 109–118. [CrossRef] [PubMed]
42. Dastangoo, S.; Hamed Mosavian, M.T.; Yeganehzad, S. Optimization of pulsed electric field conditions for sugar extraction from carrots. *Food Sci. Nutr.* **2020**, *8*, 2025–2034. [CrossRef] [PubMed]
43. Zhu, L.; Yu, B.; Chen, H.; Yu, J.; Yan, H.; Luo, Y.; Chen, D. Comparisons of the micronization, steam explosion, and gamma irradiation treatment on chemical composition, structure, physicochemical properties, and in vitro digestibility of dietary fiber from soybean hulls. *Food Chem.* **2022**, *366*, e130618. [CrossRef] [PubMed]
44. Yu, Y.; Li, Z.; Cao, G.; Li, S.; Yang, H. Effects of ball milling micronization on amino acids profile and antioxidant activities of *Polygonatum cyrtonema* Hua tuber powder. *J. Food Meas. Charact.* **2019**, *13*, 2106–2117. [CrossRef]
45. Ross, H.A.; McDougall, G.J.; Stewart, D. Antiproliferative activity is predominantly associated with ellagitannins in raspberry extracts. *Phytochemistry* **2007**, *68*, 218–228. [CrossRef]
46. Kula, M.; Majdan, M.; Głód, D.; Krauze-Baranowska, M. Phenolic composition of fruits from different cultivars of red and black raspberries grown in Poland. *J. Food Compos. Anal.* **2016**, *52*, 74–82. [CrossRef]
47. Lee, J.; Dossett, M.; Finn, C.E. Rubus fruit phenolic research: The good, the bad, and the confusing. *Food Chem.* **2012**, *130*, 785–796. [CrossRef]
48. Lee, G.E.; Kim, R.H.; Lim, T.; Kim, J.; Kim, S.; Kim, H.G.; Hwang, K.T. Optimization of accelerated solvent extraction of ellagitannins in black raspberry seeds using artificial neural network coupled with genetic algorithm. *Food Chem.* **2022**, *396*, 133712.
49. Kashchenko, N.I.; Olennikov, D.N.; Chirikova, N.K. Metabolites of Siberian raspberries: LC-MS profile, seasonal variation, antioxidant activity and, thermal stability of *Rubus matsumuranus* phenolome. *Plants* **2021**, *10*, 2317. [CrossRef]
50. Bender, A.B.B.; Speroni, C.S.; Moro, K.I.B.; Morisso, F.D.P.; dos Santos, D.R.; da Silva, L.P.; Penna, N.G. Effects of micronization on dietary fiber composition, physicochemical properties, phenolic compounds, and antioxidant capacity of grape pomace and its dietary fiber concentrate. *LWT-Food Sci. Technol.* **2020**, *117*, e108652. [CrossRef]
51. Tomas, M. Effect of dietary fiber addition on the content and in vitro bioaccessibility of antioxidants in red raspberry pure. *Food Chem.* **2022**, *375*, 131897. [CrossRef]
52. Ziemichód, A.; Różyło, R.; Dziki, D. Impact of Whole and Ground-by-Knife and Ball Mill Flax Seeds on the Physical and Sensorial Properties of Gluten Free-Bread. *Processes* **2020**, *8*, 452. [CrossRef]
53. Sobaszek, P.; Różyło, R.; Dziki, L.; Gawlik-Dziki, U.; Biernacka, B.; Panasiewicz, M. Evaluation of color, texture, sensory and antioxidant properties of gels composed of freeze-dried maqui berries and agave sugar. *Processes* **2020**, *8*, 1294. [CrossRef]

54. Southgate, D.A.T. *Determination of Food Carbohydrates*, 2nd ed.; Elsevier Applied Science: Amsterdam, The Netherlands, 1991; Chapter 3.
55. Singleton, V.L.; Orthofer, R.; Lamuela-Raventos, R.M. Analysis of total phenols and other oxidation substrates and antioxidants by means of Folin-Ciocalteu reagent. and antioxidant activity of selected plant species from the Canadian prairies. *Food Chem.* **1999**, *84*, 551–562.
56. Re, R.; Pellegrini, N.; Proteggente, A.; Pannala, A.; Yang, M.; Rice-Evans, C. Antioxidant activity applying an improved ABTS radical cation decolorization assay. *Free. Radic. Biol. Med.* **1999**, *26*, 1231–1237. [CrossRef]
57. Amarowicz, R.; Troszyńska, A.; Baryłko-Pikielna, N.; Shahidi, F. Polyphenolics extracts from legume seeds: Correlations between total antioxidant activity, total phenolics content, tannins content and astringency. *J. Food Lipids* **2004**, *11*, 278–286. [CrossRef]
58. Benzie, I.F.F.; Strain, J.J. The reducing ability of plasma as a measure of 'antioxidant power'—The FRAP assay. *Anal. Biochem.* **1996**, *239*, 70–76. [CrossRef]
59. Popov, I.; Lewin, G. Oxidants and antioxidants part B-Antioxidative homeostasis: Characterization by means of chemiluminescent technique. *Methods Enzymol.* **1999**, *300*, 437–456.

molecules

MDPI

Article

Proteases as Tools for Modulating the Antioxidant Activity and Functionality of the Spent Brewer's Yeast Proteins

Loredana Dumitrașcu, Andreea Lanciu Dorofte, Leontina Grigore-Gurgu and Iuliana Aprodu *

Faculty of Food Science and Engineering, Dunarea de Jos University of Galati, 111 Domneasca Str., 800008 Galati, Romania; loredana.dumitrascu@ugal.ro (L.D.); andreea.dorofte@ugal.ro (A.L.D.); leontina.gurgu@ugal.ro (L.G.-G.)
* Correspondence: iuliana.aprodu@ugal.ro

Abstract: The functionality of the peptides obtained through enzymatic hydrolysis of spent brewer's yeast was investigated. Hydrolysis was carried out for 4–67 h with bromelain, neutrase and trypsin. The resulting hydrolysates were characterized in terms of physical-chemical, antioxidant and techno-functional properties. The solid residues and soluble protein contents increased with the hydrolysis time, the highest values being measured in samples hydrolyzed with neutrase. Regardless of the hydrolysis time, the maximum degree of hydrolysis was measured in the sample hydrolyzed with neutrase, while the lowest was in the sample hydrolyzed with trypsin. The protein hydrolysate obtained with neutrase exhibited the highest DPPH radical scavenging activity (116.9 ± 2.9 μM TE/g dw), followed by the sample hydrolyzed with trypsin (102.8 ± 2.7 μM TE/g dw). Upon ultrafiltration, the fraction of low molecular weight peptides (<3 kDa) released by bromelain presented the highest antioxidant activity (50.06 ± 0.39 μM TE/g dw). The enzymes influenced the foaming properties and the emulsions-forming ability of the hydrolysates. The trypsin ensured the obtaining of proteins hydrolysate with the highest foam overrun and stability. The emulsions based on hydrolysates obtained with neutrase exhibited the highest viscosity at a shear rate over 10 s^{-1}. These results indicate that the investigated proteases are suitable for modulating the overall functionality of the yeast proteins.

Keywords: spent brewer's yeast; enzymatic hydrolysis; peptides; antioxidant activity; functional properties

1. Introduction

The global production of beer reached about 1.86 billion hL in 2021, out of which around 342 million hL were produced only in Europe [1,2]. The brewing industry generates about 0.4 billion tons of spent brewer's yeast (SY) each year and even more billion tons of spent grain and hops [3]. Despite their high circular economy potential, either in food, feed or for pharmaceutical and cosmetic purposes, most of the time SY is used as raw material for animal feed or discarded as waste [3]. SY has a very appealing composition as, based on dry weight, it contains high-quality proteins (45–60%), carbohydrates (15–35%), nucleic acids (4–8%), lipids (4–7%), B vitamins complex, minerals and dietary fibers [4–6]. The SY proteins contain essential amino acids in quantities that meet the FAO/WHO/UNU reference profile [7,8]. In addition, SY represents an excellent source of bioactive peptides, making it very attractive for various applications, such as for developing nutraceuticals and functional foods.

Moreover, as the global population is expected to reach about 10 billion by 2050, doubling the food demand in the next decades, the valorization of different by-products such as SY represents a sustainable, economical and healthy alternative for acquiring food-grade proteins for the increasing population [9]. Compared to other alternative food protein sources, SY has the Generally Recognized as Safe status, being available throughout the year. The initial step in the recovery of proteins and generation of peptides from SY consists in disrupting the cell walls to release the intracellular components. The disruption of the cell walls and further protein release can be achieved by using individual or various combinations of physical, chemical or enzymatic treatments. Enzymatic hydrolysis is one of the most frequently applied methods for the production of protein hydrolysates and peptides, being considered efficient, safe and relatively cheap compared to other methods [10]. However, the characteristics of the peptides usually depend on the protein chain from which they were released and the type of enzyme used for hydrolysis [11]. As reviewed by Oliveira et al. [6], endopeptidases such as Alcalase®, Corolase®, Papain® and Brauzyn®, alone or in admixture with exopeptidases (Flavourzyme® and Promod®) and other proteases (Protamex®), are most frequently used for generating protein/peptide rich extracts from SY. Many in vivo and in vitro studies performed on SY hydrolysates or isolated peptides have demonstrated antioxidant, immunomodulatory, antihyperglycemic, antihypertensive, antithrombotic, lipid-lowering, antimicrobial, antiulcer, and antiproliferative activities [12]. In particular, high antioxidant activity values of the SY extracts were reported by Podpora et al. [13], who performed autolysis assisted by enzymatic hydrolysis with papain, and Marson et al. [5], who employed enzymatic hydrolysis with Brauzyn®, Protamex® and Alcalase®.

The objective of this study was to investigate the impact of using three different proteases, namely bromelain, neutrase and trypsin, for SY proteins lysis on the antioxidant activity and technological functionality of the resulting hydrolysates. Although the potential of these enzymes for obtaining hydrolysates with increased bioactivity was demonstrated on several protein sources of animal or vegetal origin [14], there are no studies that focus on the functionality of the peptide-rich hydrolysates released by bromelain, neutrase and trypsin from the SY proteins.

2. Results and Discussion

2.1. The Effect of Enzymatic Hydrolysis on the Recovery of SRC and SPC

Hydrolysis with exogenous proteolytic enzymes represents an efficient method of cell lysis. Depending on the amount they are added, these enzymes together with endogenous enzymes accelerate leakage and recovery of intracellular compounds [6]. In the present study, the recovery of cellular compounds resulting from cell wall breakdown was measured in terms of SRC (Table 1) and SPC (Table 2). The SRC and SPC were influenced by the hydrolysis time that had a positive effect on the recovery of intracellular components. The maximum recovery of the solids from the yeast cells was obtained upon 21 h of hydrolysis, while further increase in the hydrolysis had no significant effect on the recovery of soluble solids. The type of enzyme also had a significant effect on the SRC and SPC. Compared to the control sample, after 4 h of hydrolysis, the SRC increased by 29.1%, 73.7% and 50% in the BSY, NSY and TSY samples, respectively. Among all tested enzymes, the hydrolysates produced with neutrase contained the highest SRC, regardless of the hydrolysis time. Similar results were reported in the literature, where SRC after enzymatic hydrolysis had a range of 41–61%. For example, Takalloo et al. [15] tested several extraction techniques such as autolysis, plasmolysis and enzymatic hydrolysis for the recovery of intracellular components from *Saccharomyces cerevisiae*. The authors reported that enzymatic hydrolysis resulted in the highest release of soluble solids and proteins after 48 h of hydrolysis with alcalase, when the soluble solid content increased to 52.1%. In our study, similar results were measured after 21 h of hydrolysis performed with bromelain and trypsin.

The SPC release upon SY hydrolysis varied with the enzyme and the duration of the treatment (Table 2). After 4 h of hydrolysis, the SPC of the BSY, NSY and TSY samples was 17%, 41% and 19% higher compared to the control sample. Our findings are in agreement with Marson et al. [5], who used a mixture of proteases for the 2 h hydrolysis of the spent yeast suspension and reported an increase of the protein content recovered in the extract with 16% compared to the control. Moreover, it can be seen that after 4 h of hydrolysis with neutrase the SPC was at least 1.5–1.6 times higher than in the case of bromelain and trypsin. On the other hand, it can be observed that SPC reached a plateau after 21 h of hydrolysis, regardless of the tested enzyme. Overall, regardless of hydrolysis time, the maximum SPC was obtained in the NSY sample, indicating that protein recovery was more responsive to neutrase than bromelain and trypsin. Thamnarathip et al. [16] reported a protein recovery yield on a dry basis of 31.9 ± 0.2 % for the rice bran protein samples hydrolyzed with neutrase for 8 h.

Table 1. The solid residue content (g/100 g dw) of the extracts obtained from spent brewer's yeast hydrolyzed with bromelain (BSY), neutrase (NSY) and trypsin (TSY) after different time intervals.

Sample	Hydrolysis Time (Hours)				
	4	8	21	43	67
Control	32.11 ± 2.96 ^{aD*}	32.50 ± 1.81 ^{aD}	34.74 ± 1.08 ^{aC}	33.03 ± 0.23 ^{aC}	32.47 ± 2.34 ^{aC}
BSY	41.46 ± 1.71 ^{bC}	39.44 ± 1.73 ^{cC}	53.79 ± 5.51 ^{aA}	52.58 ± 1.88 ^{aB}	52.15 ± 1.33 ^{aB}
NSY	55.80 ± 2.41 ^{bA}	55.12 ± 3.75 ^{bA}	60.69 ± 1.62 ^{aA}	60.98 ± 1.74 ^{aA}	62.18 ± 0.64 ^{aA}
TSY	48.03 ± 2.46 ^{aB}	48.33 ± 5.45 ^{aB}	53.65 ± 0.54 ^{aAB}	51.61 ± 2.34 ^{aB}	50.43 ± 1.15 ^{aB}

* Mean values that do not share the same lowercase letter ($^{a, b, c}$) for the same row are statistically significant at $p < 0.01$, based on Tukey test. Mean values that do not share the same uppercase letter ($^{A, B, C, D}$) for the same column are statistically significant at $p < 0.01$, based on Tukey test.

Table 2. The soluble protein content (g/100 g dw) of the extracts obtained from spent brewer's yeast hydrolyzed with bromelain (BSY), neutrase (NSY) and trypsin (TSY) after different time intervals.

Sample	Hydrolysis Time (Hours)				
	4	8	21	43	67
Control	22.66 ± 1.62 ^{bC*}	26.85 ± 0.00 ^{abC}	24.38 ± 0.01 ^{abD}	25.97 ± 4.94 ^{abC}	34.31 ± 3.56 ^{aC}
BSY	39.45 ± 2.52 ^{cB}	43.07 ± 3.29 ^{bcB}	51.09 ± 1.49 ^{bC}	61.56 ± 0.27 ^{aB}	67.46 ± 1.63 ^{aB}
NSY	63.24 ± 1.13 ^{cA}	68.97 ± 2.36 ^{cA}	77.29 ± 0.85 ^{bA}	85.51 ± 2.00 ^{aA}	87.65 ± 0.00 ^{aA}
TSY	41.52 ± 0.28 ^{bB}	47.68 ± 1.08 ^{bB}	57.82 ± 2.20 ^{aB}	59.70 ± 2.08 ^{aB}	59.84 ± 1.55 ^{aB}

* Mean values that do not share the same lowercase letter ($^{a, b, c}$) for the same row are statistically significant at $p < 0.01$, based on Tukey test. Mean values that do not share the same uppercase letter ($^{A, B, C, D}$) for the same column are statistically significant at $p < 0.01$, based on Tukey test.

2.2. *Degree of Hydrolysis*

DH represents an essential parameter to indicate the level of hydrolysis of proteins for obtaining peptides of different sizes and amino acid sequences [17]. The DH results obtained after yeast proteins' hydrolysis with the three different enzymes considered in the study are presented in Table 3.

Table 3. The degree of hydrolysis (%) of the proteins from spent brewer's yeast hydrolyzed with bromelain (BSY), neutrase (NSY) and trypsin (TSY) at different time intervals.

Sample	Hydrolysis Time (Hours)				
	4	8	21	43	67
Control	8.27 ± 0.69 ^{aD*}	6.99 ± 0.24 ^{bD}	6.30 ± 0.17 ^{cD}	6.49 ± 1.04 ^{bD}	7.92 ± 0.32 ^{aD}
BSY	11.50 ± 0.97 ^{cC}	9.76 ± 0.36 ^{dC}	12.87 ± 0.86 ^{cC}	16.79 ± 0.53 ^{bB}	22.55 ± 1.16 ^{aB}
NSY	22.70 ± 0.41 ^{bA}	17.55 ± 0.21 ^{cA}	26.73 ± 0.55 ^{aA}	25.74 ± 0.98 ^{aA}	26.91 ± 0.62 ^{aA}
TSY	12.52 ± 0.46 ^{dB}	12.00 ± 0.66 ^{dB}	13.62 ± 0.16 ^{cB}	14.31 ± 0.17 ^{bC}	18.05 ± 0.66 ^{aC}

* Mean values that do not share the same lowercase letter ($^{a, b, c, d}$) for the same row are statistically significant at $p < 0.01$, based on Tukey test. Mean values that do not share the same uppercase letter ($^{A, B, C, D}$) for the same column are statistically significant at $p < 0.01$, based on Tukey test.

After 4 h of hydrolysis, the DH reached significantly higher values for the samples where hydrolysis was assisted by exogenous enzymes: 11.50% in the BSY sample, 22.70% in the NSY sample and 12.52% in the TSY sample. The evolution of the DH over the 67 h of hydrolysis depended on the exogenous enzyme used for preparing the sample. For example, in the case of the BSY sample, the DH was 22.55% after 67 h of hydrolysis, almost double with respect to the value registered after 4 h; in TSY, the DH increased by about 50%, whereas in the sample hydrolyzed with neutrase, the DH increased only by about 19% (Table 3). In our study, the DH registered for hydrolysates obtained with neutrase was higher compared to the results of Thamnarathip et al. [16], who used neutrase to obtain bran protein hydrolysates and reported a DH of 8.34% after 6 h of hydrolysis. Xu et al. [18] used neutrase for hydrolyzing the casein, and after 12 h of hydrolysis, the DH was about 15%. From Table 3 it can be seen that, although during the first 4 h of hydrolysis, the DH was similar in TSY and BSY samples, until the end of the investigated hydrolysis time, bromelain was able to recognize and hydrolyze more peptide bonds than trypsin. These results suggest a better exposure of the peptide bonds cleaved by bromelain. Among all tested enzymes, and independent of the hydrolysis time, the highest DH was measured in the NSY sample and the lowest in the TSY sample, suggesting that the number of available cleavage sites for neutrase was higher than other endoproteases tested in this study. Trypsin is one of the most used enzymes for the production of bioactive peptides. Trypsin is able to recognize and cleave the peptide bonds involved at the C-terminal side, the positively charged amino acids lysine and arginine. In any case, the enzyme affinity towards these peptide bonds is reduced by the presence of acidic amino acids on either side of the cleavage site. Moreover, the presence of proline residue on the C-terminal side of the cleavage site will hinder the hydrolysis [19]. The DH reported in our study for yeast samples hydrolyzed with trypsin is similar to the results reported by Mirzaei et al. [20], where the sonication-trypsin hydrolysis of yeast suspensions resulted in a DH of 17.81% after 5 h at 37 °C. Most of the studies reporting on the hydrolysis of spent brewer yeast used Alcalase, Protamex™, Brauzyn® and Flavourzyme™. After 2 h of spent brewer's yeast hydrolysis with different enzymes (having the same enzymatic activity), Marson et al. [5] reported a DH ranging between 8.2 and 33.1%, the maximum DH being obtained through combining Protamex™ and Brauzyn®. On the other hand, the potential of using other enzymes for hydrolyzing spent brewer's yeasts has been reported in recent studies. For example, Amorin et al. [21] used autolysis (70 °C, 5 h) followed by enzymatic hydrolysis of the spent brewer's yeast using an extract of *Cynara cardunculus* and showed that the DH increased with time and enzyme concentration. The highest DH (about 30%) was reported after 4.5 h of hydrolysis with 4% *Cynara cardunculus* extract.

2.3. SDS-PAGE Analysis

The SDS-PAGE electrophoresis was conducted under reducing conditions to show that the distribution of the peptides' molecular weight in the yeast protein hydrolysates passed through ultrafiltration membranes with a cut-off of 30 kDa. As depicted in Figure 1, the hydrolysis with the three enzymes induced the formation of small peptides, most of them having a molecular weight lower than 5 kDa. In terms of band intensity, the pattern of the samples treated with neutrase (NSY line) gave a more intense spectrum compared with those hydrolyzed with bromelain (BSY line) and trypsin (TSY line). These observations are in good agreement with the DH results showing that, after 67 h, neutrase was the most aggressive in breaking down the peptide bonds within the yeast proteins' substrate (DH of 26.91%), followed by bromelain (DH of 22.55%) and trypsin (DH of 18.05%). Comparing the band intensity in Figure 1, one can observe the presence of higher amounts of peptides with molecular weights lower than 10 kDa in the BSY, NSY and TSY samples compared to the control obtained with no exogenous enzyme addition. The findings of the proteins' electrophoresis can be related to the results of the soluble proteins quantified in the yeast extracts as well. After 67 h hydrolysis, the highest soluble protein content was found in the NSY, followed by BSY, TSY and the control. The same trend can be observed in the electrophoresis results, suggesting the presence of the highest amounts of low molecular weight peptides in the NSY sample. A low-intensity smear pattern was observed in the gel for all hydrolyzed samples and was associated with peptides having a molecular weight lower than 10 kDa (Figure 1).

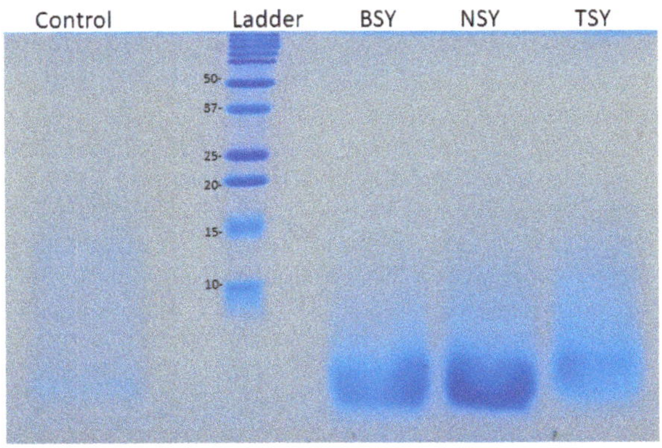

Figure 1. The SDS-PAGE profile of the spent brewer's yeast hydrolyzed for 67 h, with different exogenous enzymes, and passed through ultrafiltration membranes with the cut-off of 30 kDa. The control sample in lane 1, Ladder—Dual Xtra Standard (Bio-Rad, Hercules, CA, USA), BSY—the sample treated with bromelain, NSY—the sample treated with neutrase, and TSY—the sample treated with trypsin.

2.4. Antioxidant Activity

The potential antioxidant activity of the SY hydrolysates was measured based on $ABTS^+$ and DPPH scavenging-based assays, the results being presented in Table 4. The $ABTS^+$ scavenging activity is based on the reduction of cation-radical $ABTS^+$ into the colorless ABTS by the electron transfer of an antioxidant, while DPPH is reduced by the hydrogen atom donation of an antioxidant, changing the color of the final product from pinkish to pale yellow [22].

Table 4. The $ABTS^+$ and DPPH scavenging activity (μM TE/g dw) of the extracts obtained from spent brewer's yeast hydrolyzed with bromelain (BSY), neutrase (NSY) and trypsin (TSY) at different time intervals.

Sample	**Hydrolysis Time (Hours)**				
	4	**8**	**21**	**43**	**67**
	$ABTS^+$ scavenging activity (μM TE/g dw)				
Control	757.6 ± 0.96 ^{dC*}	774.5 ± 0.6 ^{cD}	858.5 ± 1.92 ^{bD}	1088 ± 119 ^{aC}	1024 ± 55 ^{aD}
BSY	840.6 ± 0.96 ^{eB}	845.4 ± 2.3 ^{dC}	1346 ± 28 ^{cC}	1596 ± 21 ^{bB}	1632 ± 2 ^{aB}
NSY	857.7 ± 2.13 ^{dA}	854.9 ± 1.9 ^{dB}	1564 ± 6 ^{cA}	1633 ± 7 ^{bA}	1660 ± 4 ^{aA}
TSY	855.6 ± 0.96 ^{cA}	860.4 ± 4.1 ^{cA}	1475 ± 34 ^{bB}	1573 ± 8 ^{aB}	1561 ± 44 ^{aC}
	DPPH scavenging activity (μM TE/g dw)				
Control	45.35 ± 3.17 ^{dC}	26.43 ± 0.19 ^{eD}	73.80 ± 0.19 ^{bD}	83.89 ± 1.78 ^{aD}	65.25 ± 1.8 ^{cD}
BSY	45.21 ± 3.52 ^{cC}	35.40 ± 1.38 ^{dC}	88.10 ± 0.59 ^{bC}	96.51 ± 2.57 ^{aC}	88.38 ± 0.59 ^{bC}
NSY	82.35 ± 1.98 ^{cA}	59.9 ± 4.75 ^{dA}	110.2 ± 0.59 ^{bA}	116.9 ± 2.9 ^{aA}	106.6 ± 4.9 ^{bA}
TSY	66.37 ± 2.37 ^{cB}	45.35 ± 5.15 ^{dB}	94.13 ± 0.79 ^{bB}	102.8 ± 2.7 ^{aB}	95.25 ± 2.37 ^{bB}

* Mean values that do not share the same lowercase letter ($^{a, b, c, d, e}$) for the same row are statistically significant at $p < 0.01$, based on Tukey test. Mean values that do not share the same uppercase letter ($^{A, B, C, D}$) for the same column are statistically significant at $p < 0.01$, based on Tukey test.

The results obtained using ABTS method (Table 4) showed that, after 4 h of hydrolysis, the antioxidant activity was about 11% higher in the BSY and TSY samples, and by about 13% in the NSY sample, compared to the control. Increasing the hydrolysis time generated hydrolysates with higher antioxidant activity. For all tested enzymes, a gradual increase in antioxidant activity was observed over the entire test hydrolysis time. Thus, after 67 h of hydrolysis, the antioxidant activity of the BSY and NSY samples increased by about 94%

and that of the TSY sample by about 82%. The antioxidant activity was also affected by the exogenous enzyme used for hydrolysis, the highest antioxidant activity, based on the $ABTS^+$ method, being calculated for NSY samples regardless of hydrolysis time, followed closely by samples hydrolyzed with bromelain and trypsin. Our results are in agreement with those reported by Vieira et al. [23]. These authors obtained protein hydrolysates from spent brewer grains by using several enzymatic techniques, and the sample hydrolyzed with neutrase, after ultrafiltration, showed the highest antioxidant potential.

Analyzing the results presented in Table 4, one can see that the antioxidant activity quantified using the DPPH method is always lower than values determined with the $ABTS^+$ method. These differences might be explained by the fact that the DPPH radical presents a higher stability compared to $ABTS^+$ [24] and by the differences in the reaction mechanisms standings behind the two methods. After the first 4 h of hydrolysis (Table 4), the antioxidant activity of the BSY and control samples was similar, whereas the antioxidant activity of the NSY and TSY samples was higher by 81.5% and 46.35% compared to the control. Regardless of the sample, a smaller increase of the antioxidant activity measured by the DPPH method compared to the $ABTS^+$ assay was observed with increasing the hydrolysis time. The antioxidant activity measured by the DPPH method reached a maximum after 43 h of hydrolysis. Among all tested enzymes the highest antioxidant activity was measured in the NSY sample (116.9 $\pm$ 2.9 μM TE/g dw), followed by TSY (102.8 $\pm$ 2.7 μM TE/g dw), and the lowest for the BSY sample (96.51 $\pm$ 2.57 μM TE/g dw). When compared to the control sample, the maximum increase of the antioxidant activity of about 40% was measured in the sample where hydrolysis was performed with neutrase. The presence of hydrophobic amino acids in the peptide sequences exhibits an important role in the antioxidant activity [20]. Further increase of the hydrolysis time to 67 h caused the decrease of the DPPH activity of all tested samples. The DPPH scavenging activity decrease might be attributed to the advanced hydrolysis that might affect the structure and bioactivity of the active peptides [16]. Similar results have been reported in other studies, where yeast hydrolysates showed higher antioxidant activity using the $ABTS^+$ method and lower for the DPPH [20,22,25].

In order to find out more information on the peptides contributing to the antioxidant activity, the samples obtained after 67 h of hydrolysis were subjected to ultrafiltration through membranes with different cut-offs, and the antioxidant activity of each permeate fraction was assessed. From Table 5 it can be seen that all permeate fractions were able to reduce the cation radical $ABTS^+$. In the case of the permeates with peptides having molecular weights lower than 30 kDa, the highest $ABTS^+$ scavenging activity was measured for the BSY sample (1223 $\pm$ 10 μM TE/g dw), followed closely by NSY (1203 $\pm$ 4 μMl TE/g dw) and TSY (1086 $\pm$ 23 μM TE/g dw). On the other hand, for the same fraction, when antioxidant activity was evaluated by the DPPH method, the strongest scavenging activity was measured in the TSY sample (71.08 $\pm$ 4.22 μM TE/g dw) and the lowest in the BSY sample (60.15 $\pm$ 7.53 μM TE/g dw).

Regarding permeates with peptides having molecular weights lower than 10 kDa, the highest antioxidant activity based on $ABTS^+$ scavenging activity was recorded for NSY (1286 $\pm$ 10 μM TE/g dw), and for BSY (85.10 $\pm$ 5.3 μM TE/g dw) when using the DPPH method. Other authors reported that increased antioxidant potential of the 10 kDa fraction resulted after the hydrolysis with neutrase [23]. Moreover, the fraction concentrating peptides with molecular weight lower than 10 kDa exerted a protective effect against free-radical-induced cytotoxicity in Caco-2 and HepG2 cell lines.

When analyzing fractions concentrating the lowest size peptides (molecular weight < 3 kDa), one can see that the strongest scavenging activity by using $ABTS^+$ assay was obtained for samples hydrolyzed with neutrase and bromelain (Table 5). However, when using the DPPH assay, the highest antioxidant activity was measured in the sample hydrolyzed with bromelain (50.06 $\pm$ 0.39 μM TE/g dw) and the lowest in the sample hydrolyzed with neutrase (27.63 $\pm$ 0.30 μM TE/g dw). Bromelain shows preferences for peptide bonds involving glutamic acid, aspartic acid, lysine or arginine residues at the N-terminal, cleaving particularly the peptide chain at arginine–alanine and alanine–glutamic acid bonds. Therefore, one can assume that in the <3 kDa fraction, these amino acids are present in higher amounts, contributing to the increased antioxidant activity of this fraction [26].

Table 5. The $ABTS^+$ and DPPH scavenging activity of the extracts obtained from spent brewer's yeast hydrolyzed for 67 h with bromelain (BSY), neutrase (NSY) and trypsin (TSY) and passed through ultrafiltration membranes with the cut-offs of 30, 10 and 3 kDa.

Sample	Membrane Cut-Off		
	30 kDa	**10 kDa**	**3 kDa**
	$ABTS^+$ scavenging activity (μM TE/g dw)		
Control	483 ± 7 ^{bD*}	656 ± 109 ^{aC}	511 ± 119 ^{aD}
BSY	1223 ± 10 ^{aA}	1208 ± 61 ^{aA}	1126 ± 34 ^{bB}
NSY	1203 ± 4 ^{aB}	1286 ± 10 ^{aA}	1139 ± 17 ^{bA}
TSY	1086 ± 23 ^{aC}	1088 ± 8 ^{aB}	742 ± 42 ^{bC}
	DPPH scavenging activity (μM TE/g dw)		
Control	43.05 ± 2.98 ^{aC}	40.25 ± 3.42 ^{aD}	27.07 ± 1.75 ^{bC}
BSY	60.15 ± 7.53 ^{bB}	85.10 ± 5.30 ^{aA}	50.06 ± 0.39 ^{cA}
NSY	64.36 ± 1.57 ^{bB}	68.00 ± 1.98 ^{aB}	27.63 ± 0.30 ^{cC}
TSY	71.08 ± 4.22 ^{aA}	61.83 ± 2.21 ^{bC}	36.60 ± 0.39 ^{cB}

* Mean values that do not share the same lowercase letter ($^{a, b, c}$) for the same row are statistically significant at $p < 0.01$, based on Tukey test. Mean values that do not share the same uppercase letter ($^{A, B, C, D}$) for the same column are statistically significant at $p < 0.01$, based on Tukey test.

Recent studies showed the potential of using bromelain for obtaining hydrolysates with antioxidant and antimicrobial activity. Selamassakul et al. [26] showed that hydrolysis of brown rice protein with bromelain produced low molecular weight peptides that could be used to enhance the biological activity of foods, whereas Ghanbari et al. [17] found that the 7 h hydrolysis of sea cucumber with bromelain generated peptides with antimicrobial activity. Mirzaei et al. [20] used trypsin to obtain hydrolysates from yeasts, and the fraction including peptides with molecular weight < 3 kDa presented ACE inhibitory activities as well as strong DPPH and $ABTS^+$ scavenging activities of 179.24 ± 4.8 μM TE/mg protein and 4653.36 ± 5 μM TE/mg protein, respectively. The results were attributed to the total content of hydrophobic amino acids with aromatic or branched side chains at each of the C-terminal tripeptide positions.

Analyzing the results presented in Tables 4 and 5, it can be seen that, after separation through membranes with various cut-offs, the antioxidant activity of the resulting fractions was lower compared to the initial hydrolysates. Our observation is in agreement with Mirzaei et al. [20], who indicated that the overall antioxidant activity of the peptides released from *Saccharomyces cerevisiae* proteins is the result of combined actions of fractions with molecular weights < 3 kDa and 5–10 kDa.

2.5. Color Coordinates

The color coordinates measured on the hydrolyzed yeast slurries after the thermal inactivation of the enzymes are presented in Table 6. In the case of the control sample, the hydrolysis time had no significant effect on the luminosity coordinate. On the other hand, regardless of the exogenous enzyme used for hydrolysis, L* decreased with increasing hydrolysis duration ($p < 0.01$). The highest decrease of L* was measured in the NSY sample, followed by the BSY sample, indicating a darker color of the hydrolysates produced with neutrase and bromelain (Table 6). Moreover, similar to the results reported by Marson et al. [5], in our study, L* was negatively correlated with the soluble protein content (BSY − R2 = 0.84; NSY − R2 = 0.91; TSY − R2 = 0.94; $p < 0.01$), an indication that darker samples contain higher protein levels. The a* coordinate was influenced by both the exogenous enzyme and hydrolysis time (Table 6). For the control, BSY and NSY samples, a* coordinate increased with increasing hydrolysis time, indicating the tendency towards the red of these samples. For the TSY sample, a* coordinate decreased from 5.83 ± 0.05 to 4.98 ± 0.03, with an increasing time of hydrolysis. After 4 h of hydrolysis, the highest b* coordinate was measured for NSY and the lowest for the TSY sample (Table 6). On the other hand, after 67 h of hydrolysis, the highest increase was measured for the control sample. From Table 6 it can be seen that similar to a* values, b* values for the TSY sample increased after 8 h of hydrolysis, reaching a maximum of 12.94 ± 0.02. Similar results were reported by Bertolo et al. [27]. For control yeast suspension, the authors reported a b* value of 16.42 ± 0.02. Based on the above-mentioned results, it can be concluded that the enzymes used in this study for the yeast proteins' hydrolysis exerted a significant effect on the color of the hydrolysates.

Table 6. The influence of the exogenous enzyme used for the spent brewer's yeast hydrolysis (BSY—bromelain, NSY—neutrase, and TSY—trypsin) and hydrolysis time on the color parameters of the hydrolyzed yeast slurries.

Sample	Hydrolysis Time (Hours)				
	4	8	21	43	67
			L*		
Control	79.97 ± 0.23 ^{Aa*}	78.88 ± 0.14 ^{cA}	80.01 ± 0.19 ^{aA}	79.24 ± 0.16 ^{bA}	79.38 ± 0.50 ^{aA}
BSY	77.52 ± 0.20 ^{Ab}	75.87 ± 0.10 ^{bB}	74.66 ± 0.39 ^{cB}	72.21 ± 0.46 ^{dB}	69.83 ± 0.18 ^{eC}
NSY	71.53 ± 0.26 ^{Ad}	69.68 ± 0.11 ^{bD}	66.88 ± 0.24 ^{cD}	66.76 ± 0.29 ^{cD}	66.01 ± 0.20 ^{dD}
TSY	75.86 ± 0.33 ^{aC}	73.34 ± 0.10 ^{bC}	71.83 ± 0.37 ^{cC}	71.15 ± 0.34 ^{cC}	71.01 ± 0.27 ^{dB}
			a* coordinate		
Control	5.97 ± 0.05 ^{dC}	5.99 ± 0.01 ^{dD}	6.39 ± 0.03 ^{cC}	7.16 ± 0.02 ^{bB}	7.53 ± 0.01 ^{aB}
BSY	6.6 ± 0.01 ^{dB}	6.98 ± 0.02 ^{cB}	6.62 ± 0.02 ^{dB}	7.00 ± 0.03 ^{bC}	7.22 ± 0.02 ^{aC}
NSY	6.78 ± 0.03 ^{eA}	7.36 ± 0.03 ^{dA}	7.5 ± 0.03 ^{cA}	7.53 ± 0.06 ^{bA}	7.79 ± 0.02 ^{aA}
TSY	5.83 ± 0.05 ^{bD}	6.40 ± 0.02 ^{aC}	5.42 ± 0.03 ^{cD}	4.80 ± 0.03 ^{eD}	4.98 ± 0.03 ^{dD}
			b* coordinate		
Control	12.93 ± 0.01 ^{eC}	13.21 ± 0.01 ^{dC}	13.91 ± 0.15 ^{cC}	15.15 ± 0.00 ^{bA}	16.29 ± 0.01 ^{aA}
BSY	13.53 ± 0.06 ^{dB}	13.46 ± 0.01 ^{eB}	14.39 ± 0.08 ^{cA}	14.57 ± 0.05 ^{bB}	14.84 ± 0.04 ^{aB}
NSY	14.38 ± 0.07 ^{aA}	13.69 ± 0.02 ^{cA}	14.24 ± 0.09 ^{bB}	14.49 ± 0.09 ^{aC}	14.38 ± 0.07 ^{bC}
TSY	12.65 ± 0.04 ^{bD}	12.94 ± 0.02 ^{aD}	12.15 ± 0.01 ^{cD}	12.1 ± 0.03 ^{cD}	12.68 ± 0.02 ^{bD}

* Mean values that do not share the same lowercase letter ($^{a, b, c, d, e}$) for the same row are statistically significant at $p < 0.01$, based on Tukey test. Mean values that do not share the same uppercase letter ($^{A, B, C, D}$) for the same column are statistically significant at $p < 0.01$, based on Tukey test.

2.6. Technological Functionality of the Yeast Protein Hydrolysates

The influence of the exogenous enzyme-assisted hydrolysis on the technological functionality of the yeast proteins was determined by assessing the foaming properties and rheological behavior of the emulsions.

The foaming ability of the yeast protein hydrolysates was determined upon incorporating air into the samples at three different homogenization speeds, and the results are presented in Table 7. The foaming capacity of all samples increased with the homogenization speed. Regardless of the homogenization speed used for obtaining the foams, the samples prepared with exogenous enzymes exhibited significantly higher foaming capacity compared to the control ($p < 0.01$). As one can see in Figure 1, the exogenous enzyme-assisted hydrolysis released higher amounts of peptides with low molecular weights, contributing to the foaming capacity of the samples. As indicated by Liang et al. [28], who studied the effect of the controlled pepsin-assisted hydrolysis on the foaming properties of soy proteins, the significantly higher foaming ability of the hydrolysates might be due to the presence of the low molecular weight peptides with amphiphilic properties and more flexible structure, which are absorbed faster at the gas-water interface. The enzyme-assisted hydrolysis allows better exposure of the proteins' hydrophobic patches which establish contacts with the gas phase, while the hydrophilic groups tend to interact with the liquid phase [29]. The lower foaming capacity of the control sample might be associated with the lower degree of hydrolysis (Table 1), which explains the higher abundance of the large molecular weight proteins and aggregates in the dispersion subjected to foaming, therefore resulting in slower diffusion to the gas-water interface. The highest overrun values were observed in the case of yeast proteins hydrolysate prepared with trypsin (FC of 85.0–132.5%). Among the enzyme-assisted hydrolyzed samples, NSY exhibited the lowest foaming ability, most probably as the result of the higher DH value (Table 3), resulting in higher amounts of peptides with very low molecular weights. A previous study of Van der Ven et al. [30], dealing with whey protein hydrolysates, indicated that the peptide fractions with 3–5 kDa have better foaming properties compared with the fraction having larger (over 20 kDa) or very small (<3 kDa) peptides.

Table 7. Foaming properties of the extracts obtained from spent brewer's yeast hydrolyzed for 67 h with bromelain (BSY), neutrase (NSY) and trypsin (TSY).

Sample	Foaming Capacity, %			Foam Stability, %		
	5000 rpm	7000 rpm	9000 rpm	5000 rpm	7000 rpm	9000 rpm
Control	$32.5 \pm 3.5^{cC*}$	62.5 ± 3.5^{dB}	92.5 ± 3.5^{cA}	46.4 ± 5.1^{cA}	58.0 ± 0.5^{cA}	56.7 ± 1.7^{dA}
BSY	75.0 ± 0.0^{aC}	100.0 ± 0.0^{bB}	122.5 ± 3.5^{abA}	81.7 ± 2.4^{abA}	86.3 ± 1.8^{bA}	83.7 ± 0.5^{bA}
NSY	55.0 ± 7.1^{bC}	87.5 ± 3.5^{cB}	120.0 ± 0.0^{bA}	68.3 ± 2.4^{bB}	81.4 ± 2.8^{bA}	76.0 ± 1.5^{cAB}
TSY	85.0 ± 0.0^{aB}	125.0 ± 0.0^{aA}	132.5 ± 3.5^{aA}	85.3 ± 4.2^{aA}	98.0 ± 2.8^{cA}	94.4 ± 2.5^{aA}

* Mean values that do not share the same lowercase letter ($^{a, b, c, d}$) for the same row are statistically significant at $p < 0.01$, based on Tukey test. Mean values that do not share the same uppercase letter ($^{A, B, C}$) for the same column are statistically significant at $p < 0.01$, based on Tukey test.

The foam stability over 30 min of storage at room temperature ranged between 56.7 and 98.0% in case of the samples prepared with exogenous enzymes, which is significantly higher compared to the control (FS of 46.4–58.0%). The interactions established between the peptides released through enzyme hydrolysis appear to contribute to the formation of a cohesive and flexible film around the gas bubbles [29]. The higher foam destabilization tendency and liquid drainage registered for the control sample are due to the poor viscoelastic and mechanical properties of the film, which is prone to easy rupture, leading to larger air bubbles and coalescence phenomenon [28]. The homogenization speed exerted no significant influence on the foam stability (Table 7). The only exception concerns the NSY sample, which exhibited significantly lower FS when foaming at a lower homogenization speed ($p < 0.01$).

The ability of the yeast protein hydrolysates to form viscoelastic layers on the surface of the oil droplets was further assessed by determining the rheological behavior of the emulsions prepared with a volume fraction of sunflower oil of 50%. The rheological measurements on the emulsions under flow conditions indicated the shear stress increase over the entire shear rate domain considered in the study. The apparent viscosity of the emulsions measured at particular shear rate values are presented in Table 8. Regardless of the protein hydrolysate used to prepare the emulsion, the viscosity values varied depending on the applied shear rate. The highest apparent viscosity values were measured at lower shear rates for the emulsions based on control and BSY hydrolysates. On the other hand, at high shear rates no important differences in terms of the apparent viscosity were registered among the investigated emulsion samples (Table 8).

Rheological measurements under low amplitude oscillatory conditions, during the strain sweep test at a constant frequency of 1 Hz were first conducted to assess the LVR of the emulsions. The critical strain (γc) values which mark the limit of the LVR, beyond which the emulsions no longer exhibit linear viscoelastic behavior, are presented in Table 8. Similar upper limits of the LVR of 0.5–1% were previously reported by Vasilean et al. [31] for the soy protein emulsions prepared with sunflower, canola and palm oils.

Table 8. Rheological properties of the emulsions prepared with extracts obtained from spent brewer's yeast hydrolyzed for 67 h with bromelain (BSY), neutrase (NSY) and trypsin (TSY).

Sample	Stepped Flow Test		Strain Sweep Test	
	Viscosity (Pa·s) at Shear Rate of 1 s^{-1}	Viscosity (Pa·s) at Shear Rate of 10 s^{-1}	Critical Strain (γc), %	Yield Strain, (γy), %
Control	0.182 ± 0.002 [a]*	0.025 ± 0.001 [a]	0.36 ± 0.06 [b]	3.17 ± 0.01 [b]
BSY	0.178 ± 0.005 [a]	0.023 ± 0.002 [a]	1.43 ± 0.23 [a]	25.22 ± 0.21 [a]
NSY	0.109 ± 0.013 [b]	0.032 ± 0.002 [a]	0.36 ± 0.06 [b]	1.00 ± 0.01 [c]
TSY	0.105 ± 0.002 [b]	0.022 ± 0.007 [a]	0.32 ± 0.01 [b]	-

* Mean values that do not share the same lowercase letter ([a, b, c]) for the same column are statistically significant at $p < 0.01$, based on Tukey test.

The oscillatory tests based on the progressive increase of the deformation (%) also allowed the identification of the emulsion flow threshold, corresponding to the point where the phase inversion occurs, and the flowing process is considered initiated [32]. The yield strain values (γy) were recorded when the G″ values, corresponding to the viscous component of the sample, exceed the G′ values corresponding to the elastic component. The emulsions exhibited different responses to the applied strain. Except for the emulsion prepared with trypsin-assisted hydrolyzed extract, all samples exhibited a solid-like behavior, with G′ prevailing over the G″, up to strain values depending on the enzyme. The highest yield strain value of 25.22% was registered in the case of emulsion prepared with BSY, whereas the lowest (γy of 1.00%) was in the case of the NSY-based emulsion. The emulsions prepared with yeast protein hydrolysates obtained with trypsin presented higher values of G″ compared to G′ throughout the entire scanned deformation range.

Frequency sweep tests were further run within the LVR at strain values below the critical strain determined for each tested emulsion (Table 8). The results of the frequency sweep test, in terms of the evolution of the complex modulus (G*), are presented in Figure 2.

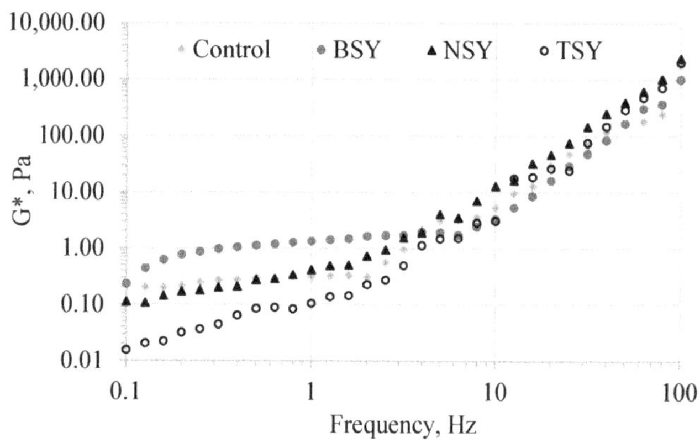

Figure 2. Evolution of the complex modulus (G*) during the frequency sweep tests on the emulsions prepared with an extract obtained from spent brewer's yeast hydrolyzed for 67 h with bromelain (BSY), neutrase (NSY) and trypsin (TSY).

The complex modulus provides information on the overall resistance to the deformation of the tested emulsions, integrating both the recoverable (elastic) and the non-recoverable (viscous) component. The values of the complex modulus represent a direct measure of the rigidity of the analyzed samples when exposed to stress below the yield stress. At low-frequency values, which provide information on the slow-motion flow behavior on a long time scale, the emulsions prepared with the yeast extract hydrolyzed with bromelain exhibited the highest G′ and G″ values and, consequently, the highest complex modulus (Figure 2), suggesting that the network structure of the emulsion is stronger compared to the other tested samples. The lowest complex modulus values were registered for the TSY-based emulsions, which exhibited liquid-like behavior, being weakly flocculated. At high-frequency values, showing the fast motion behavior at short timescales, the highest complex modulus values were registered for the TSY- and NSY-based emulsions.

3. Materials and Methods

Dried SY with a protein concentration of 41.96% was kindly provided by a beer factory from Ploiesti, Romania. Bromelain was provided by Carl Roth (Karlsruhe, Germany), Neutrase 5.0 BG by Novo Nordisk (Bagsværd, Denmark) and trypsin by Merck (Darmstadt, Germany), whereas o-phthaldialdehyde (OPA), N-acetyl-cysteine (NAC), sodium dodecyl sulfate (SDS), 2,2-diphenyl-1-picryl-hydrazyl (DPPH) and 2,2′-azinobis(3-ethylbenzothiazoline-6-sulphonate) (ABTS) were purchased from Sigma Aldrich (Burlington, MA, USA). All other reagents were of analytical grade.

3.1. Enzymatic Hydrolysis of SY Suspensions

Dried SY was suspended in distilled water to achieve a solid concentration of 12% (*w*/*w*) and homogenized at 15,000 rpm for 10 min (Ultra Turrax® IKA T18 basic and S18N-19G dispersing tool, IKA-Werke GmbH and Co. KG, Staufen, Germany). The resulting suspension with a pH of 6.5 was pretreated for 1 h at 75 °C under continuous stirring at 150 rpm. The suspension was then cooled to 50 °C and the pH was adjusted to 7.0. The concentration of the enzymes added to the yeast slurry, with respect to the solid content of the suspension, was as follows: bromelain 0.5% (BSY), neutrase 1% (NSY) and trypsin 1% (TSY). The hydrolysis was carried out for 67 h at 55 °C, while continuously shaking the samples at 100 rpm. The control sample was prepared with no enzyme addition. At the end of the hydrolysis step, the pH was adjusted to 7.0 and the enzymes were inactivated by heating the hydrolysates at 95 °C for 5 min. Cell debris was removed upon separation as residue through centrifugation at 14,000 rpm for 10 min. The liquid extract, which retained all soluble peptide fractions, was further freeze-dried (CHRIST Alpha 1–4 LD plus, Osterode am Harz, Germany).

3.2. Separation of Peptide from SY Hydrolysates

The supernatant resulting after centrifugation was passed through ultrafiltration membranes in an Amicon® stirred cell model (Merck KGaA, Darmstadt, Germany) with a cut-off of 30 kDa, 10 kDa and 3 kDa, respectively. The antioxidant activity of the resulting supernatant was measured by using DPPH and $ABTS^+$ assay.

3.3. Characterization of SY Hydrolysates

3.3.1. Soluble Residue Content (SRC) and Soluble Protein Content (SPC)

The SRC was measured on the supernatant fraction by heating the samples in a Memmert UNB 400 oven (Memmert GmbH+Co. Kg, Schwabach, Germany) to constant weight [33]. The SPC was measured on the supernatant fraction by using the Lowry method, as previously reported by Dumitrascu et al. [33].

3.3.2. Degree of Hydrolysis (DH)

The evaluation of the degree of hydrolysis was performed according to OPA spectrophotometric assay, as described by Spellman et al. [34]. Fresh OPA reagent (100 mL) was obtained by mixing 10 mL of 50 mM OPA in methanol, 10 mL of 50 mM NAC, 5 mL of SDS (20 %, *w*/*v*) and 75 mL of 0.1 M borate buffer (pH = 9.5). The reaction was initiated by mixing 400 µL of samples with 3 mL of OPA reagent, and after 15 min the mixture absorbance was measured at 340 nm (A_{sample}). A blank sample was prepared by replacing the sample with distilled water. The standard sample was obtained by adding 400 µL serine standard to 3 mL of OPA solution and measuring the absorbance at 340 nm ($A_{standard}$). The DH was calculated as indicated by Nielsen et al. [35] (Equations (1)–(3)):

$$DH\ (\%) = h/h_{tot} \cdot 100 \quad (1)$$

where h represents the number of hydrolyzed bonds, and h_{tot} = 7.5 is the total number of peptide bonds per protein equivalent.

$$h = (\text{serine-NH}_2 - \beta)/\alpha \quad (2)$$

$$\text{serine-NH}_2 = (A_{sample} - A_{blank})/(A_{standard} - A_{blank}) \cdot 0.9516\ \text{meq/L} \cdot d\ 100/X \cdot P \quad (3)$$

where, β was considered 0.4, α was considered 1, d is the dilution, X is the grams of the sample and P is the protein % in the sample.

3.3.3. The Proteins Gel Electrophoresis

The peptides obtained upon ultra-filtration through the membrane with a cut-off of 30 kDa were analyzed by sodium dodecyl sulfate—polyacrylamide gel electrophoresis (SDS-PAGE). The concentration of the polyacrylamide was 15% in the case of the resolving gel (pH 8.8) and 4.5% in the case of the stacking gel (pH 6.8). A 30% acrylamide/bis-acrylamide ratio of 37.5:1, with a 2.7% crosslinker (Bio-Rad, Hercules, CA, USA), was used. Each sample (control, BSY, NSY, and TSY) was suspended in 4 × Laemmli sample buffer (Bio-Rad, Hercules, CA, USA) and treated under reducing conditions using β-mercaptoethanol, according to the manufacturer's recommendations. After thermal treatment at 95 °C for 5 min in a water bath, followed by centrifugation for 3 min at 5.000 rpm, a volume of 20 μL of each sample was loaded into the wells. The electrophoresis was run at a constant voltage of 100 V for 100 min. The protein bands were fixed for 1 h by immersing the gels in a 40% methanol solution and 10% acetic acid, stained for 40 min in a 0.1% *w*/*v* Coomassie brilliant blue R-250 solution (Bio-Rad, Hercules, CA, USA) and de-stained for 30 min in 10% *v*/*v* acetic acid. Afterwards, the gel was maintained in MilliQ water for 24 h and photographed with a Canon PowerShot G16 digital camera (Canon Inc., Tokyo, Japan).

3.3.4. Antioxidant Activity Assays

The antioxidant activity of the spent brewer's yeast protein hydrolysates and of the peptide mixtures separated through ultrafiltration with membranes having cut-offs of 30 kDa, 10 kDa and 3 kDa was evaluated using the methods based on DPPH and $ABTS^+$ radicals-scavenging activity, as detailed in Dumitrascu et al. [33]. The antioxidant activity was expressed as μM Trolox Equivalent (TE)/g dw yeast.

3.3.5. Color Coordinates

The color coordinates were measured by using the Chroma Meter CR-410 (Konica Minolta Sensing Americas Inc., Ramsey, NJ, USA). The color coordinates were expressed considering the CIELab scale, where L* represents the luminosity (0—darkest black to 100—brightest white), a* represents the green (negative values)/red (positive values) colors and b* represents blue (negative values)/yellow (positive values) colors [5]. The measurements were performed on the yeast protein hydrolysates, after enzyme inactivation and prior to centrifugation.

3.3.6. Foaming Properties

The foaming properties of the yeast protein hydrolysates were determined using the methods described by Liang et al. [28], with slight modifications. A volume (V_0) of 50 mL protein solution of 6% (*w*/*v*) was subjected to foaming in a graduated cylinder, using the Ultra Turax® IKA T18 basic homogenizer with the S18N-19G dispersing tool (IKA-Werke GmbH and Co. KG, Staufen, Germany). The volume of the foam generated after 2 min of homogenization at three different speed values of 5000, 7000 and 9000 rpm (V_f) was used to calculate the foaming capacity (FC) as follows:

$$FC\ (\%) = (V_f - V_0)/V_0 \times 100 \quad (4)$$

The foam collapse was observed over 30 min of storage at room temperature. The volume of the foam measured after 30 min (V_{30}) was used to calculate the foam stability (FS) as follows:

$$FS\ (\%) = (V_{30} - V_0)/(V_f - V_0) \times 100 \quad (5)$$

3.3.7. Rheological Properties of the Emulsions

Emulsions were prepared by homogenizing for 5 min at a speed of 15,000× *g* (Ultra Turrax® IKA T18, IKA-Werke GmbH and Co. KG, Staufen, Germany) the 1:1 (*v*/*v*) mixtures consisting of protein suspensions of 6% concentration (*w*/*v*) and sunflower oil (Spornic, Prutul SA, Galati, Romania).

The obtained emulsions were immediately used for rheological measurements at 20 °C, using a controlled-stress rheometer (AR2000ex, TA Instruments Ltd., New Castle, DE, USA) and a cone–plate geometry (cone angle of 2° and diameter of 40 mm). A closing gap of 1000 µm was selected for all measurements, and the edges were covered with mineral oil to avoid moisture loss.

The linear viscoelastic region (LVR) of the emulsions was first identified by running a strain sweep test in the oscillating strain domain of 0.1–100%, at a frequency of 1 Hz. The storage modulus (G′), loss modulus (G″) and complex modulus (G*) were further registered in the frequency domain of 0.1 to 100 Hz while running a frequency sweep test at constant strain within the LVR. The stepped flow was finally applied to measure the steady shear viscosities while raising the shear rate in the 0.1–100 s^{-1} domain. The results were analyzed by means of TA Rheology Advantage Data Analysis Software V 4.8.3. (TA Instruments, New Castle, DE, USA).

3.4. Statistical Analysis

The results are expressed as mean value followed by standard deviation. The significant differences between samples were assessed using one-way ANOVA, whereas post hoc analysis at $p < 0.01$ was performed with Tukey test for post hoc analysis. Correlations were determined by Pearson test. Minitab 19 (Minitab LLC, State College, PA, USA) software was employed to perform the statistical analysis tests.

4. Conclusions

The antioxidant properties and technological functionality of the yeast protein hydrolysates obtained using bromelain, neutrase and trypsin were investigated. When hydrolysis of the spent brewer's yeast was carried out with neutrase, higher soluble solids and protein contents were released compared to the samples hydrolyzed with bromelain and trypsin. The luminosity of the hydrolysates was correlated with the soluble protein content. Regardless of the exogenous enzyme used for preparing the yeast protein hydrolysates, the maximum antioxidant activity assessed using the DPPH method was obtained after 43 h of hydrolysis. Regarding the peptides with a molecular weight lower than 3 kDa, the highest antioxidant activity was obtained in the case of the sample hydrolyzed with bromelain and the lowest for the samples prepared with neutrase. The technological functionality of the yeast proteins hydrolysates was estimated based on the foaming properties and rheological behavior of the emulsions. The results suggested that the exogenous enzyme-assisted hydrolysis is a promising approach for improving the foaming properties of the yeast proteins. The hydrolysates prepared with trypsin exhibited the best foaming properties. The rheological measurements indicated that the stability of the emulsions highly depends on the enzyme used for protein hydrolysis. Additional studies will be conducted to investigate in detail the bioactivity and cytotoxicity of the peptides obtained through enzymatic hydrolysis of the spent brewer's yeast. Moreover, the functionality of the yeast proteins hydrolysates in complex matrices, such as the value-added food products, will be considered as well.

Author Contributions: Conceptualization, L.D. and I.A.; methodology, L.D., L.G.-G. and I.A.; software, L.D. and I.A.; validation L.D., A.L.D. and L.G.-G.; formal analysis, L.D. and I.A.; investigation, L.D., A.L.D., L.G.-G. and I.A.; resources, L.D., and I.A.; data curation, L.D.; writing—original draft preparation, L.D., A.L.D., L.G.-G. and I.A.; writing—review and editing, L.D. and I.A.; supervision, L.D. and I.A.; project administration, L.D.; funding acquisition, L.D. All authors have read and agreed to the published version of the manuscript.

Funding: This work was supported by a grant from the Ministry of Research, Innovation and Digitization, CNCS—UEFISCDI, project number PN-III-P1-1.1-TE-2021-0459, within PNCDI III.

Institutional Review Board Statement: Not applicable.

Informed Consent Statement: Not applicable.

Data Availability Statement: Not applicable.

Conflicts of Interest: The authors declare no conflict of interest.

References

1. Beer Production Worldwide from 1998 to 2021. Available online: https://www.statista.com/statistics/270275/worldwide-beer-production/#:~:text=In%202021%2C%20the%20global%20beer%20production%20amounted%20to,production%20are%20China%2C%20the%20United%20States%20and%20Brazil (accessed on 10 March 2023).
2. European Beer Trends Statistics Report—2022 Edition. Available online: https://brewersofeurope.org/uploads/mycms-files/documents/publications/2022/european-beer-trends-2022.pdf (accessed on 10 March 2023).
3. Moates, G.; Sweet, N.; Bygrave, K.; Waldron, K. Top 20 Food Waste Streams. REFRESH. 2016. Available online: https://eu-refresh.org/sites/default/files/D6_9_Waste_Streams_Final.pdf (accessed on 7 February 2022).
4. Feldmann, H. *Yeast: Molecular and Cell Biology*, 2nd ed.; Wiley-VCH Verlag GmbH & Co., KgaA: Weinheim, Germany, 2012; pp. 5–24.
5. Marson, G.V.; de Castro, R.J.S.; Machado, M.T.D.C.; da Silva Zandonadi, F.; Barros, H.D.D.F.Q.; Maróstica Júnior, M.R.; Sussulini, A.; Hubinger, M.D. Proteolytic enzymes positively modulated the physicochemical and antioxidant properties of spent yeast protein hydrolysates. *Proc. Biochem.* **2020**, *91*, 34–45. [CrossRef]
6. Oliveira, A.S.; Ferreira, C.; Pereira, J.O.; Pintado, M.E.; Carvalho, A.P. Valorisation of protein-rich extracts from spent brewer's yeast (*Saccharomyces cerevisiae*): An overview. *Biomass Conv. Bioref.* **2022**, *2022*, 1–23. [CrossRef]
7. Caballero-Córdoba, G.M.; Sgarbieri, V.C. Nutritional and toxicological evaluation of yeast (*Saccharomyces cerevisiae*) biomass and a yeast protein concentrate. *J. Sci. Food Agric.* **2000**, *80*, 341–351. [CrossRef]
8. San Martin, D.; Ibarruri, J.; Iñarra, B.; Luengo, N.; Ferrer, J.; Alvarez-Ossorio, C.; Bald, C.; Gutierrez, M.; Zufía, J. Valorisation of brewer's spent yeasts' hydrolysates as high-value bioactive molecules. *Sustainability* **2021**, *13*, 6520. [CrossRef]
9. Dumitrascu, L. Non-animal protein sources and resources- new strategies of valorization into value-added ingredients. In *Food, Nutrition and Environment: Position Papers in Central European Space*; Raspor, P., Ed.; Croatian Academy of Engineering: Zagreb, Croatia, 2022; pp. 129–138.
10. Nwachukwu, I.D.; Aluko, R.E. Structural and functional properties of food protein-derived antioxidant peptides. *J. Food Biochem.* **2019**, *43*, e12761. [CrossRef]
11. Tacias-Pascacio, V.G.; Morellon-Sterling, R.; Siar, E.; Tavano, O.; Berenguer-Murcia, Á.; Fernandez-Lafuente, R. Use of Alcalase in the production of bioactive peptides: A review. *Int. J. Biol. Macromol.* **2020**, *165*, 2143–2196. [CrossRef]
12. Ribeiro-Oliveira, R.; Martins, Z.E.; Sousa, J.B.; Ferreira, I.M.P.L.V.O.; Diniz, C. The health-promoting potential of peptides from brewing by-products: An up-to-date review. *Trends Food Sci. Technol.* **2021**, *118*, 143–153. [CrossRef]
13. Podpora, B.; Świderski, F.; Sadowska, A.; Rakowska, R.; Wasiak-Zys, G. Spent brewer's yeast extracts as a new component of functional food. *Czech J. Food Sci.* **2016**, *34*, 554–563. [CrossRef]
14. Ulug, S.K.; Jahandideh, F.; Wu, J. Novel technologies for the production of bioactive peptides. *Trends in Food Sci. Technol.* **2021**, *108*, 27–39. [CrossRef]
15. Takalloo, Z.; Nikkhah, M.; Nemati, R.; Jalilian, N.; Sajedi, R.H. Autolysis, plasmolysis and enzymatic hydrolysis of baker's yeast (*Saccharomyces cerevisiae*): A comparative study. *World J. Microbiol. Biotechnol.* **2020**, *36*, 68. [CrossRef]
16. Thamnarathip, P.; Jangchud, K.; Nitisinprasert, S. Identification of peptide molecular weight from rice bran protein hydrolysate with high antioxidant activity. *J. Cereal Sci.* **2016**, *69*, 329–335. [CrossRef]
17. Ghanbari, R.; Ebrahimpour, A.; Hamid, A.A.; Ismail, A.; Saari, N. Actinopyga lecanora hydrolysates as natural antibacterial agents. *Int. J. Mol. Sci.* **2012**, *13*, 16796–16811. [CrossRef] [PubMed]
18. Xu, W.; Kong, B.; Zhao, X. Optimization of some conditions of neutrase-catalyzed plastein reaction to mediate ACE-inhibitory activity in vitro of casein hydrolysate prepared by neutrase. *J. Food Sci. Technol.* **2014**, *51*, 276–284. [CrossRef] [PubMed]
19. O'Connor, J.O.; Garcia-Vaquero, M.; Meaney, S.; Tiwari, B.K. Bioactive peptides from algae: Traditional and novel generation strategies, structure-function relationships, and bioinformatics as predictive tools for bioactivity. *Mar. Drugs* **2022**, *20*, 317. [CrossRef] [PubMed]

20. Mirzaei, M.; Mirdamadi, S.; Ehsani, M.R.; Aminlari, M.; Hosseini, E. Purification and identification of antioxidant and ACE-inhibitory peptide from *Saccharomyces cerevisiae* protein hydrolysate. *J. Func. Foods.* **2015**, *19*, 259–268. [CrossRef]
21. Amorim, M.; Marques, C.; Pereira, J.O.; Guardão, L.; Martins, M.J.; Osório, H.; Moura, D.; Calhau, C.; Pinheiro, H.; Pintado, M. Antihypertensive effect of spent brewer yeast peptide. *Process Biochem.* **2019**, *76*, 213–218. [CrossRef]
22. Oliveira, A.S.; Pereira, J.O.; Ferreira, C.; Faustino, M.; Durão, J.; Pintado, M.E.; Carvalho, A.P. Peptide-rich extracts from spent yeast waste streams as a source of bioactive compounds for the nutraceutical market. *Innov. Food Sci. Emerg. Technol.* **2022**, *81*, 103148. [CrossRef]
23. Vieira, E.F.; Dias, D.; Carmo, H.; Ferreira, I. Protective ability against oxidative stress of brewers' spent grain protein hydrolysates. *Food Chem.* **2017**, *228*, 602–609. [CrossRef]
24. Marecek, V.; Mikyska, A.; Hampel, D.; Cejka, P.; Neuwirthova, J.; Malachova, A.; Cerkal, R. ABTS and DPPH methods as a tool for studying antioxidant capacity of spring barley and malt. *J. Cereal Sci.* **2017**, *73*, 40–45. [CrossRef]
25. Guo, H.; Guo, S.; Liu, H. Antioxidant activity and inhibition of ultraviolet radiation-induced skin damage of selenium-rich peptide fraction from selenium-rich yeast protein hydrolysate. *Bioorg. Chem.* **2020**, *105*, 104431. [CrossRef]
26. Selamassakul, O.; Laohakunjit, N.; Kerdchoechuen, O.; Yang, L.; Maier, C.S. Isolation and characterisation of antioxidative peptides from bromelain-hydrolysed brown rice protein by proteomic technique. *Proc. Biochem.* **2018**, *70*, 179–187. [CrossRef] [PubMed]
27. Bertolo, A.P.; Biz, A.P.; Kempka, A.P.; Rigo, E.; Cavalheiro, D. Yeast (*Saccharomyces cerevisiae*): Evaluation of cellular disruption processes, chemical composition, functional properties and digestibility. *J. Food Sci. Technol.* **2019**, *56*, 3697–3706. [CrossRef] [PubMed]
28. Liang, G.; Chen, W.; Qie, X.; Zeng, M.; Qin, F.; He, Z.; Chen, J. Modification of soy protein isolates using combined pre-heat treatment and controlled enzymatic hydrolysis for improving foaming properties. *Food Hydrocoll.* **2020**, *105*, 105764. [CrossRef]
29. Moll, P.; Salminen, H.; Griesshaber, E.; Schmitt, C.; Weiss, J. Homogenization improves foaming properties of insoluble pea proteins. *J. Food Sci.* **2022**, *87*, 4622–4635. [CrossRef]
30. Van der Ven, C.; Gruppen, H.; de Bont, D.B.; Voragen, A.G. Correlations between biochemical characteristics and foam-forming and-stabilizing ability of whey and casein hydrolysates. *J. Agric. Food Chem.* **2002**, *50*, 2938–2946. [CrossRef]
31. Vasilean, I.; Aprodu, I.; Vasilean, I.; Patrașcu, L. Pulse flour based emulsions—The effect of oil type on technological and functional characteristics. *Stud. Univ. Babes-Bolyai Chem.* **2018**, *63*, 199–214. [CrossRef]
32. Pătraşcu, L.; Banu, I.; Vasilean, I.; Aprodu, I. Effect of gluten, egg and soy proteins on the rheological and thermo-mechanical properties of wholegrain rice flour. *Food Sci. Technol. Int.* **2017**, *23*, 142–155. [CrossRef]
33. Dumitrascu, L.; Lanciu (Dorofte), A.; Aprodu, I. A preliminary study on using ultrasounds for the valorization of spent brewer's yeast. *Ann. Univ. Dunarea Jos Galati Fascicle VI Food Technol.* **2022**, *46*, 141–153. [CrossRef]
34. Spellman, D.; Mc Evoy, E.; O'Cuinn, G.; FitzGerald, R.J. Proteinase and exopeptidase hydrolysis of whey protein: Comparison of the TNBS, OPA and pH stat methods for quantification of degree of hydrolysis. *Int. Dairy J.* **2003**, *13*, 447–453. [CrossRef]
35. Nielson, P.M.; Petersen, D.; Dambmann, C. Improved method for determining food protein degree of hydrolysis. *J. Food Sci.* **2001**, *66*, 642–646. [CrossRef]

molecules

MDPI

Article

Bioactive Potential of Elderberry (*Sambucus nigra* L.): Antioxidant, Antimicrobial Activity, Bioaccessibility and Prebiotic Potential

Ioana Mariana Haş [1], Bernadette-Emőke Teleky [2,3], Katalin Szabo [2,3], Elemer Simon [2,3], Floricuta Ranga [3], Zorița Maria Diaconeasa [3], Anamaria Lavinia Purza [4], Dan-Cristian Vodnar [2,3,*], Delia Mirela Tit [4,*] and Maria Nițescu [5]

1 Doctoral School of Biomedical Sciences, University of Oradea, 410087 Oradea, Romania
2 Institute of Life Sciences, University of Agricultural Sciences and Veterinary Medicine, 400372 Cluj-Napoca, Romania
3 Department of Food Science and Technology, University of Agricultural Sciences and Veterinary Medicine, 400372 Cluj-Napoca, Romania
4 Department of Pharmacy, Faculty of Medicine and Pharmacy, University of Oradea, 410028 Oradea, Romania
5 Department of Preclinical–Complementary Sciences, University of Medicine and Pharmacy "Carol Davila", 050474 Bucharest, Romania
* Correspondence: dan.vodnar@usamvcluj.ro (D.-C.V.); dtit@uoradea.ro (D.M.T.)

Abstract: Due to its abundance of physiologically active ingredients, one of the oldest medicinal herbs, elderberry (EB) *Sambucus nigra* L., is beneficial for both therapeutic and dietary purposes. This study determined the bioaccessibility of the phenolic compounds and the prebiotic potential of the polyphenols from freeze-dried EB powder (FDEBP), along with the antioxidant and antimicrobial activities of this extract. The most significant phenolic compounds in black EB are represented by anthocyanins (41.8%), predominating cyanidin-sambubiosides and cyanidin-glucosides (90.1% of the identified anthocyanins). The FRAP assay obtained the highest antioxidant activity value (185 ± 0.18 μmol Fe^{2+}/g DW). The most sensitive to the antimicrobial activity of the extract was proven to be *Staphylococcus aureus*, and *Pseudomonas aeruginosa* had the lowest minimum inhibitory concentration of 1.95 mg/mL. To determine the prebiotic potential of the polyphenols, the cell growth of five probiotic strains (*Lactobacillus plantarum*, *L. casei*, *L. rhamnosus*, *L. fermentum* and *Saccharomyces boulardii*) was tested. The influence on cell growth was positive for all five probiotic strains used. Overall, the most significant increase ($p < 0.05$) was recorded at 1.5% FDEBP, on *L. casei* with a growth index (GI) of 152.44%, very closely followed by GI at 0.5% and 1% concentrations. The stability of the total phenolic compounds through simulated gastronitestinal digestion was increased (93%), and the bioaccessibility was also elevated (75%).

Keywords: elderberry; polyphenols; bioaccessibility; gastrointestinal digestion; prebiotic potential

1. Introduction

Since prehistoric times, plants have been used for food and as remedies [1]. In recent years, the interest in food supplements and functional foods, which contain bioactive plant-based compounds, has continuously increased. Numerous scientific studies were carried out using different methods and revealed the benefits of products rich in bioactive compounds [2,3] for preventing and treating certain diseases, including cardiovascular and metabolic diseases, pathologies responsible for millions of deaths worldwide [4–7].

An important source of bioactive compounds, relatively little studied compared to other berries, is black elderberry (EB), *Sambucus nigra* L.—also known as elder, black elder, European elder, European black elder, and common elder; it is part of the family Adoxaceae, genus Sambucus, being a common species. It has three subspecies: *S. nigra* ssp. *canadensis*, *S. nigra* ssp. *Cerulea* and *S. nigra* ssp. *nigra* L., the latter being the European elder [8–10]. Since ancient times, black EB has been used as a natural remedy in the treatment of various ailments: EB was used in the treatment of colds and flu, constipation or for its diuretic and anti-inflammatory effects [10]; elderflowers were used as a remedy for various respiratory and skin conditions, joint pain or for the diuretic effect [11]; even the elder leaves were used to treat various skin diseases [12].

The existence of phenolic constituents, which have a significant antioxidant effect and can therefore remove free radicals and combat oxidative stress, a factor contributing to the deterioration of the human body and the emergence of several diseases, is mainly responsible for the therapeutic properties of EB [13,14]. Through the wide variety of polyphenols contained, anthocyanins, flavonols, phenolic acids, and proanthocyanidins, EBs have proven their multiple beneficial effects, showing cardiovascular protection [15], antidiabetic properties [16], the ability to counteract obesity and metabolic dysfunction [17], antiviral and antibacterial activity [18], antioxidant capacity [19], antitumor potential [20], antidepressant action [21], and more recently, the prebiotic effect [22].

Phenolic compounds can deteriorate as a consequence of being exposed to light, oxygen, enzymatic activity, unfavorable pH condition, temperature, water, and metal ions; thus, their positive qualities may be modified [23]. For better comprehension and assessment of the potential biological characteristics of phenolic compounds, it is essential to confirm their stability and absorption in the digestive tract. Phenolic components are subjected to physiochemical alteration (due to pH, temperature, and digestive enzymes) in the gastrointestinal environment [22]. Yet, the bioaccessibility of dietary phenolic compounds during gastrointestinal digestion (GID) determines their therapeutic effects [24].

From this perspective, in order to evaluate the biological properties, along with the aspects regarding the chemical composition, this study determined the bioaccessibility of the phenolic compounds after GID and the prebiotic potential of the polyphenols from black EB from the spontaneous flora of Romania. Additionally, the antioxidant and antimicrobial action was tested. As far as we know, this is the first study of this type on black EB. The results provide new insights into advancing knowledge and research opportunities for the development of new nutraceutical or adjunctive strategies that use this product's bioactive potential.

2. Results

2.1. Antioxidant Activity Analysis

The antioxidant activity of the EB extract (*S. nigra* L.) was analyzed with the help of four different assay methods (DPPH, ABTS, FRAP, and CUPRAC), and the results are presented in Table 1. The lowest value was obtained by ABTS, and the highest with the FRAP assay.

Table 1. Antioxidant capacity of *S. nigra* L. measured by different complementary assays.

Assay Method	Antioxidant Activity
DPPH (μmol TE/g DW)	104.35 ± 0.22
ABTS (μmol TE/g DW)	30.36 ± 0.18
FRAP (μmol Fe^{2+}/g DW)	185 ± 0.18
CUPRAC (μmol TE/g DW)	52.3 ± 0.11

Values are expressed as mean values ± SD, n = 3; TE—Trolox equivalents; DW—Dry weight.

2.2. Antimicrobial Activity Assay

The antimicrobial activity of the lyophilized EB powder extract was evaluated on a total of seven strains containing Gram-positive and Gram-negative bacteria and yeasts. The lyophilized black EB powder extract shows antimicrobial activity on all tested microorganisms. The most sensitive to the activity of the extract was proven to be *S. aureus*, *P. aeruginosa* and the two yeasts, the lowest minimum inhibitory concentration (MIC) being 1.95 mg/mL. However, in the case of *S. enterica* and both strains of *E. coli*, the MIC was weaker in comparison with the other tested strains, as it presented an MIC of 3.91 mg/mL. The results are presented in Table 2.

Table 2. Minimum inhibitory concentration against *Staphylococcus aureus*, *Salmonella enterica*, *Escherichia coli* (25922 and 8739), *Pseudomonas aeruginosa*, *Candida albicans*, and *C. parapsilosis*.

Tested Strain	*S. aureus* 25923	*S. enterica* 6017	*E. coli* 25922	*E. coli* 8739	*P. aeruginosa* 27853	*C. albicans* 10231	*C. parapsilosis* 22019
FDEBME * (mg/mL)	1.95 ± 0.1	3.91 ± 0.2	3.91 ± 0.2	3.91 ± 0.2	1.95 ± 0.1	1.95 ± 0.1	1.95 ± 0.1
Gentamicin (μg/mL)	≤0.098	≤0.098	≤0.098	12.5 ± 0.5	12.5 ± 0.5	12.5 ± 0.5	12.5 ± 0.5

* FDEBME—freeze-dried elderberry methanolic extract.

2.3. Qualitative and Quantitative Analysis of the Extracts by HPLC-DAD-ESI-MS, before and after GID

The high-performance liquid chromatography (HPLC-DAD-ESI-MS) analysis of the extract from the powder obtained from lyophilized EB revealed the presence of 12 polyphenolic compounds belonging to the subclasses: anthocyanins, flavonols, hydroxycinnamic acids, and hydroxybenzoic acid derivatives.

Quantitative data show that the most significant amount of phenolic compounds in black EB is represented by anthocyanins, precisely 41.8%, predominating cyanidin-sambubiosides and cyanidin-glucosides, the two compounds constituting 90.1% of the identified anthocyanins. As for flavonols, they represented 25.5% of the total phenolic compounds in lyophilized EB. The HPLC-DAD analysis revealed the presence of quercetin derivatives (94%) and kaempferol. Rutin is the most present compound of this class, representing 75.7% of the total flavonols and 19.27% of the total phenolic compounds identified. Hydroxycinnamic acids were present in a proportion of 18.6%, and hydroxybenzoic acid derivatives in a proportion of 14.1%.

The total content of polyphenols in the analyzed extract was 41.28 mg/g of lyophilized EB powder, of which consisted 17.25 mg of anthocyanins, 10.51 mg of flavonols, 7.69 mg of hydroxycinnamic acid derivatives and 5.83 mg of hydroxybenzoic acid derivatives (Table 3).

2.4. The Bioaccessibility of Phenolic Compounds of Sambucus nigra L. Fruits during Simulated Digestion

The calculation of bioaccessibility was carried out according to the formula presented by Stefănescu et al. [25], which is:

$$BI\ (\%) = (\text{Phenolic content after gastrointestinal digestion (in vitro)}/\text{Phenolic content before digestion}) \times 100 \quad (1)$$

Then, the individual phenolic compound content was assessed HPLC-DAD-ESI-MS in the gastric and intestinal phases. The results are presented in Table 3, and as can be seen, the anthocyanin content, such as cyanidin-diglucoside and cyanidin glucoside, decreased from 1.20 ± 0.07 and 15.56 ± 0.19 to 1.03 ± 0.09 and 8.14 ± 0.08 after SIF; cyanidin was only detected before digestion. Only hydroxybenzoic acids presented an increase throughout digestion, for instance, hydroxybenzoic acid from 3.49 ± 0.05 to 5.31 ± 0.11, and protocatechuic acid from 2.34 ± 0.11 to 7.79 ± 0.15.

The bioaccessibility (Table 4) of the bioactive compounds from EB can be perceived as the amount of the compound released inside the intestinal tract and available for assimilation. Before and after digestion, four compounds were detected, with a final bioaccessibility of 74.54 ± 5.7%.

Table 3. The in vitro effect of gastrointestinal digestion on the phenolic content of EB mg/g.

Peak	Rt (min)	UV λmax (nm)	[M + H]+ (m/z)	Compound	Subclass	BD	SGF	SIF
1	3.81	270	139	Hydroxybenzoic acid	Hydroxybenzoic acid	3.49 ± 0.05	5.32 ± 0.14	5.31 ± 0.11
2	9.63	528, 280	611	Cyanidin-diglucoside	Anthocyanin	1.20 ± 0.07	1.14 ± 0.09	1.03 ± 0.09
			743	Cyanidin-sambubioside-glucoside				
3	10.12	295	155	Protocatechuic acid	Hydroxybenzoic acid	2.34 ± 0.11	7.32 ± 0.13	7.79 ± 0.15
4	11.09	529, 280	449	Cyanidin-glucoside	Anthocyanin	15.56 ± 0.19	13.80 ± 0.23	8.14 ± 0.08
			581	Cyanidin-sambubioside				
5	12.91	323	355	5-Caffeoylquinic acid (Chlorogenic acid)	Hydroxycinnamic acid	1.50 ± 0.10	1.48 ± 0.14	1.28 ± 0.07
6	13.6	322	181	Caffeic acid	Hydroxycinnamic acid	1.27 ± 0.09	1.18 ± 0.10	0.84 ± 0.09
7	14.06	530, 280	287	Cyanidin	Anthocyanin	0.49 ± 0.03	N.D.	N.D.
8	14.47	356, 256	611	Kaempferol-diglucoside	Flavonol	0.67 ± 0.05	0.57 ± 0.04	0.45 ± 0.01
9	15.59	332	369	Feruloyquinic acid	Hydroxycinnamic acid	4.92 ± 0.11	N.D.	N.D.
10	15.88	360, 255	611	Quercetin-rutinoside (Rutin)	Flavonol	7.9 ± 0.09	6.30 ± 012	5.42 ± 0.16
11	16.57	360, 255	465	Quercetin-glucoside	Flavonol	1.26 ± 0.08	1.15 ± 0.09	0.51 ± 0.08
12	21.91	360, 255	303	Quercetin	Flavonol	0.63 ± 0.03	N.D.	N.D.
				Total phenolics		41.27 ± 0.15	38.26 ± 0.21	30.76 ± 0.17

Values are expressed as mean values ± SD, n = 3; BD—before digestion, SGF—simulated gastric fluid, SIF—simulated intestinal fluid; N.D.—not determined.

2.5. The Prebiotic Potential of the Phenolic Compounds of Sambucus nigra L. Fruits

The prebiotic potential of the freeze-dried EB powder (FDEBP) was tested on the following probiotic strains: *L. plantarum, L. casei, L. rhamnosus, L. fermentum* and *S. boulardii*, at three different concentrations: 0.5%, 1%, and 1.5% (*w*/*v*). The cell growth of the probiotic strains was tested after inoculation and after 24 h incubation period. The results are shown in Figure 1. For the control media, we used glucose as a carbon source and the difference of $\log_{10}$ CFU/mL between incubation and inoculation was expressed as 100% GI. The influence on cell growth was most prominent in the case of *L. casei*, which yielded significant positive growth for all tested concentrations of FDEBP, with the highest results of 152.44 % GI ($p < 0.05$) recorded at 1.5% FDEBP. Moreover, positive growth was also recorded for *L. rhamnosus* and *L. plantarum*, with the highest results recorded at 141.36% ($p < 0.05$) GI with 1.5% and GI of 133.31% ($p < 0.05$) at 0.5%, respectively. In the case of *L. fermentum*, the only tested concentration that yielded a positive GI of 115.04% ($p < 0.05$) was at 0.5% FDEBP concentration. However, the least growth influence was recorded by the *S. boulardi* strain for which the 1.5% FDEBP yielded a significant GI of 107.22%. Overall, the results showed that most of the probiotic tested strains presented a positive growth influence with FDEBP as a carbon source in comparison with glucose; thus, the tested samples exhibited a prebiotic potential.

Table 4. The bioaccessibility of EB extract.

Compound	Bioaccesibility (%)
Anthocyanins	53.17 ± 1.5
Flavonols	60.64 ± 3.8
Hydroxycinnamic acids	27.59 ± 2.1
Hydroxybenzoic acid	224.64 ± 5.8
Total phenolics	74.54 ± 5.7

Values are expressed as mean values ± SD.

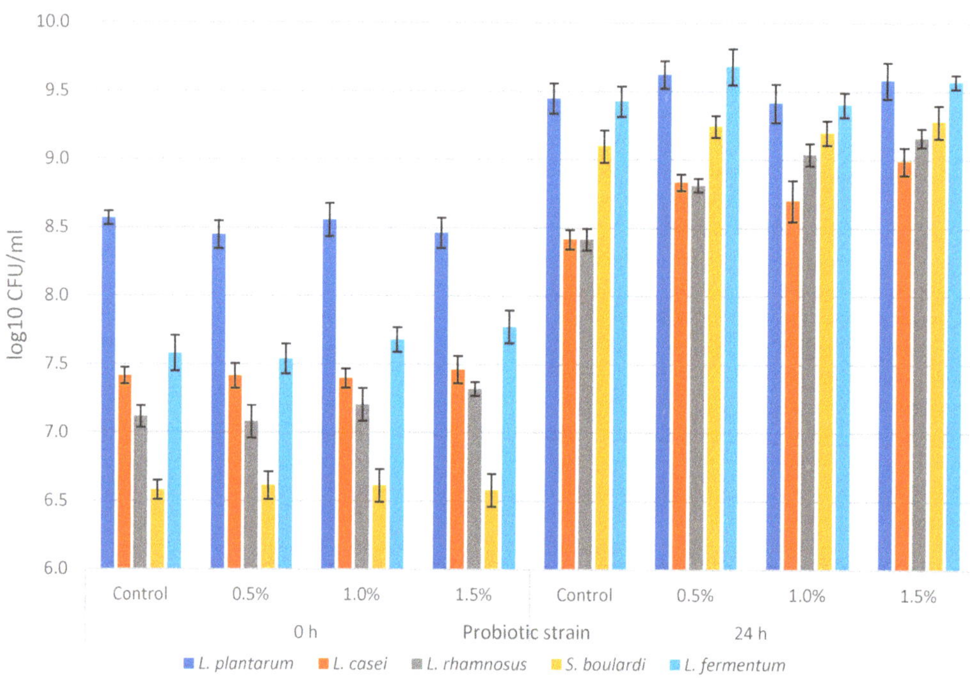

Figure 1. Cell viability of probiotic strains after inoculation and after 24 h (n = 3).

3. Discussion

The most significant bioactive substances found in EB in relatively high concentrations are polyphenols, which are recognized for their free radical scavenging (antioxidant) action [26]—the proprieties of each active compound being strongly correlated and induced by its unique structure [27]. As a result, most of the studies carried out on elderberries have evaluated their phenolic compound content, and thus are recognized for their multiple beneficial effects on health. There are fewer studies that follow up what happens to these bioactive compounds during digestion, and the present study proposed the quantitative and qualitative evaluation of phenolic compounds through HPLC after simulating digestion. Bioaccessibility can be defined as the amount of an ingested nutrient that is available for absorption in the gut after digestion [28] and it is an essential aspect to follow—it is known that phenolic compounds are unstable under certain conditions, and GID involves pH and temperature variations, contact with digestive enzymes, etc. [25,29].

The main polyphenols found in EB, according to published data, are chlorogenic acid, cryptochlorogenic acid, neochlorogenic acid, kaempferol-3-glucoside (astragaline), kaempferol-3-rutinoside, isorhamnetin-3-rutinoside, quercetin-3-glucoside (isoquercitrin), quercetin-3-rutinoside (rutin), quercetin. Rutin is the main flavonoid found in this plant, while EB also contains minor levels of astragaline and isoquercitrin [26]. Additionally, in our study, we identified the main phenolic compounds to be cyanidin-glucoside and cyanidin-sambubioside followed by rutin and the other compounds identified cyanidin-diglucoside, cyanidin-sambubioside-glucoside, hydroxybenzoic acid, protocatechuic acid, chlorogenic acid, caffeic acid, cyaniding, kaempferol-diglucoside, feruloyquinic acid, quercetin-glucoside and quercetin.

EBs present strong anti-inflammatory characteristics that are linked to their significant antioxidant properties. Compared to similar studies, such as by Imenšek et al. [30], who also analyzed the antioxidant activity from specific hybrids of *S. nigra*, the DPPH (96 ± 14 μmol TE/g DW) and FRAP (208 ± 25 μmol TE/g DW) results were similar; however, the ABTS values (130 ± 14 μmol TE/g DW) were higher than in our study. This study also showed the effect of maturation on the antioxidant activity of the selected plants, which indicated a growing pattern.

Data from the specialized literature attribute the antimicrobial effect to tannins and triterpenes, as well as to peptides and oligosaccharides that are present in black EB [31]. The authors of a recent study, however, draw attention to the fact that the antimicrobial activity is due to the combination of bioactive compounds from the black EB extract, rather than to certain compounds considered individually [9]. The antimicrobial activity of EB extracts has been demonstrated in several previous studies on several Gram-positive and Gram-negative bacterial strains from the following genera: *Staphylococcus*, *Pseudomonas*, *Enterococcus*, *Escherichia*, *Streptococcus*, *Klebsiella*, *Bacillus*, *Corynebacterium*, *Proteus* [9,32–34]. Mohammadsadeghi et al. showed that EB extract had an inhibitory effect on the development of some *Candida* species, including *Candida albicans* [32]. In none of the existing examples in the scientific literature, however, were identified studies on the antimicrobial effect on *Candida parapsilosis*, a pathogenic agent causing fungal diseases associated with significantly increased morbidity and mortality [35,36]. The present antimicrobial activity analysis results reveal a significant antimicrobial potential of EB at a concentration between 1.95 mg/mL and 3.91 mg/mL lyophilized EB powder extract. Furthermore, to our knowledge, it is the first study to demonstrate the antimicrobial effect on *C. parapsilosis*.

The bioaccessibility testing of phenolic compounds was carried out using the updated in vitro static digestion method, developed by the INFOGEST working group [37]. The in vitro digestion methods have proven to be an efficient and useful solution in anticipating the effects of in vivo digestion [38]. Anthocyanins appear to be the most unstable polyphenolic compounds during GID. In the gastric phase, the amount decreases, compared to the amount found in the extract, by 13.4%, and after intestinal digestion, the bioaccessibility of anthocyanins is 53.2%. The biggest decrease was recorded for cyanidin-glucosides and cyanidin-sambubiosides, respectively, while cyanidin was not detected by HPLC. Moreover, the data from the scientific literature mention the instability of anthocyanins in an alkaline environment [39], an aspect that is also confirmed in our study. In the intestinal tract, their hydrolysis or degradation takes place, forming phenolic acids and aldehydes. Variations at the B ring level in the structure of anthocyanins determine the degradation of cyanidin and the formation of protocatechuic acid [40]. This is one of the explanations for the spectacular changes in the amount of hydroxybenzoic acid derivatives, the subclass that became the most predominant after digestion—the total amount increased by 225% in the intestinal phase compared to the amount in the extract. The bioaccessibility of protocatechuic acid was 332.34%. Hydroxybenzoic acid showed an increase of 152.23%, the hypothesis being that HA acid can be generated either as a degradation product of anthocyanins, or as a metabolite [41,42]. Derivatives of benzoic acid have demonstrated their cardiovascular protective effects, action against cancer and obesity along with the inhibition of the inflammatory response in inflammatory bowel diseases [43–45].

Regarding flavonols, their presence decreases in the gastric phase compared to time 0 by 23.6%, and at the end of the intestinal phase, 60.6% of the initial amount remains. Quercetin is the flavonol that was not detected by HPLC in any of the two gastrointestinal phases. The elderberry extract contained hydroxycinnamic acids in a proportion of 18.6% of the total phenolic compounds; at the end of the gastric phase, they represented 6.95%, and 6.90% in the intestinal phase. Numerous studies carried out on polyphenols have shown a higher stability of phenolic compounds in the gastric phase with their degradation in the intestinal tract [46–48]. In our study, the stability of the total phenolic compounds was high (93%), and the bioaccessibility in the intestinal phase was 75%.

As there is a degree of complementarity between bioaccessibility, the prebiotic effect and overall health benefits of the bioactive compounds [49], we also monitored in the study the prebiotic potential of the phenolic compounds from black elderberry.

Prebiotics are defined as "a substrate that is selectively utilized by host microorganisms conferring a health benefit" [50], and according to the newest definition and recent studies, polyphenols were shown as a prebiotic substrate [51]. It seems that the relationship between polyphenols-intestinal microbiota is mutual: phenolic compounds can modulate the intestinal microbiota and, at the same time, microorganisms have the ability to modulate the activity of polyphenols [52]. Various preclinical studies have shown that dietary polyphenols have a prebiotic effect, stimulating the growth of different beneficial microorganisms [53–55]. This effect was attributed especially to anthocyanins, proanthocyanidins and catechins [56]. A previous longitudinal intervention study followed the prebiotic properties of a purified extract from EB, observing a major change in microbial diversity immediately after initiating the administration of the extract. Furthermore, in some study participants, the relative abundance of *Akkermansia* spp. increased even after supplementation was completed [22].

As far as we know, our study is among the first studies that investigated the prebiotic potential of phenolic compounds from black EB, specifically on *L. plantarum, L. casei, L. rhamnosus, L. fermentum and S. boulardii.* The prebiotic potential was proven on all five strains, withal, more in vitro and in vivo studies are necessary to support the findings, and to implicitly demonstrate the prebiotic potential of these compounds and their possible health-effects. However, through the results obtained regarding the prebiotic potential of the tested bacterial strains and the antimicrobial effect on *C. parapsilosis*, the current study offers new research perspectives unexplored until now, which can increase the use of bioactive potential of black EB fruits.

4. Materials and Methods

4.1. Plant Material

The present study used black *Sambucus nigra* L. fructus (elderberry) from the spontaneous flora of Romania, Bihor County (46°43′36.5″ N, 21°54′32.4″ E). The species were identified in the Pharmaceutical Botany department of the Faculty of Medicine and Pharmacy, Oradea University, Romania.

The EBs were harvested in September 2022, frozen at −20 °C and then lyophilized, using a Telstar Lyo Quest 55 plus lyophilizer (Azbil Group, Terrassa, Spain) at a temperature of −55 °C and pressure of 0.001 mbar for 72 h. Then, the lyophilized fruits were transformed by grinding and sieving into a fine powder and kept in the dark until the determinations were made.

4.2. Methanolic Extraction

The fine powder obtained from lyophilized EB (0.5 g) was extracted with 10 mL of methanol acidified with 1% hydrochloric acid of concentration 37% by vortexing (Heidolph Reax top, Heidolph Instruments, Schwabach, Germany) for 1 min, then sonication in an ultrasonic bath (Elmasonic E 15 H, Elma Schmindbauer, Singen, Germany) for 15 min and centrifugation (10,000 rpm for 10 min at 24 °C) in an Eppendorf AG 5804 centrifuge (Eppendorf, Hamburg, Germany). The extraction process was repeated until the complete decoloration of the sample was achieved. At the end of each extraction, the supernatant was filtered through a 0.45 μm Chromafil Xtra nylon filter (Macherey-Nagel, Duren, Germany) and combined in a flask. The obtained extract was brought to dryness by evaporating the solvent with a rotary evaporator (Rotavapor R-124, Buchi, Flawil, Switzerland) and brought back into a known volume of methanol (mL solvent retook). The extract solution was used for identifying and quantifying phenolic compounds from the lyophilized EB extract, using high-performance liquid chromatography (HPLC), respectively, for the determination of antioxidant and antimicrobial activity.

4.3. Qualitative and Quantitative Determinations of Phenolic Compounds Phenolic Compounds from Freeze-Dried Elderberry Extract

To identify and quantify the phenolic compounds from the lyophilized EB powder extract, an HPLC-DAD-ESI-MS system was used, consisting of an Agilent 1200 HPLC with a UV–vis detector (DAD) coupled to a mass detector (MS) with a single quadrupole Agilent 6110 (Agilent Technologies, Santa Clara, CA, USA). For the separation of phenolic compounds, the Kinetex XB C18 column (Phenomenex, Torrance, CA, USA) was used, having, as mobile phases, water + 0.1% acetic acid (solvent A), and acetonitrile + 0.1% acetic acid (solvent B), at a temperature of 25 °C, for 30 min, with a flow rate of 0.5 mL/min. The elution program was as follows: 5% B (0 min); 5% B (0–2 min); 5–40% B (2–18 min); 40–90% B; 90% B (20–24 min); 90–5% B (24–25 min); 5% B (25–30 min). For MS fragmentation, the ESI (+) ionization module was used, with a scan range between 120 and 1200 m/z, capillary voltage of 3000 V, at a temperature of 350 °C, and nitrogen flow of 7 L/min. The spectral values were recorded for all peaks in the 200–600 nm range. Phenolic compounds were identified at 280 nm, 340 nm and 520 nm. The results were analyzed using the Agilent ChemStation software (Rev B.02.01 SR2, Palo Alto, CA, USA). The phenolic compounds in the EB extract were identified considering the retention intervals; UV–vis absorption spectra and mass spectra were recorded for each peak. For the quantification of phenolic compounds, calibration curves were made with standard substances. Thus, for the quantification of the identified anthocyanins, a calibration curve was made with Cyanidin (R^2 = 0.9951). For the quantification of hydroxybenzoic acids, the calibration curve was made with gallic acid (R^2 = 0.9978), hydroxycinnamic acids were quantified as a chlorogenic acid equivalent (R^2 = 0.9937), and flavonols as rutin equivalent (R^2 = 0.9981).

4.4. Antioxidant Activity Assay

The antioxidant activity of the EB extract was tested using four complementary methods: DPPH, FRAP, ABTS and CUPRAC.

The DPPH (2,2-diphenyl-1-picrylhydrazyl) test was based on the ability of the compound to donate an electron (H^+) from the structure to the DPPH radical. For the determination, the protocol previously reported by Brand-Williams et al. [57] was applied, which is the most frequently used in studies,. To summarize, lyophilized EB extract (35 µL) was mixed with 250 µL of the DPPH solution (0.02 mg/mL) and incubated for 30 min in the dark; then, the absorbance was measured at 517 nm. The resulting data were expressed as a micromole Trolox equivalent (µmol TE)/g sample.

The antioxidant method of neutralizing the ABTS radical is based on the reaction between ABTS [2,20-azino-bis(3-ethylbenzothiazoline-6-sulfonic acid)] and a compound with antioxidant activity. This reaction causes a decrease in absorbance [58]. The determination was made according to the protocol described by Arnao et al. [59], adjusted to be suitable for the 96-well microplates. In short, 20 µL of the lyophilized EB extract was mixed with 170 µL of ABTS and incubated for 6 min in the dark, then the absorbance was measured at 734 nm using a BioTek microplate reader (Synerg y HT, BioTek Instruments, Winooski, VT, USA). The resulting data were expressed as a micromole µmol TE/g sample.

The FRAP (Ferric Reducing Antioxidant Power) method is a colorimetric method that quantifies the ability of compounds with antioxidant activity to reduce (Fe^{3+}) to (Fe^{2+}) [60]. The assay was performed according to the protocol described by Benzie and Strain [61]. A total of 20 µL of the sample extract was added to 180 µL of the FRAP reagent. After an incubation time of 3 min, the absorbance was measured at 593 nm. The antioxidant potential was expressed as the µM Fe^2 equivalent/g sample.

The CUPRAC test (cupric ion reducing antioxidant capacity) is a spectrophotometric technique which measures the antioxidant capacity of a compound, based on the ability of antioxidants to reduce (Cu^{2+}) to (Cu^{+}) [62]. The determination was performed based on the protocol described by Apak et al. [63]. The results were expressed as the μmol TE/g sample.

4.5. Antimicrobial Capacities

The following seven standard strains were tested: *Escherichia coli* ATCC 25922, *E. coli* ATCC 8739, *Staphylococcus aureus* ATCC 29213, *Pseudomonas aeruginosa* ATCC 27853, *Salmonella enterica* NCTC 6017, *Candida albicans* ATCC 10231 and *C. parapsilosis* ATCC 22019. They were all acquired from American Type Culture Collection (ATCC), VA, USA. The microorganisms were grown on a specific medium, Tryptic Soy agar (M1968, HiMedia Laboratories, Pvt. Ltd., Thane, India) for *E. coli* ATCC 8739, *P. aeruginosa* and both *Candida* strains, and on Mueller–Hinton agar (Oxoid Ltd., Basingstoke, Hampshire, England) for the others, within the Food Biotechnology Laboratory of the University of Agricultural Sciences and Veterinary Medicine Cluj-Napoca, Romania. The plates were incubated for 24 h at 37 °C for bacteria and 30 °C for yeasts, respectively. Bacterial and yeast morphology were confirmed by optical microscopy [64].

For each tested strain, several colonies cultivated on agar plates (Oxoid Ltd., Basingstoke, Hampshire, UK) were transferred in a sterile saline solution (8.5 g/L NaCl) and adjusted to match the turbidity of McFarland 0.5 standard which corresponded to $1.5–3 \times 10^8$ CFU/mL. Then, bacterial suspensions were serially diluted 10-fold in a ratio of 1:9 in sterile serum, and 10^5 CFU/mL solutions were added to each microplate well.

The minimum inhibitory concentration (MIC) was determined using the resazurin microtiter plate-based antibacterial assay [65–67]. A total of 100 μL of sterile specific growth broth medium was added to the wells of a 96-well microplate. Then, 100 μL of lyophilized EB methanolic extract (FDEBME) was added in the first well, and serial 11-fold dilutions were made in the subsequent wells of each row by transferring 100 μL from well to well. The surplus of 100 μL in the last well of the row was discarded. Then, 10 μL of appropriate inoculum was added to all wells. The positive control was gentamicin (0.4 mg/mL in saline solution). The extracts' solvent solution (methanol: H_2O 1:1) was added as a negative control. The microplates were incubated for 20–22 h at 37 °C or 30 °C, respectively, and then 20 μL of the 0.2 mg/mL resazurin aqueous solution was added in all wells. The microplates were subjected to a subsequent two-hour incubation. After this period, resazurin (a blue non-fluorescent dye) was oxidized to resorufin (fluorescent pink) wherever the wells contained viable bacterial cells. Thus, the concentration in the last well on each row that remained blue was considered to completely inhibit bacterial growth, the MIC. The assay was run in triplicate. The results are expressed as the mean ± standard deviation.

4.6. Static In Vitro Digestion of the S. nigra Samples

The updated static in vitro digestion method, developed by the INFOGEST working group, was used to simulate the GID of the samples. The protocol extensively described by Brodkorb et al. [37] is based on sequential oral, gastric and intestinal digestion. In contrast, parameters such as electrolytes, enzymes, bile, pH, dilution, and digestion time are established on available physiological data. The samples (microcapsules and lyophilized powder of *S. nigra* fruits) were subjected to a three-stage in vitro digestion process, mimicking the conditions of the mouth, stomach and small intestine. Due to the absence of starch in the matrix, the oral phase was conducted without amylase.

The samples (2 g) were diluted with 3 mL of water to achieve the proper consistency and were further diluted 1:1 (*wt/wt*) with simulated oral fluid (SOF) to achieve a swallowable bolus with a paste-like consistency. The SOF was composed of electrolyte solutions KCl, KH_2PO_4, $NaHCO_3$, NaCl, $MgCl_2 \cdot 6H_2O$, $(NH_4)_2CO_3$, alongside $CaCl_2(H_2O)_2$ and water. Further, the oral bolus was mixed with 10 mL of the simulated gastric fluid (SGF). The SGF was composed of electrolyte solutions KCl, KH_2PO_4, $NaHCO_3$, $MgCl_2 \cdot 6H_2O$, $(NH_4)_2CO_3$, alongside $CaCl_2(H_2O)_2$ solution (0.3 M), porcine pepsin (2000 U/mL in the final digestion mixture), and water. The pH of the samples was adjusted to 3 by adding HCl (1 M), and the mixture was homogenized and incubated for 2 h in a shaking incubator (New Brunswick Innova 44, Eppendorf AG, Hamburg, Germany). For the intestinal phase, the samples were mixed with 20 mL of pre-warmed simulated intestinal fluid (SIF) to achieve a final ratio of 1:1 (*v*/*v*). The SIF was composed of electrolyte solutions KCl, KH_2PO_4, $NaHCO_3$, NaCl, $MgCl_2 \cdot 6H_2O$, alongside $CaCl_2(H_2O)_2$, the bile extract solution (10 mM in total digesta) and pancreatic enzymes (100 U/mL). The pH was set to 7 using NaOH (1 M), and the mixture was homogenized and incubated at 37 °C for 2 h in a shaking incubator (95 rpm). After the process was complete, 1 mL of the samples was filtered and further analyzed by HPLC to establish the bioaccessibility index.

4.7. The Prebiotic Potential of Phenolic Compounds of Sambucus nigra L. Fruits

To determine the prebiotic potential of the FDEBP, we tested its influence on the cell growth for the following probiotic strains: *L. plantarum* ATCC 14917, *L. casei* ATCC 393, *L. rhamnosus* LMG 25626, *L. fermentum* CECT5716 Lc40 and *S. boulardii* MYA 796, which were all acquired from the American Type Culture Collection (ATCC, Manassas, VA, USA).

The probiotic strains were obtained in freeze-dried powder form. They were activated in test tubes containing 10 mL of Man–Rogosa–Sharpe (MRS) broth (1.10661, Merck, Rahway, NJ, USA) for *Lactobacillus* strains and Potato Dextrose (PD) broth (GM403, HIMEDIA) for probiotic yeast *S. boulardii*. The tubes were incubated aerobically for 18–24 h at 37 °C in case of bacteria and at 30 °C in case of the yeast. The grown microorganisms were propagated further (10% *v*/*v*) in 100 mL flasks with 45 mL of fresh sterile media, that were used as inoculum after incubation in same conditions [68].

For the prebiotic potential assay, freeze-dried EB powder was used as the carbon source in the formulation of MRS broth and Potato Dextrose broth in three different concentrations: 0.5%, 1% and 1.5% (*w*/*v*). The control media consisted of MRS or Potato Dextrose broth with glucose as a carbon source. For the experiment, 50 mL of media in 100 mL flasks, containing either 0.5%, 1% or 1.5% of FDEBP or glucose as a carbon source, was autoclaved at 121 °C for 15 min. After that, the media was cooled to room temperature and inoculated in sterile conditions with 10% (*v*/*v*) inoculum. The inoculum used in the assay was obtained by overnight culturing in MRS broth or PD broth, respectively, for each strain. The flasks were incubated in the Heidolph 1000 shaking incubator (Heidolph Instruments, Schwabach, Germany) with 150 rpm for 24 h at 37 °C or 30 °C, respectively, in aerobic conditions.

The cell growth of the probiotic strains was tested after inoculation and after a 24 h incubation period by using the pour plate method for *Lactobacillus* strains and the spread plate method for the yeast of the serially diluted samples. Roughly 1 mL of the sample was serially diluted in 9 mL of the sterile serum solutions (0.85% *w*/*v* NaCl solution), and 1 mL of the tested dilution was pipetted in a sterile Petri dish over which 15–20 mL of semi-molten agar was poured, followed by mixing and cooling in the case of the pour plate method. For the spread plate method, 100 µL of the tested dilution was pipetted on a solidified agar plate and spread with a Driglaski spatula until complete absorption. The plates were then incubated at appropriate temperatures for 24–48 h in aerobic conditions, after which the grown colonies were counted. The experiments were run in triplicate, and the results were expressed as the mean colony-forming units CFU/mL ± standard deviation ($n = 3$). The prebiotic potential was calculated as follows: growth index GI (%) = (sample $\log_{10}$ CFU/mL 24 h − 0 h)/(control $\log_{10}$ CFU/mL 24 h − 0 h) × 100 for each strain. The difference between 24 h incubation and inoculation viability of each control was expressed as 100% GI.

5. Conclusions

The results show that the black elder, from the spontaneous flora of Romania, presents a high antioxidant and antimicrobial potential. Polyphenols showed a significant bioaccessibility index. During gastrointestinal digestion, anthocyanins were the most unstable polyphenolic compounds, while hydroxybenzoic acid derivatives increased significantly in the intestinal phase, compared to the amount in the extract. Moreover, this study provided, for the first time, results regarding the prebiotic potential of elderberries on *L. plantarum*, *L. casei*, *L. rhamnosus*, *L. fermentum* and *S. boulardii*, opening new directions for research and exploration of these berries. For future perspectives, in vivo studies should also be carried out in order to confirm the health benefits of elderberry.

Author Contributions: Conceptualization, I.M.H. and D.M.T.; methodology, I.M.H., B.-E.T., K.S., E.S., F.R. and Z.M.D.; software, F.R. and D.-C.V.; validation, B.-E.T., K.S., D.M.T. and D.-C.V.; formal analysis, B.-E.T.; investigation, E.S., K.S., M.N., A.L.P. and F.R.; resources, I.M.H., M.N. and D.-C.V.; data curation, B.-E.T., D.M.T. and K.S.; writing—original draft preparation, I.M.H., A.L.P. and B.-E.T.; writing—review and editing, I.M.H., D.M.T. and D.-C.V.; visualization, D.M.T. and D.-C.V.; supervision, D.M.T. and D.-C.V.; project administration, D.M.T. and D.-C.V.; funding acquisition, D.M.T. and D.-C.V. All authors have read and agreed to the published version of the manuscript.

Funding: This research was funded by the European Union's Horizon 2020 research and innovation program under the Marie Skłodowska-Curie grant agreement No. 101007783—FRIETS.

Institutional Review Board Statement: Not applicable.

Informed Consent Statement: Not applicable.

Data Availability Statement: Data are contained within the article.

Acknowledgments: The authors would like to thank the whole team from the Department of Food Science, University of Agricultural Sciences and Veterinary Medicine, Cluj-Napoca for their continued support.

Conflicts of Interest: The authors declare no conflict of interest.

Sample Availability: Samples of the compounds are available from the authors.

References

1. Martău, G.A.; Teleky, B.-E.; Odocheanu, R.; Soporan, D.A.; Bochis, M.; Simon, E.; Vodnar, D.C. *Vaccinium* Species (Ericaceae): Phytochemistry and Biological Properties of Medicinal Plants. *Molecules* **2023**, *28*, 1533. [CrossRef]
2. Khuntia, A.; Martorell, M.; Ilango, K.; Bungau, S.G.; Radu, A.F.; Behl, T.; Sharifi-Rad, J. Theoretical evaluation of Cleome species' bioactive compounds and therapeutic potential: A literature review. *Biomed. Pharmacother.* **2022**, *151*, 113161. [CrossRef]
3. Pallag, A.; Bungau, S.; Tit, D.M.; Jurca, T.; Sirbu, V.; Honiges, A.; Horhogea, C. Comparative study of polyphenols, flavonoids and chlorophylls in *Equisetum arvense* L. populations. *Rev. Chim.* **2016**, *67*, 530–533.
4. Behl, T.; Bungau, S.; Kumar, K.; Zengin, G.; Khan, F.; Kumar, A.; Kaur, R.; Venkatachalam, T.; Tit, D.M.; Vesa, C.M.; et al. Pleotropic Effects of Polyphenols in Cardiovascular System. *Biomed. Pharmacother.* **2020**, *130*, 110714. [CrossRef]
5. Szabo, K.; Teleky, B.-E.; Ranga, F.; Roman, I.; Khaoula, H.; Boudaya, E.; Ltaief, A.B.; Aouani, W.; Thiamrat, M.; Vodnar, D.C. Carotenoid Recovery from Tomato Processing By-Products through Green Chemistry. *Molecules* **2022**, *27*, 3771. [CrossRef] [PubMed]
6. Shen, Y.; Zhang, N.; Tian, J.; Xin, G.; Liu, L.; Sun, X.; Li, B. Advanced approaches for improving bioavailability and controlled release of anthocyanins. *J. Control. Release* **2022**, *341*, 285–299. [CrossRef] [PubMed]
7. Teleky, B.E.; Martău, G.A.; Ranga, F.; Pop, I.D.; Vodnar, D.C. Biofunctional soy-based sourdough for improved rheological properties during storage. *Sci. Rep.* **2022**, *12*, 17535. [CrossRef] [PubMed]
8. Sambucus, L. Taxonomic Serial No.: 35315. Integrated Taxonomic Information System. Available online: http://www.itis.gov (accessed on 14 February 2023).
9. Przybylska-Balcerek, A.; Szablewski, T.; Szwajkowska-Michałek, L.; Swierk, D.; Cegielska-Radziejewska, R.; Krejpcio, Z.; Suchowilska, E.; Tomczyk, Ł.; Stuper-Szablewska; Stuper-Szablewska, K. Sambucus nigra Extracts–Natural Antioxidants and Antimicrobial Compounds. *Molecules* **2021**, *26*, 2910. [CrossRef] [PubMed]
10. Sidor, A.; Gramza-Michałowska, A. Advanced research on the antioxidant and health benefit of elderberry (*Sambucus nigra*) in food—a review. *J. Funct. Foods* **2015**, *18*, 941–958. [CrossRef]
11. Ho, G.T.T.; Zou, Y.F.; Aslaksen, T.H.; Wangensteen, H.; Barsett, H. Structural characterization of bioactive pectic polysaccharides from elderflowers (*Sambuci flos*). *Carbohydr. Polym.* **2016**, *135*, 128–137. [CrossRef] [PubMed]
12. Fazio, A.; Plastina, P.; Meijerink, J.; Witkamp, R.F.; Gabriele, B. Comparative analyses of seeds of wild fruits of *Rubus* and *Sambucus* species from Southern Italy: Fatty acid composition of the oil, total phenolic content, antioxidant and anti-inflammatory properties of the methanolic extracts. *Food Chem.* **2013**, *140*, 817–824. [CrossRef] [PubMed]
13. Kashi, D.S.; Shabir, A.; Da Boit, M.; Bailey, S.J.; Higgins, M.F. The Efficacy of Administering Fruit-Derived Polyphenols to Improve Health Biomarkers, Exercise Performance and Related Physiological Responses. *Nutrients* **2019**, *11*, 2389. [CrossRef] [PubMed]

14. Pascuta, M.S.; Vodnar, D.C. Nanocarriers for sustainable active packaging: An overview during and post COVID-19. *Coatings* **2022**, *12*, 102. [CrossRef]
15. Festa, J.; Singh, H.; Hussain, A.; Da Boit, M. Elderberry extract inhibits tumour necrosis factor induced monocyte adhesion to endothelial cells via modulation of the NF-κB pathway. *Cardiovasc. Res.* **2022**, *118*, cvac066-172. [CrossRef]
16. Salvador, Â.C.; Król, E.; Lemos, V.C.; Santos, S.A.O.; Bento, F.P.M.S.; Costa, C.P.; Almeida, A.; Szczepankiewicz, D.; Kulczyński, B.; Krejpcio, Z.; et al. Effect of elderberry (*Sambucus nigra* L.) extract supplementation in STZ-induced diabetic rats fed with a high-fat diet. *Int. J. Mol. Sci.* **2017**, *18*, 13. [CrossRef]
17. Zielińska-Wasielica, J.; Olejnik, A.; Kowalska, K.; Olkowicz, M.; Dembczyński, R. Elderberry (*Sambucus nigra* L.) fruit extract alleviates oxidative stress, insulin resistance, and inflammation in hypertrophied 3T3-L1 adipocytes and activated RAW 264.7 macrophages. *Foods* **2019**, *8*, 326. [CrossRef]
18. Mocanu, M.L.; Amariei, S. Elderberries—A Source of Bioactive Compounds with Antiviral Action. *Plants* **2022**, *11*, 740. [CrossRef]
19. Domínguez, R.; Zhang, L.; Rocchetti, G.; Lucini, L.; Pateiro, M.; Munekata, P.E.S.; Lorenzo, J.M. Elderberry (*Sambucus nigra* L.) as potential source of antioxidants. Characterization, optimization of extraction parameters and bioactive properties. *Food Chem.* **2020**, *330*, 127266. [CrossRef]
20. Ma, X.; Ning, S. Cyanidin-3-glucoside attenuates the angiogenesis of breast cancer via inhibiting STAT3/VEGF pathway. *Phyther. Res.* **2019**, *33*, 81–89. [CrossRef]
21. Mahmoudi, M.; Ebrahimzadeh, M.A.; Dooshan, A.; Arimi, A.; Ghasemi, N.; Fathiazad, F. Antidepressant activities of *Sambucus ebulus* and *Sambucus nigra*. *Eur. Rev. Med. Pharmacol. Sci.* **2014**, *18*, 3350–3353. [PubMed]
22. Reider, S.; Watschinger, C.; Längle, J.; Pachmann, U.; Przysiecki, N.; Pfister, A.; Zollner, A.; Tilg, H.; Plattner, S.; Moschen, A.R. Short- and Long-Term Effects of a Prebiotic Intervention with Polyphenols Extracted from European Black Elderberry—Sustained Expansion of *Akkermansia* spp. *J. Pers. Med.* **2022**, *12*, 1479. [CrossRef]
23. Cao, H.; Saroglu, O.; Karadag, A.; Diaconeasa, Z.; Zoccatelli, G.; Conte-Junior, C.A.; Gonzalez-Aguilar, G.A.; Ou, J.; Bai, W.; Zamarioli, C.M.; et al. Available technologies on improving the stability of polyphenols in food processing. *Food Front.* **2021**, *2*, 109–139. [CrossRef]
24. Sánchez-Velázquez, O.A.; Mulero, M.; Cuevas-Rodríguez, E.O.; Mondor, M.; Arcand, Y.; Hernández-Álvarez, A.J. In vitro gastrointestinal digestion impact on stability, bioaccessibility and antioxidant activity of polyphenols from wild and commercial blackberries (*Rubus* spp.). *Food Funct.* **2021**, *12*, 7358–7378. [CrossRef]
25. Ștefănescu, B.E.; Nemes, S.A.; Teleky, B.E.; Călinoiu, L.F.; Mitrea, L.; Martău, G.A.; Szabo, K.; Mihai, M.; Vodnar, D.C.; Crișan, G. Microencapsulation and Bioaccessibility of Phenolic Compounds of *Vaccinium* Leaf Extracts. *Antioxidants* **2022**, *11*, 674. [CrossRef]
26. Młynarczyk, K.; Walkowiak-Tomczak, D.; Łysiak, G.P. Bioactive properties of *Sambucus nigra* L. As a functional ingredient for food and pharmaceutical industry. *J. Funct. Foods* **2018**, *40*, 377–390. [CrossRef] [PubMed]
27. Glevitzky, I.; Dumitrel, G.A.; Glevitzky, M.; Pasca, B.; Otrisal, P.; Bungau, S.; Cioca, G.; Pantis, C.; Popa, M. Statistical analysis of the relationship between antioxidant activity and the structure of flavonoid compounds. *Rev. Chim.* **2019**, *70*, 3103–3107. [CrossRef]
28. Chiang, Y.C.; Chen, C.L.; Jeng, T.L.; Lin, T.C.; Sung, J.M. Bioavailability of cranberry bean hydroalcoholic extract and its inhibitory effect against starch hydrolysis following in vitro gastrointestinal digestion. *Food Res. Int.* **2014**, *64*, 939–945. [CrossRef]
29. Szabo, K.; Teleky, B.-E.; Ranga, F.; Simon, E.; Pop, O.L.; Babalau-Fuss, V.; Kapsalis, N.; Vodnar, D.C. Bioaccessibility of microencapsulated carotenoids, recovered from tomato processing industrial by-products, using in vitro digestion model. *LWT-Food Sci. Technol.* **2021**, *152*, 112285. [CrossRef]
30. Imenšek, N.; Kristl, J.; Šumenjak, T.K.; Ivančič, A. Antioxidant activity of elderberry fruits during maturation. *Agriculture* **2021**, *11*, 555. [CrossRef]
31. Hearst, C.; Mccollum, G.; Nelson, D.; Ballard, L.M.; Millar, B.C.; Goldsmith, C.E.; Rooney, P.J.; Loughrey, A.; Moore, J.E.; Rao, J.R. Antibacterial activity of elder (*Sambucus nigra* L.) flower or berry against hospital pathogens. *J. Med. Plants Res.* **2010**, *4*, 1805–1809. [CrossRef]
32. Mohammadsadeghi, S.; Malekpour, A.; Zahedi, S.; Eskandari, F. The antimicrobial activity of elderberry (*Sambucus nigra* L.) extract against gram positive bacteria, gram negative bacteria and yeast. *Res. J. Appl. Sci.* **2013**, *8*, 240–243.
33. Krawitz, C.; Mraheil, M.A.; Stein, M.; Imirzalioglu, C.; Domann, E.; Pleschka, S.; Hain, T. Inhibitory activity of a standardized elderberry liquid extract against clinically-relevant human respiratory bacterial pathogens and influenza A and B viruses. *BMC Complement. Altern. Med.* **2011**, *11*, 16. [CrossRef] [PubMed]
34. Konečná, M.; Sedlák, V.; Tkáčiková, L.; Kšonžeková, P.; Mydlárová-Blaščáková, M.; Gruľová, D.; Gaľová, J.; Gogaľová, Z.; Babejová, A.; Vašková, H.; et al. Inhibition of the growth of gram-negative bacteria by anthocyanins of berries fruits. *Sci. Bull. Uzhhorod Univ. Biol. Ser.* **2019**, 42–47. [CrossRef]
35. Mitrea, L.; Ranga, F.; Fetea, F.; Dulf, F.V.; Rusu, A.; Trif, M.; Vodnar, D.C. Biodiesel-derived glycerol obtained from renewable biomass-A suitable substrate for the growth of *Candida zeylanoides* yeast strain ATCC 20367. *Microorganisms* **2019**, *7*, 265. [CrossRef]
36. Trofa, D.; Gácser, A.; Nosanchuk, J.D. *Candida parapsilosis*, an emerging fungal pathogen. *Clin. Microbiol. Rev.* **2008**, *21*, 606–625. [CrossRef] [PubMed]
37. Brodkorb, A.; Egger, L.; Alminger, M.; Alvito, P.; Assunção, R.; Ballance, S.; Bohn, T.; Bourlieu-Lacanal, C.; Boutrou, R.; Carrière, F.; et al. INFOGEST static in vitro simulation of gastrointestinal food digestion. *Nat. Protoc.* **2019**, *14*, 991–1014. [CrossRef]

38. Bohn, T.; Carriere, F.; Day, L.; Deglaire, A.; Egger, L.; Freitas, D.; Golding, M.; Le Feunteun, S.; Macierzanka, A.; Menard, O.; et al. Correlation between in vitro and in vivo data on food digestion. What can we predict with static in vitro digestion models? *Crit. Rev. Food Sci. Nutr.* **2018**, *58*, 2239–2261. [CrossRef] [PubMed]
39. Bermúdez-Soto, M.J.; Tomás-Barberán, F.A.; García-Conesa, M.T. Stability of polyphenols in chokeberry (*Aronia melanocarpa*) subjected to in vitro gastric and pancreatic digestion. *Food Chem.* **2007**, *102*, 865–874. [CrossRef]
40. Del Bò, C.; Ciappellano, S.; Klimis-Zacas, D.; Daniela, M.; Claudio, G.; Riso, P.; Porrini, M. Anthocyanin absorption, metabolism, and distribution from a wild blueberry-enriched diet (*Vaccinium angustifolium*) is affected by diet duration in the sprague-dawley rat. *J. Agric. Food Chem.* **2010**, *58*, 2491–2497. [CrossRef]
41. Woodward, G.; Kroon, P.; Cassidy, A.; Kay, C. Anthocyanin stability and recovery: Implications for the analysis of clinical and experimental samples. *J. Agric. Food Chem.* **2009**, *57*, 5271–5278. [CrossRef]
42. Kay, C.D.; Mazza, G.; Holub, B.J. Anthocyanins exist in the circulation primarily as metabolites in adult men. *J. Nutr.* **2005**, *135*, 2582–2588. [CrossRef] [PubMed]
43. Choi, J.H.; Lee, H.J.; Kim, Y.S.; Yeo, S.H.; Kim, S. Effects of *Maclura tricuspidata* (Carr.) Bur fruits and its phytophenolics on obesity-related enzymes. *J. Food Biochem.* **2020**, *44*, e13110. [CrossRef] [PubMed]
44. Han, X.; Li, M.; Sun, L.; Liu, X.; Yin, Y.; Hao, J.; Zhang, W. p-Hydroxybenzoic Acid Ameliorates Colitis by Improving the Mucosal Barrier in a Gut Microbiota-Dependent Manner. *Nutrients* **2022**, *14*, 5383. [CrossRef] [PubMed]
45. Wang, X.N.; Wang, K.Y.; Zhang, X.S.; Yang, C.; Li, X.Y. 4-Hydroxybenzoic acid (4-HBA) enhances the sensitivity of human breast cancer cells to adriamycin as a specific HDAC6 inhibitor by promoting HIPK2/p53 pathway. *Biochem. Biophys. Res. Commun.* **2018**, *504*, 812–819. [CrossRef]
46. McDougall, G.J.; Dobson, P.; Smith, P.; Blake, A.; Stewart, D. Assessing potential bioavailability of raspberry anthocyanins using an in vitro digestion system. *J. Agric. Food Chem.* **2005**, *53*, 5896–5904. [CrossRef]
47. Gil-Izquierdo, A.; Zafrilla, P.; Tomás-Barberán, F.A. An in vitro method to simulate phenolic compound release from the food matrix in the gastrointestinal tract. *Eur. Food Res. Technol.* **2002**, *214*, 155–159. [CrossRef]
48. Liu, G.; Ying, D.; Guo, B.; Cheng, L.J.; May, B.; Bird, T.; Sanguansri, L.; Cao, Y.; Augustin, M. Extrusion of apple pomace increases antioxidant activity upon: In vitro digestion. *Food Funct.* **2019**, *10*, 951–963. [CrossRef]
49. Precup, G.; Pocol, C.B.; Teleky, B.-E.; Vodnar, D.C. Awareness, Knowledge, and Interest about Prebiotics–A Study among Romanian Consumers. *Int. J. Environ. Res. Public Health* **2022**, *19*, 1208. [CrossRef]
50. Hill, C.; Guarner, F.; Reid, G.; Gibson, G.R.; Merenstein, D.J.; Pot, B.; Morelli, L.; Canani, R.B.; Flint, H.J.; Salminen, S.; et al. Expert consensus document: The international scientific association for probiotics and prebiotics consensus statement on the scope and appropriate use of the term probiotic. *Nat. Rev. Gastroenterol. Hepatol.* **2014**, *11*, 506–514. [CrossRef]
51. Plamada, D.; Vodnar, D.C. Polyphenols—Gut Microbiota Interrelationship: A Transition to a New Generation of Prebiotics. *Nutrients* **2022**, *14*, 137. [CrossRef]
52. Fuke, N.; Nagata, N.; Suganuma, H.; Ota, T. Regulation of gut microbiota and metabolic endotoxemia with dietary factors. *Nutrients* **2019**, *11*, 2277. [CrossRef]
53. Mitrea, L.; Nemes, S.-A.; Szabo, K.; Teleky, B.-E.; Vodnar, D.-C. Guts Imbalance Imbalances the Brain: A Review of Gut Microbiota Association With Neurological and Psychiatric Disorders. *Front. Med.* **2022**, *9*, 81324. [CrossRef] [PubMed]
54. Simon, E.; Călinoiu, L.F.; Mitrea, L.; Vodnar, D.C. Probiotics, prebiotics, and synbiotics: Implications and beneficial effects against irritable bowel syndrome. *Nutrients* **2021**, *13*, 2112. [CrossRef] [PubMed]
55. Li, J.; Wu, T.; Li, N.; Wang, X.; Chen, G.; Lyu, X. Bilberry anthocyanin extract promotes intestinal barrier function and inhibits digestive enzyme activity by regulating the gut microbiota in aging rats. *Food Funct.* **2019**, *10*, 333–343. [CrossRef] [PubMed]
56. Rodríguez-Daza, M.C.; Pulido-Mateos, E.C.; Lupien-Meilleur, J.; Guyonnet, D.; Desjardins, Y.; Roy, D. Polyphenol-Mediated Gut Microbiota Modulation: Toward Prebiotics and Further. *Front. Nutr.* **2021**, *8*, 689456. [CrossRef]
57. Teleky, B.-E.; Mitrea, L.; Plamada, D.; Nemes, S.A.; Călinoiu, L.-F.; Pascuta, M.S.; Varvara, R.-A.; Szabo, K.; Vajda, P.; Szekely, C.; et al. Development of Pectin and Poly(vinyl alcohol)-Based Active Packaging Enriched with Itaconic Acid and Apple Pomace-Derived Antioxidants. *Antioxidants* **2022**, *11*, 1729. [CrossRef]
58. Tena, N.; Martín, J.; Asuero, A.G. State of the art of anthocyanins: Antioxidant activity, sources, bioavailability, and therapeutic effect in human health. *Antioxidants* **2020**, *9*, 451. [CrossRef]
59. Gerasimenko, I.; Sheludko, Y.; Unger, M.; Stöckigt, J.; Arnao, M.B. Estimation of free radical-quenching activity of leaf pigment extracts. *Phytochem. Anal.* **2001**, *12*, 138–143. [CrossRef]
60. Munteanu, I.G.; Apetrei, C. Analytical methods used in determining antioxidant activity: A review. *Int. J. Mol. Sci.* **2021**, *22*, 3380. [CrossRef]
61. Benzie, I.F.F.; Strain, J.J. Ferric reducing/antioxidant power assay: Direct measure of total antioxidant activity of biological fluids and modified version for simultaneous measurement of total antioxidant power and ascorbic acid concentration. *Methods Enzymol.* **1999**, *299*, 15–27. [CrossRef] [PubMed]
62. Gulcin, İ. Antioxidants and antioxidant methods: An updated overview. *Arch. Toxicol.* **2020**, *94*, 651–715. [CrossRef] [PubMed]
63. Apak, R.; Güçlü, K.; Özyürek, M.; Karademir, S.E. Novel total antioxidant capacity index for dietary polyphenols and vitamins C and E, using their cupric ion reducing capability in the presence of neocuproine: CUPRAC method. *J. Agric. Food Chem.* **2004**, *52*, 7970–7981. [CrossRef]

64. Stefănescu, B.-E.; Călinoiu, L.F.; Ranga, F.; Fetea, F.; Mocan, A.; Vodnar, D.C.; Crisan, G. The Chemical and Biological Profiles of Leaves from Commercial Blueberry Varieties. *Plants* **2020**, *9*, 1193. [CrossRef] [PubMed]
65. Semeniuc, C.A.; Pop, C.R.; Rotar, A.M. Antibacterial activity and interactions of plant essential oil combinations against Gram-positive and Gram-negative bacteria. *J. Food Drug Anal.* **2017**, *25*, 403–408. [CrossRef]
66. Bogdan, M.A.; Bungau, S.; Tit, D.M.; Zaha, D.C.; Nechifor, A.C.; Behl, T.; Chambre, D.; Lupitu, A.I.; Copolovici, L.; Copolovici, D.M. Chemical profile, antioxidant capacity, and antimicrobial activity of essential oils extracted from three different varieties (Moldoveanca 4, vis magic 10, and alba 7) of *Lavandula angustifolia*. *Molecules* **2021**, *26*, 4381. [CrossRef]
67. Vică, M.L.; Glevitzky, M.; Tit, D.M.; Behl, T.; Heghedűş-Mîndru, R.C.; Zaha, D.C.; Ursu, F.; Popa, M.; Glevitzky, I.; Bungău, S. The antimicrobial activity of honey and propolis extracts from the central region of Romania. *Food Biosci.* **2021**, *41*, 101014. [CrossRef]
68. Mitrea, L.; Călinoiu, L.-F.; Precup, G.; Bindea, M.; Rusu, B.; Trif, M.; Ferenczi, L.-J.; Ştefănescu, B.-E.; vodnar, D.C. Inhibitory Potential of *Lactobacillus plantarum* on *Escherichia coli*. *Bull. Univ. Agric. Sci. Veter-Med. Cluj-Napoca. Food Sci. Technol.* **2017**, *74*, 99–101. [CrossRef]

molecules

MDPI

Review

Blackthorn—A Valuable Source of Phenolic Antioxidants with Potential Health Benefits

Oana-Raluca Negrean [1], Anca Corina Farcas [1,*], Oana Lelia Pop [1,2,*] and Sonia Ancuta Socaci [1,3]

1. Department of Food Science, University of Agricultural Science and Veterinary Medicine of Cluj-Napoca, 400372 Cluj-Napoca, Romania
2. Molecular Nutrition and Proteomics Lab, CDS3, Life Science Institute, University of Agricultural Science and Veterinary Medicine of Cluj-Napoca, 400372 Cluj-Napoca, Romania
3. Life Science Institute, University of Agricultural Science and Veterinary Medicine of Cluj-Napoca, 400372 Cluj-Napoca, Romania

* Correspondence: anca.farcas@usamvcluj.ro (A.C.F.); oana.pop@usamvcluj.ro (O.L.P.)

Abstract: *Prunus spinosa* L. fruit, commonly known as blackthorn, is a rich source of bioactive compounds, including flavonoids, anthocyanins, phenolic acids, vitamins, minerals, and organic acids, which exhibit significant antioxidant and antibacterial properties. Notably, flavonoids such as catechin, epicatechin, and rutin have been reported to have protective effects against diabetes, while other flavonoids, including myricetin, quercetin, and kaempferol, exhibit antihypertensive activity. Solvent extraction methods are widely used for the extraction of phenolic compounds from plant sources, owing to their simplicity, efficacy, and broad applicability. Furthermore, modern extraction techniques, such as microwave-assisted extraction (MAE) and ultrasound-assisted extraction (UAE), have been employed to extract polyphenols from *Prunus spinosa* L. fruits. This review aims to provide a comprehensive analysis of the biologically active compounds found in blackthorn fruits, emphasizing their direct physiological effects on the human body. Additionally, the manuscript highlights the potential applications of blackthorn fruits in various industries, including the food, cosmetics, pharmaceutical, and functional product sectors.

Keywords: antioxidant; antibacterial properties; bioactivities; blackthorn; *Prunus spinosa* L.

1. Introduction

Widely cultivated in New Zealand, Tasmania, and eastern North America, *Prunus spinosa* L., known as blackthorn or sloe berry, is a small to medium-sized thorny tree native to Europe, Western Asia, and Northern Africa [1]. The tiny, globose blackthorn fruits have an astringent flavor and a deep purple color. They are usually used to make jams, drinks, and supplements and can be preserved and used to make herbal teas [1–3].

Blackthorn contains a variety of bioactive polyphenolic compounds, including phenolic acids (neo-chlorogenic and caffeic derivatives), flavonoids (rutin, catechins, procyanidins, quercetin, kaempferol), and anthocyanins [4–10], nor-isoprenoid glycosides, and A-type proanthocyanidins [11] along as well as ascorbic acid, carbohydrates, macro and microminerals [2,12,13].

Flavonoids, particularly anthocyanins, are known for their potential health benefits due to their antioxidant properties. Studies have suggested that consuming foods with high antioxidant levels, such as flavonoids, can lower the risk of developing various diseases. Therefore, including anthocyanin-rich foods in one's diet can have a significantly positive impact on human health [14,15].

According to ethnopharmacological investigations, blackthorn fruit has been used to treat external inflammatory conditions in the mouth and throat, diabetes, pneumonia, and diarrhea [16–18]. For many years, humans have used the leaves and fruits of blackthorn in traditional medicine as a laxative and the flowers as a natural vermicidal treatment [19]. It was also mentioned that due to its various bioactive components, a different part of the blackthorn plant has historically been used to stimulate and regulate the menstrual flow and function and treat leucorrhoea, but also for their analgesic, antispasmodic, anti-edema, and antidysenteric properties [20].

Furthermore, a series of recent researches have shown that blackthorn aqueous or alcoholic extracts have significant antioxidant, antibacterial (against both Gram-positive and Gram-negative bacteria), anti-inflammatory activities, and more than that, they have cancer cell growth-inhibiting effects [8,19,21–24].

Figure 1 illustrates the main applications of blackthorn across diverse domains, alongside their potential health-promoting properties, which will be discussed in detail in the following sections.

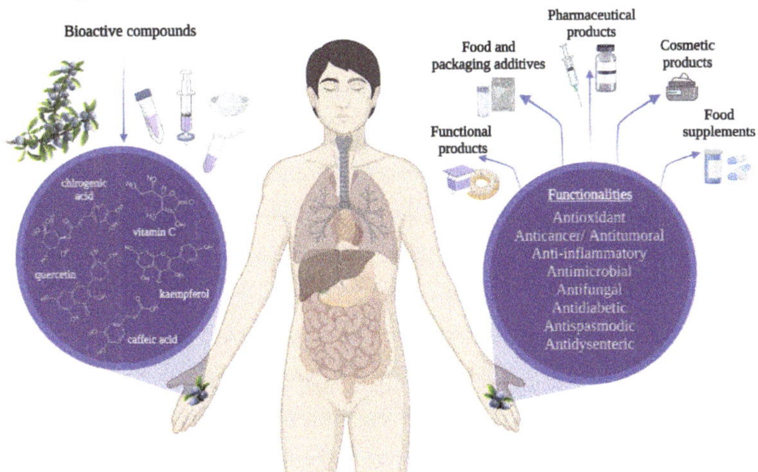

Figure 1. Blackthorn applications and potential health properties.

Blackthorn extracts may be used in various areas, including the food industry, as natural colorants and as preservatives because of the polyphenols' antioxidant and antibacterial properties [8,25–27]. For example, blackthorn fruits added to ice cream samples improved qualities such as color, gumming structure, appearance, and general acceptability [28,29]. Blackthorn fruits are recognized as an excellent source of vitamin C, with numerous studies reporting a value of around 25 mg 100 g^{-1} fw [3,30,31]. Moreover, the plant possesses significant quantities of macro and microelements such as calcium (19.859–34.234 µg/g dw), potassium (1202.822–18,706.98 µg/g dw), iron (3.399–16.18 µg/g dw), sodium (530.11 µg/g dw), zinc (0.350–1.803 µg/g dw), manganese (1.868–4.58 µg/g dw), and magnesium (8.574–11.827 µg/g dw) [2,12]. Băbălău-Fuss et al., 2021 determined that blackthorn oil has oleic acid as a majority fatty acid. Additionally, it was shown in this study that blackthorn oil presents a total concentration of mono and polyunsaturated fatty acids of 80.74% [32]. Because they have a high content of beta-sitosterol, vanillin, and gamma-tocopherols, cold press oils may have various applicability in food, health, and pharmaceutical industries [33,34]. The blackthorn extracts are also suitable for cosmetic use because of their photoprotective effects [35]. At wavelengths between 200 and 400 nm, phenolic compounds present a significant role in UV ray absorption [36].

Another application of blackthorn fruit extracts is in the food industry. Mandic et al. (2018) developed a study on incorporating extract of blackthorn fruits in the natural casing of Kranjska sausages. This study showed that the blackthorn extracts, both aqueous and ethanol, succeeded in reducing the number of lactic acid bacteria from the surface of vacuum-packed sausages that were stored at 4 °C for 60 days [37].

In this paper, our objective is to conduct a comprehensive analysis of the biologically active compounds present in blackthorn fruits, highlighting their primary physiological effects on the human body. Additionally, we aim to emphasize the beneficial properties of blackthorn fruits in their potential application across various industries, including the food, cosmetic, and pharmaceutical sectors, as well as in the development of functional products.

The Methodological Approach

The methodology employed in this bibliographic research is a narrative review, which entails conducting a thorough search of pertinent literature related to the research topic, followed by an evaluative assessment of the identified studies. The aim is to synthesize the existing evidence to highlight the potential applications of blackthorn. The review draws information from academic databases, specifically PubMed, Scopus, and Web of Science, while excluding grey literature sources such as conference proceedings, institutional repositories, and government reports. The studies were selected based on their relevance to the research topic, the quality of their design and methodology, and the availability of data. Studies that did not meet these criteria, such as those not relevant to the research topic, lacking a control group, or not written in English, were excluded. The search strategy utilized a combination of keywords such as "blackthorn", "*Prunus spinosa*", "antioxidants", "flavonoids", "polyphenols", "health benefits", "anti-inflammatory", "anti-cancer", and "immune system", along with Boolean operators.

2. Bioactive Compounds

The blackthorn fruit is a rich source of components with antioxidant and antibacterial characteristics, including flavonoids, anthocyanins, phenolic acids, vitamins, minerals, and organic acids [29,38]. Among these components, polyphenols are particularly noteworthy for their strong contribution to the antioxidant capacity of fruits and vegetables compared to other antioxidant compounds, such as vitamins C and E or carotenoids [39–41]. Phenolic phytochemicals are also of major impact on defense responses, including anti-aging, anti-inflammatory, antioxidant, and antiproliferative effects [42]. Therefore, the abundance of polyphenols and other antioxidants in blackthorn fruit highlights its potential as a valuable resource for preventing oxidative stress and related health issues.

The most common flavonoids in these fruits are quercetin and its glycosides, such as rutin, which are found in higher concentrations in the peel [43]. In previous studies, substances such as quercetin, cyanidin, caffeic acids, catechin, rutin, epicatechin, kaempferol, gallic acid, chlorogenic acid, syringic acid, vanillic acid, ferulic acid, p-coumaric acid were found in blackthorn [2,10,19,31,44]. Gallic acid has powerful anti-oxidative, anti-inflammatory, antibacterial, antiviral, anti-melanogenic, antimutagenic, and anticancer properties [45–47]. The most representative phenolic compounds found in blackthorn fruits and their quantities are displayed in Figure 2.

It is well known that quercetin glucoside bioavailability is significantly higher than quercetin rutinoside bioavailability, indicating that the small intestine actively absorbs the glucosides [44]. Due to its complete structural functionality and high radical scavenging capacity, quercetin functions as a potent antioxidant [48,49]. Caffeoylquinic acid and quercetin were the dominant compounds in Marcetic et al.'s (2022) study of phenols in blackthorn fruit extracts [50]. The primary phenolic acids found in blackthorn fruits are neochlorogenic and chlorogenic acid (3-caffeoylquinic acid) [43,51–55]. The only flavonoid found by Najgebauer-Lejko et al. (2021) in blackthorn puree was myricetin. They also remarked that chlorogenic (144.98 mg/100 g DW) and caffeic acids (100.39 mg/100 g DW) were the most abundant compounds among phenolic acids [56].

Previously, researchers from Northeastern Portugal discovered that among three samples of wild fruits, strawberry tree, blackthorn, and wild rose, blackthorn fruits were the only ones that contained phenolic acids from the hydroxycinnamic acid derivate subgroup. Blackthorn fruits had the highest levels of phenolic acids (29.78 mg/100 g), flavone/ols (57.48 mg/100 g), and anthocyanins (100.40 lg/100 g), despite the absence of flavan-3-ols compounds [5].

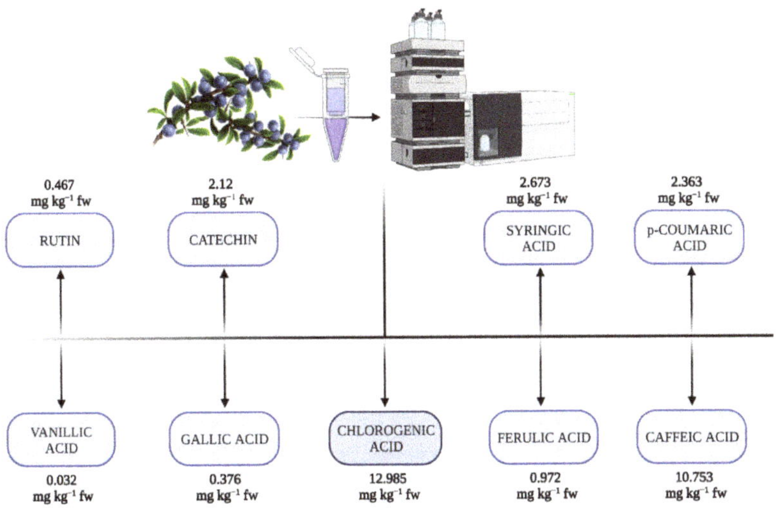

Figure 2. The main phenolic compounds from blackthorn fruits (adapted from [31]).

Mechchate et al. (2021) reported that catechin, epicatechin, and rutin were found to be potent antihypertensive flavonoid agents [57]. In addition, rutin, myricetin, quercetin, and kaempferol protect against diabetes [58]. In isolated arteries, quercetin and similar flavonoids have a vasodilation action that is both endothelium-dependent and independent [59,60]. Blackthorn fruits have an astringent flavor because of the tannins that are present there. Like many other polyphenols, tannins have various biological effects, including anti-inflammatory, anti-carcinogenic, anti-microbial, and protective of the cardiovascular system [61–63].

The colorful fruits are a rich source of antioxidant compounds such as phenolic pigments and anthocyanins, which have numerous health benefits [64,65]. The high anthocyanin concentration gives blackthorn (*Prunus spinosa* L.) fruit its dark blue hue. Anthocyanins are a sizable subclass of flavonoids that produce vibrant shades ranging from scarlet to blue or red to orange [66]. Cyanidin is the main anthocyanin found in stone fruits belonging to the Prunus genus. Anthocyanins' advantages for cardiovascular health have been connected to their capacity to fend off oxidative stress [67]. Katanic Stankovic et al. (2022) found a substantial amount of anthocyanins (461.27 mg cyanidin-3-glucoside eq/kg fresh fruit) in an ethyl acetate extract of Serbian blackthorn fruits, equivalent to aronia. The HPLC analysis was used to identify phenolic acids and anthocyanins from investigated extracts [68]. Sabatini et al. (2020) took into account the hypothesis that the great efficacy in the anti-microbial effect of blackthorn fruits may be due to anthocyanins or the combined action of bioactive compounds [69]. The anthocyanin components cyanidin-3-*O*-glucoside, cyanidin-3-*O*-rutinoside, and peonidin-3-*O*-glucoside were found in the aqueous extract by Veličković et al. (2014), while phenolic acids (neochlorogenic and caffeic acid) and flavonoids were present in the ethanol and ethanol-aqueous extracts (myricetin and quercetin) [10]. Mineral or organic acid-containing solvents are commonly used to extract anthocyanins from plant organs [70]. Anthocyanins are likely substantial contributors to the antioxidant activities of blackthorn leaves, as evidenced by correlations between quantities of phenols, flavonoids, and anthocyanins and values obtained for antioxidant activity established in a previous study [71].

According to Wang et al. (2022), anthocyanins could prevent the growth of dangerous bacteria by disrupting respiratory metabolism and inhibiting gene expression. Furthermore, by stimulating vital enzyme activities, the presence of anthocyanins may promote the growth of probiotics [72].

Pinacho et al. (2015) conducted a study to evaluate the phenolic compounds of blackthorn and their antioxidant capacity, specifically examining the impact of in vitro digestion. The researchers found that branches of the plant exhibited a higher antioxidant capacity compared to other plant components (leaves and fruits). To investigate the effect of digestion, an ethanolic extract of blackthorn branches was subjected to in vitro oral, gastric, and intestinal digestion. Results indicated that no significant changes occurred during oral or gastric digestion, suggesting the stability of the phenolic compounds under these conditions. However, intestinal digestion resulted in extensive degradation of the chemicals, likely due to the alkaline conditions in the small intestine. The researchers observed the formation of various structural forms with significant antioxidant activity, some of which were previously unknown or unrecognized. Based on their findings, the researchers concluded that the blackthorn extracts could potentially serve as a valuable resource for preventing oxidative stress-related illnesses [73].

3. Extraction Methods

The chemical nature of the phenolic compounds extracted from plant parts, the sample matrix structure, the time of extraction, the storage condition, and the presence of any interfering substances are influencing more or less the extraction efficiency. Plant phenolic extracts are a mixture of different classes of phenols that are selectively soluble in different solvents [74]. Because of their ease of use, efficiency, and broad applicability, solvent extractions are the most used procedures to extract phenolic compounds from plant sources and free them from the vacuolar structures where they are found [75]. The type of solvent used and how polar it is, the temperature and length of the extraction process, as well as the chemical and physical characteristics of the materials being extracted, all affect chemical extraction [75]. For phenolic extraction, typical solvents include water, acetone, methanol, ethanol, *N*,*N*-dimethylformamide (DMF), or combinations of these substances in water. Their extraction efficiency has been studied due to polarity differences [76,77]. Methanol, for example, has some disadvantages, such as potentially hazardous effects on human health. Its residues may remain in the final product, requiring additional time-consuming purification steps that influence the final cost of the process. On the other hand, if for extraction are used pure organic solvents such as benzoic or cinnamic acid, extraction may not be completed [70].

Additionally, modern extraction methods have recently been used to facilitate the isolation of polyphenolic compounds (accelerated solvent extraction, microwave-assisted extraction, and supercritical fluid extraction) [78]. The main techniques currently used for the extraction of phenolic compounds from blackthorn, as well as the factors that influence the recovery yield, are summarized in Figure 3.

High-performance liquid chromatography (HPLC) and ultra-high-performance liquid chromatography (UHPLC) are the most frequently used methods for polyphenol separation. Liquid chromatography is often combined with mass spectrometry (LC-MS) for more selective and reliable compound identification, as reported by Rajbhar et al. (2015). Also, thin-layer chromatography (TLC), capillary electrophoresis (CE), and gas chromatography (GC) can be applied for separating individual polyphenolics from fruit samples [74,79].

Besides the importance of quantification of individual polyphenols, total phenolic content, and methods for determining antioxidant activity are important. The Folin-Ciocalteu assay is used to determine total phenolic content (TPC), with the calibration curve typically constructed using gallic acid as a standard [65,79].

Quantifying polyphenolic compounds is a challenging and complex process that requires careful consideration of various factors. In order to better understand the extraction of phenols from blackthorn, several studies have been conducted to evaluate the working parameters and extraction yield. Therefore, some studies related to the extraction of phenols from blackthorn, as well as the working parameters and the extraction yield are summarized in Table 1.

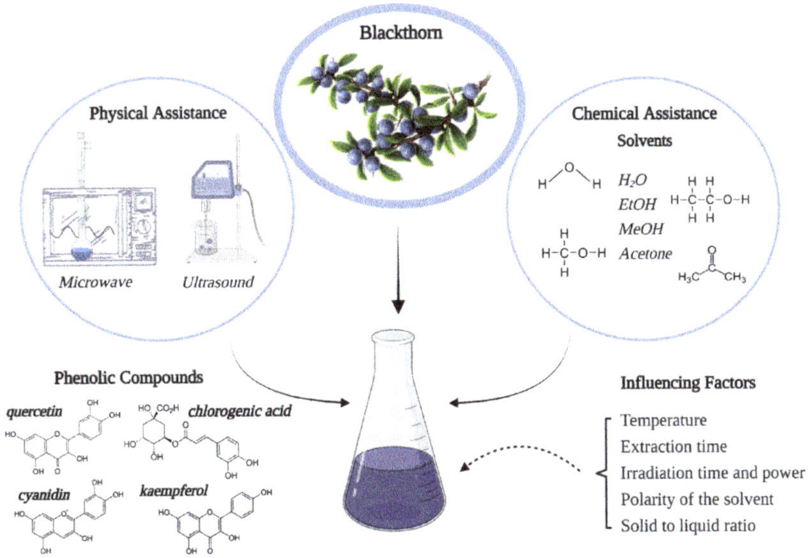

Figure 3. The main extraction techniques and factors affecting recovery yield of phenolic compounds from blackthorn.

Modern extraction methods such as microwave-assisted extraction (MAE) and ultrasound-assisted extraction (UAE) are often used to increase the extraction yield of polyphenols from plants. UAE can improve mass transfer, cell disruption, penetration, and capillary effects [74,80]. In addition, MAE's benefits require little or no solvent, reduce extraction time, and are environmentally friendly [81,82].

Table 1. The extraction of phenols from different blackthorn parts and the applied protocols.

Plant Part	Conditions and Effectiveness of Extraction	Antioxidant Content			References
		TPC	TFC	TAC	
SOLVENT EXTRACTION					
Leaves	Distilled water Yield: 13.65%	142.40 ± 3.82 mg GAE/g dw	36.28 ± 0.41 mg QE/g dw	-	[71]
	Ethanol 96% Yield: 9.14%	116.63 ± 1.62 mg GAE/g dw	45.52 ± 0.9 mg QE/g dw	-	
	Acetone Yield: 4.36%	181.19 ± 1.70 mg GAE/g dw	80.10 ± 0.00 mg QE/g dw	-	
Fruit	Distilled water Yield: 18.45%	23.19 ± 2.52 mg GAE/g dw	2.96 ± 0.22 mg QE/g dw	14.00 μg/g dw	[83]
	Ethanol 96% Yield: 11.09%	19.98 ± 1.28 mg GAE/g dw	3.07 ± 0.27 mg QE/g dw	9.00 μg/g dw	
	Acetone Yield: 9.4%	26.78 ± 4.44 mg GAE/g dw	2.89 ± 0.36 mg QE/g dw	23.00 μg/g dw	
	Methanol/Water (50:50, pH 2) Acetone/Water (70:30)	37.97 ± 0.30 mg GAE/g fw	2.26 ± 0.15 mg RUE/g fw	25.85 ± 1.51 mg pelargonidin 3-glucoside/g fw	[30]

Table 1. *Cont.*

Plant Part	Conditions and Effectiveness of Extraction	Antioxidant Content			References
Flower	Methanol/Water (7:3, *v/v*)	206.07 ± 10.86 mg GAE/g dw	125.12 ± 0.55 mg/g dw	45.13 ± 2.38 mg CYE/g dw	[84]
	Diethyl ether Yield: 1.23 g dw	464.57 ± 20.57 mg GAE/g dw	490.63 ± 8.16 mg/g dw	49.5 ± 2.23 mg CYE/g dw	
	Ethyl acetate Yield: 4.00 g dw	584.07 ± 12.98 mg GAE/g dw	325.53 ± 4.23 mg/g dw	109.43 ± 3.71 mg CYE/g dw	
	n-butanol Yield: 4.86 g dw	296.57 ± 3.28 mg GAE/g dw	241.27 ± 4.74 mg/g dw	46.6 ± 1.14 mg CYE/g dw	
	Water residue Yield: 13.08	64.6 ± 1.93 mg GAE/g dw	1.88 ± 0.04 mg/g dw	12.43 ± 0.25 mg CYE/g dw	
		MICROWAVE-ASSISTED EXTRACTION			
Flowers	Ethanol 1 min 60 °C	54.45 ± 0.12 mg GAE/g dw Borije	1.547 ± 0.001 mg QE/g dw Vareš	0.339 ± 0.063 mg CGE/g dw Trnovo	[2]
Leaves	Ethanol 1 min 60 °C	17.78 ± 0.10 mg GAE/g dw Borije	0.479 ± 0.001 mg QE/g dw Vareš	1.353 ± 0.060 mg CGE/g dw Trnovo	
Fruits	Ethanol 1 min 60 °C	6.87 ± 0.01 mg GAE/g dw Borije	0.149 ± 0.001 mg QE/g dw Vareš	0.746 ± 0.092 mg CGE/g dw Trnovo	
		ULTRASOUND-ASSISTED EXTRACTION			
Fruit	Ethanol 75% Ultrasound bath: 240 W, 35 kHz 25 °C for 30 min	25.9 ± 0.2 mg GAE/g fw	5.09 ± 0.12 mg RUE/g fw	0.16 ± 0.001 mg Mv-3- glc/g fw	[68]
	Ethanol 40% Ultrasound bath: 95 W, 35 kHz 67 °C for 10 min	1.02 mg GAE/g dw	-	-	[35]
	Ethanol Ultrasound bath: 20 min	4.116 ± 0.003 mg GAE/g dw Borije	0.064 ± 0.001 mg QE/g dw Vareš	1.258 ± 0.029 mg CGE/g dw Trnovo	[2]
Flowers	Ethanol Ultrasound bath: 20 min	24.41 ± 0.03 mg GAE/g dw Borije	0.677 ± 0.001 mg QE/g dw Vareš	0.718 ± 0.058 mg CGE/g dw Trnovo	
Leaves	Ethanol Ultrasound bath: 20 min	8.31 ± 0.03 mg GAE/g dw Borije	0.282 ± 0.001 mg QE/g dw Vareš	1.364 ± 0.03 mg CGE/g dw Trnovo	

TPC—Total Phenolic Content; TFC—Total Flavonoid Content; TAC—Total Anthocyanin Content;. GAE—gallic acid equivalent; QE—quercetin equivalent; CYE—cyanidine chloride equivalents; CGE—cyanidin-3-glucoside equivalents; RUE—rutin equivalent; Mv-3-glc—malvidin-3-glucoside; dw—dry weight; fw—fresh weight.

The studies detailed in Table 1 highlight that ethanol is the most commonly used solvent to extract chemical compounds from different parts of blackthorn. However, it has been demonstrated that phenolic compounds are extracted more effectively in acetone than in ethanol. For example, the fruit TPC of the acetone extract was 26.78 ± 4.44 mg GAE/g, while for the ethanol extract was 19.98 ± 1.28 mg GAE/g [83]. On the other hand, despite the higher TPC in the fruit acetone extract, the ethanolic extract showed higher in vitro antioxidant and antibacterial properties, as well as a potential antidiabetic action. Additionally, the total phenolic content determined in the ethanolic extract using conventional methods showed a higher content (19.98 ± 1.28 mg GAE/g [83]) than microwave-assisted extraction (6.87 ± 0.01 mg GAE/g [2]).

4. The Potential Effects of Bioactive Compounds

4.1. Antioxidant Activity

Antioxidants are commonly integrated into preservation procedures to enhance stability by mitigating the deleterious impact of reactive oxygen species on products as well as in the human body [85,86]. Anthocyanins, which are found in blackthorn, comprise one of the main classes of water-soluble flavonoids that contribute to flavonoids' antioxidant activities [87]. Wang and Lin (2000) discovered a strong relationship between antioxidant capacity (oxygen radical absorbance capacity, ORAC), anthocyanins (pH differential method), and total phenols (spectrophotometric method, using Folin-Ciocalteu reagent) [88]. Fraternale et al. (2009) researched the antioxidant activity of blackthorn fruit juice. They compared the antioxidant activity of blackthorn juice with grape juice discovering that the antioxidant of blackthorn juice was higher than the antioxidant activity presented by grape juice (55.13 mg/g DW, respectively 1.15 mg/g DW) [89]. After evaluation of the Folin–Ciocalteu method, FRAP, and ABTS assays, blackthorn extracts showed significant high correlations between antioxidant activities and total phenolic compounds, supporting the theory that polyphenols are the primary contributor to the antioxidant activity of blackthorn fruits [30].

Opriș et al. (2021) studied the antioxidant capacity of blackthorn fruits to integrate the concentrated extracts into the sunscreen formula. The highest content of polyphenols was extracted with ethanol at 40%. Ethanol extract presented antioxidant and UV-protective activities. Blackthorn fruit extracts were applied to two different samples from western and central Serbia. The first sample had a considerably greater total phenolic component content as well as stronger antioxidant activity in DPPH, ABTS, and FRAP assays. However, there were no significant differences between the extracts, implying that, in addition to anthocyanins, other compounds contribute to lipid peroxidation inhibition [35,50]. A recent study on blackthorn leaf extract showed that acetone extract was the most effective in scavenging DPPH and ABTS free radicals (44.57 and 16.12 μg mL^{-1}), with acetone revealing the best antioxidant activity. Furthermore, acetone leaf extract was richer in phenols, flavonoids, and anthocyanins than water and ethanol, proving that phenol content and antioxidant activity are closely related [71].

In recent times, another investigation has been conducted to examine the impact of preservation techniques and storage conditions on the antioxidative efficacy of blackthorn phytochemicals. The freeze-drying method showed a significant contribution to increasing the concentration of colored compounds. Although, in terms of antioxidant compound content and bioactivity, lyophilization provided the most favorable storage parameters. Further analysis revealed that the ability of phenolics to reduce iron ions in frozen samples increased during storage [90].

In a study by Ruiz-Rodríguez, De Ancos et al. (2014), the bioactive substances of blackthorn and hawthorn fruits, including vitamin C (ascorbic and dehydroascorbic acids), were evaluated for their antioxidant potential. The results indicated that blackthorn fruits had a total vitamin C content of 11.27 mg/100 g^{-1} fw, with dehydroascorbic acid being the primary contributor. In contrast, hawthorn fruits had a higher total vitamin C content of 30.35 mg/100 g^{-1} fw. Blackthorn fruits also exhibited a higher antioxidant capacity than hawthorn fruits, with the DPPH values indicating a strong correlation with vitamin C and phenolic components. Vitamin C levels in blackthorn fruits were found to be consistent with other study findings, with 20.86 mg/100 g in frozen and 23.84 mg/100 g in fresh raw material. Furthermore, freezing and frozen storage did not significantly alter the antioxidant content of blackthorn fruits [3].

4.2. Anti-Microbial and Antifungal Activities

Consumer demand for healthy food, free of synthetic additives has grown recently, generating interest in using natural anti-microbial agents. Blackthorn is used as a medicine by people in countries like Turkey and Serbia to treat bronchial asthma and nephritis [89]. Gündüz (2013) studied the anti-microbial activity of blackthorn purees (fresh and processed) on *Salmonella* spp., this Gram-negative bacteria being reduced about 3.94 and 2.37 log units in fresh and processed purees at 25 °C for 30 min [26]. In accordance, Smullen et al. (2007) demonstrated that extracts containing polyphenols were effective in inhibiting *Streptococcus* mutants, the blackthorn skin extract showing a minimum inhibitory concentration (MIC) of 2 mg/mL [91].

For a more detailed antibacterial and antifungal activity, Veličković I. et al. (2020) tested the different blackthorn extracts on various microorganisms such as *Bacillus cereus*, *Stapylococcus aureus*, *Listeria monocytogenes*, *Escherichia coli*, *Aspergillus niger*, *Penicillium funiculosum*. According to the findings, the ethanol extract was more effective against the tested bacteria, whereas the aqueous extract demonstrated superior antifungal properties [83]. In 2021, the same research group analyzed the blackthorn leaf extract, doing similar tests for antifungal and antibacterial activities using the microdilution method. The pathogenic bacteria examined responded better to the ethanolic extract, especially *E. cloacae* and *B. cereus* [26,71,91]. In contrast to antibacterial activity, the aqueous sample's antimycotic characteristics were marginally better but still inferior to the positive control (ketoconazole). In both cases, the ethanol and aqueous extracts were chosen due to their non-toxicity. Also, the ethanol extract was richer in flavonoids, which may explain the efficient anti-microbial activity. Furthermore, ethanolic fruit extract obtained by Velickovic et al. (2014) showed anti-microbial activity, using the disc diffusion method, against all tested microorganisms (*S. aureus, E. coli, Salmonella abony*, and *Pseudomonas aeruginosa*) except *Bacillus cereus* [10]. Kumarasamy et al. (2004) confirmed the effectiveness of methanol extracts of blackthorn seeds against *Lactobacillus plantarum, Staphylococcus aureus*, and *Citrobacter freundii*, by using a 96-well microplate-based broth dilution assay [11]. The best result among aerial parts, studied by Dedić et al. (2021), was the leaf extract obtained using MAE, with an inhibition zone of 21.67 mm against *Bacillus subtilis*. However, the fruits from the Trnovo region showed the best results against all the bacterial strains. The aerial parts extracts had the most effective anti-microbial activity against *B. subtilis, S. aureus, E. faecalis*, and *P. aeruginosa* (agar well diffusion method), as well as antifungal activity. The positive controls were streptomycin (10 μg, Oxoid) and respectively antimycotic nystatin (10 μg, Oxoid) [2].

4.3. *Antidiabetic Effect*

Millions of people worldwide suffer from a chronic metabolic condition known as diabetes [22,92]. Based on natural substances, researchers are working to develop a new medication [22]. Because of their variety of benefits, safety, tolerance, and affordability, medicinal herbs like blackthorn may function as an alternative to conventional antidiabetic medications [93]. Genotypes of North Serbian blackthorn, rich in polyphenols with biological activities, were associated with antidiabetic and anti-proliferative effects and antioxidant properties [22]. Also, Stanković et al. (2022) examined the possible anti-diabetic activities of blackthorn ethanolic extracts by testing their ability to inhibit α-glucosidase. As a result, blackthorn extract showed a significant glucosidase inhibitory activity (129.46 ± 0.73). Furthermore, compared to standard anti-diabetic drugs (acarbose), blackthorn extract had a higher IC50 value [34,83].

Moreover, Crnić et al. (2021) examined how blackthorn flower extract affected the glycaemic balance in C57BL/6 mice that were neither normoglycemic nor alloxan-induced hyperglycemic. The data confirmed that short-term (in minutes) or long-term (for 10 days) oral consumption of blackthorn flower extract can slightly but significantly raise blood sugar concentration in metabolically healthy (normoglycemic) mice. Consuming blackthorn flower extract for 10 days reduced blood glucose levels, in C57BL/6 mice, it improved glucose tolerance, boosted insulin production, and reduced serum α-amylase activity, which may have hyperglycemic protective benefits [94].

4.4. *Anti-Inflammatory Effect*

In past studies, different scientists [4,9,10,89,95–97] have demonstrated a significant antioxidant activity of blackthorn extracts by 2,2-diphenyl-1-picrylhydrazyl test, knowing the fact that fruits with a high capacity of radical scavenging are associated with low occurrences of degenerative diseases (inflammation and arthritis) [98]. Blackthorn fruit ethanolic extract was investigated for its wound-healing capacity by Coppari et al. (2021) [99]. The human umbilical vein endothelial cells were mechanically scratched within T25 tissue culture flasks and then subjected to evaluation using a phase contrast microscope. The anti-inflammatory features of fruit extract improved wound healing closure by 70%. Magiera et al. (2022) also studied the anti-inflammatory capacity of blackthorn extracts.

The fruits from Poland were high in polyphenols, anthocyanins, and flavonols, compounds that contribute to antioxidant and anti-inflammatory capacity. The hydroalcoholic extract showed biological effects such as the secretion of key anti-inflammatory factors (IL-10). These findings may help explain the historic use of blackthorn fruits in treating gastrointestinal chronic inflammatory illnesses, given that natural polyphenols accumulate to a significant degree in the intestines [23].

4.5. Anticancer/Antitumoral Effects

A valuable source of phenolic compounds is blackthorn flower, which contains kaempferol, quercetin, kaempferol 3-*O*-L-arabinofuranoside, quercetin 3-*O*-L-arabinofuranoside, kaempferol 3-*O*-L-ramnopyranoside, kaempferol 7-*O*-L-ramnopyranoside, kaempferol 3-*O*-D-x [7,73]. These polyphenols have strong anti-oxidant and pro-oxidant qualities, and they may be used to treat cancer either as a preventative measure (antioxidants) or as a cancer cell killer (prooxidants) [100]. Red grape, blackberry, black cherry, black currant, elderberry, blackthorn, and plum fruit peel polyphenolic extracts led to caspase-dependent cell death in breast cancer MCF-7 cells, which was connected to an increase in oxidative stress and resulted in the release of pro- and anti-apoptotic mitochondrial proteins from the Bcl-2 family [101].

Although Kello et al.'s (2017) focused on the efficiency of blackthorn extracts on breast cancer cells, Murati et al. (2019) exposed human liver cancer cells (HepG2) to various doses of blackthorn flower extract (10–200 g/mL) and measured cytotoxic activity using the neutral red and kenacid blue techniques after 24, 48, and 72 h of incubation to determine its anticancer properties. Significant inhibition of Hep G2 cellular proliferation was observed at concentrations exceeding 50 g/mL, indicating that the blackthorn flower extract may have detrimental effects on human liver cancer cells. The extract's impact was characterized by apoptosis and necrosis, which were most likely a result of heightened oxidative stress levels [101,102].

According to Condello et al. (2019), the extract of blackthorn fruits combined with a nutraceutical activator complex is well tolerated by mice, inhibits the growth of colorectal cancer, and should be considered for incorporation into multi-drug protocols for colon cancer treatment [103]. More specifically, their research demonstrated the anticancer efficacy of Trigno M, an extract of blackthorn enriched in phenolic acids, flavonoids, and anthocyanins, in combination with the nutraceutical activator complex (NAC), in 2D, 3D, on the reduction of tumor growth, through an in vivo colorectal cancer testing protocol [103].

4.6. Other Studies

Balta et al. (2019) investigated tartrazine toxicity development in albino Wistar rats and the protective action of blackthorn fruits. When mixed with benzoates, the synthetic food pigment tartrazine (also known as Yellow 5 or E102) can cause hyperactivity syndrome in children combined with agitation, confusion, and rhinitis [104,105]. The use of tartrazine in groups I, II, and III of rats, led to lesions in all rodents' kidneys, spleens, and livers, according to histopathological analysis [105]. Tartrazine produced histological abnormalities that led to substantial liver tissue lesions and changes in blood parameters [106]. In this sense, blackthorn powder showed a good protective effect on blood parameters but not offering any discernible benefits for the organs [105].

In another study, Balta V. et al. (2021) examined the extracts' individual phenolic components for their bioavailability and absorption. They investigated the effects of phenolics from an ethanol-water extract of blackthorn flowers given orally to C57/BL6 mice for 28 days at doses of 25 mg total phenolics/kg body weight. The gut, liver, and kidney phenolic contents were measured using the UPLC-MS/MS technique after 1, 7, 14, 21, and 28 days of extract administration. After consuming blackthorn flower extract for at least three weeks, the number of phenolics in mice tissues considerably increased. In vivo, the extract promoted anti-oxidative defense pathways in an organ-specific manner and possessed significant bioactive properties. The primary components in the gut were

3-O-feruloylquinic acid, 4-O-p-coumaroylquinic acid, kaempherol pentoside, and quercetin rhamnoside, whereas the main components in the liver and kidneys were quercetin 3-O rutinoside, ferulic acid, and 4-O-p-coumaroylquinic acid [107].

Compared to Balta's study (2021), Dikic et al. (2022) focused on the functionality and bioavailability of polyphenols in blackthorn flower extract (PSE), studying the benefic effects of the extract on the brain. In this regard, a dose of 25 mg polyphenols/kg body weight, was administrated to experimental animals for 28 days. The brain was found to contain a total of 68.7% polyphenols from blackthorn extract. For 59.1% of the chemicals found in the brain, higher (p0.05) C_{max}/AUClast values were observed in the PSE treatment group compared to the control group, showing reasonably good bioaccumulation in the brain. The substances that were most abundant in PSE were not always the ones that were most bioabsorbable in the brain. In contrast to phenolic acids, quercetins, or epigallocatechin-3-gallate, kaempferol were not considerably dispersed. On the 28th day, the chemicals with the highest concentrations were 4-p-coumaroylquinonic acid, (-)-epicatechin, quercetin-3-*O*-rutinoside, quercetin-rhamnoside, kaempherol-3-rutinoside, and quercetin-3-glucoside. Researchers concluded that the compounds they had tested would be good candidates for developing or testing "neuro-nutriceuticals"—polyphenol combinations that are specifically targeted at the brain [108].

5. Applications

Since ancient times, people have valued blackthorn for treating some stomach affections and the flu. In recent years, they have become more significant as researchers examine them and look for potential applications in various industries. According to Fraternale et al. (2009), these fruits may be useful to the food industry, particularly in the sector that develops dietary supplements and nutraceuticals with enhanced bioactive capabilities [89]. In a recent study, Marcetic et al. (2022) investigated the potential prebiotic effect of blackthorn fruits, noting that the findings are in agreement with those of Milutinovic et al. (2021), who claimed that plants rich in polyphenols have prebiotic activity with a significant impact on *Saccharomyces boulardii* yeast [50,109]. The fact that anthocyanins can prevent the proliferation of harmful bacteria by downregulating gene expression, altering metabolic enzymes, and affecting respiratory metabolism, was also confirmed in a recent study conducted by Wang et al. (2022) [72].

The research conducted by Najgebauer-Lejko et al. (2021) was to obtain probiotic yogurt with 10% sweetened purees of elderberries, sea buckthorn, and blackthorn and assess their chemical composition, acidity, polyphenol, and anthocyanin content. The antioxidant strength and antiradical power were also evaluated. Their high amount of polyphenols, particularly anthocyanins, elderberry, and blackthorn significantly increased the antioxidant capacity of probiotic yogurts. As a result, both prototypes have great potential in acting as natural colorants, the final products turning in a deep purple and stable color. On the other hand, blackthorn puree might be a particularly excellent source of dietary fiber [56].

Blackthorn extract rich in phenolic acids and anthocyanins was encapsulated by Blagojevic et al. (2022) using halloysite and maltodextrin. In this way, the release of bioactive compounds was achieved in a controlled manner using yogurt as a support food, and their bioavailability was significantly improved. Also, these findings suggest that using blackthorn extract and its halloysite and maltodextrin encapsulates form can be a sustainable direction to sustain the development of new functional foods, dietary supplements, and nutraceuticals [110].

Other recent research has highlighted the importance of blackthorn fruit extract in products with solar protection properties. Scientists attempted to obtain various blackthorn polyphenol-rich extracts using intensive treatments such as sonication and test them in sunscreen composition [35,36]. Because blackthorn extract has demonstrated strong antioxidant properties; its in vitro sun protection factor and photostability in the developed sunscreen formulation were also investigated. They concluded that the optimized blackthorn extract can be effectively incorporated into cosmetic formulations increasing its photoprotective properties, and the optimized product remaining stable at the end of the stability study (120 min). These promising results open up new possibilities for the use of blackthorn extract in cosmetic products

Mandic et al. (2018) investigated the effect of blackthorn fruit (*Prunus spinosa* L.) extract on the quality of Kranjska sausage. The results showed that vacuum-packed sausage's quality and shelf life were greatly improved when natural casings were treated with an aqueous or ethanol extract of blackthorn fruits. The number of lactic acid bacteria on the exterior surfaces of Kranjska sausage filled in casings that had previously been treated with an aqueous or ethanol extract of blackthorn fruits decreased over 60 days of storage in vacuum packs. They concluded that adding blackthorn fruit extract to the sausage filling, where its impact would be more noticeable, would significantly increase the extract's antioxidant function [37].

For many years, the industrial production of natural-based colorants has consisted primarily of obtaining colorant-rich extracts through the use of conventional heat-assisted extraction with water as a solvent, followed by several drying steps. Procedures used to extract anthocyanins such as ultrasound, microwave, and supercritical fluid-assisted extraction techniques have caught the attention of industrials and researchers recently. Leichtweis et al.'s (2019) objective was to investigate the composition of blackthorn anthocyanins and promote the commercial value of these wild fruits through the development of an anthocyanin-based coloring extract. Moreover, their study aimed to optimize the extraction of anthocyanin compounds from fruit epicarp using methods such as ultrasound extraction and heat-assisted extraction. Ultrasound extraction was shown to be the most effective method, presenting 18.17 mg/g anthocyanin content. By examining extracts rich in anthocyanins that have potential use as natural colorants in several industrial domains, their study helped value blackthorn's wild fruits. Overall, a workable environmentally friendly method was developed that might be used for small-scale study for industrial manufacturing of colorants based on blackthorn anthocyanins [27].

Furthermore, blackthorn fruit extract was evaluated as natural purple colorant used in doughnut icing. Initially, icing presented a dark purple tone, but after 24 h, the product showed a decrease in color intensity. Despite their color decrease, the antioxidant and antimicrobial activities presented a significant increase, ranking blackthorn extracts as promising candidates as natural food preservatives and colorants [25].

6. Conclusions

Blackthorn may be considered a valuable source of bioactive compounds that can maintain human health by preventing or shortening the convalescence period for many disorders due to their antioxidant, anti-inflammatory, antiproliferative, antiviral, and antibacterial properties. Since it has been shown to have health-promoting qualities, blackthorn is now attracting more attention. The potential positive and toxicological effects on the human body are still of great curiosity, even though blackthorn has been well-examined for its content in biologically active compounds. Nevertheless, it still needs to be clarified how to identify the substances specifically responsible for blackthorn's antibacterial, antiviral, and anticancer properties.

Additionally, the research must be focused on their use as a viable and sustainable alternative to medications, considering consumers' preferences for introducing more effective plant-based medications to the market and environmentally friendly manufacturing practices. Likewise, consumer demands for safe food free of chemicals have grown, which would make blackthorn anthocyanin extracts excellent as natural food ingredients, enhancing the color and the stability of the products. Therefore, it is imperative to re-evaluate blackthorn fruits as a significant source of safe and cost-effective antioxidants. This plant has the potential to be an excellent source of functional foods, nutritional supplements, and ingredients that prevent lipid oxidation in fat-rich foods. Moreover, as a future perspective, due to already proven antibacterial and antifungal properties, blackthorn extracts could represent an interesting research topic in the development and optimization of innovative and intelligent packaging solutions.

Author Contributions: Conceptualization, A.C.F., O.L.P. and S.A.S.; investigation, data curation, writing—original draft preparation, O.-R.N., O.L.P. and A.C.F.; supervision, A.C.F., S.A.S. and O.L.P.; project administration, O.L.P. and A.C.F.; funding acquisition, A.C.F. and O.L.P. All authors have read and agreed to the published version of the manuscript.

Funding: This work was supported by grants from the Romanian Ministry of Education and Research, CCCDI—UEFISCDI, projects PN-III-P4-ID-PCE-2020-2306 and PN-III-P4-ID-PCE-2020-2126, within PNCDI III.

Data Availability Statement: Not applicable.

Conflicts of Interest: The authors declare no conflict of interest.

References

1. Popescu, I.; Caudullo, G. Prunus spinosa in Europe: Distribution, habitat, usage and threats. In *European Atlas of Forest Tree Species*; San-Miguel-Ayanz, J., de Rigo, D., Caudullo, G., Houston Durrant, T., Mauri, A., Eds.; Publication Office of the European Union: Luxembourg, 2016; p. 145.
2. Dedić, A.; Dţudţević-Čančar, H.; Alispahić, A.; Tahirović, I.; Muratović, E. In-Vitro antioxidant and antimicrobial activity of aerial parts of *Prunus Spinosa* l. growing wild in Bosnia and Herzegovina. *Int. J. Pharm. Sci. Res.* **2021**, *12*, 3643–3653.
3. Sikora, E.; Bieniek, M.I.; Borczak, B. Composition and antioxidant properties of fresh and frozen stored blackthorn fruits (*Prunus spinosa* L.). *Acta Sci. Pol. Technol. Aliment.* **2013**, *12*, 365–372.
4. Ganhao, R.; Estévez, M.; Kylli, P.; Heinonen, M.; Morcuende, D. Characterization of selected wild Mediterranean fruits and comparative efficacy as inhibitors of oxidative reactions in emulsified raw pork burger patties. *J. Agric. Food Chem.* **2010**, *58*, 8854–8861. [CrossRef] [PubMed]
5. Guimaraes, R.; Barros, L.; Duenas, M.; Carvalho, A.M.; Queiroz, M.J.; Santos-Buelga, C.; Ferreira, I.C. Characterisation of phenolic compounds in wild fruits from Northeastern Portugal. *Food Chem.* **2013**, *141*, 3721–3730. [CrossRef] [PubMed]
6. Irizar, A.C.; Fernandez, M.F.; González, A.G.; Ravelo, A.G. Constituents of *Prunus spinosa*. *J. Nat. Prod.* **1992**, *55*, 450–454. [CrossRef]
7. Olszewska, M. Flavonoids from the Leaves of *Prunus spinosa* L. *Pol. J. Chem.* **2002**, *76*, 967–974.
8. Pozzo, L.; Russo, R.; Frassinetti, S.; Vizzarri, F.; Arvay, J.; Vornoli, A.; Casamassima, D.; Palazzo, M.; Croce, C.M.D.; Longo, V. Wild Italian *Prunus spinosa* L. Fruit Exerts In Vitro Antimicrobial Activity and Protects Against In Vitro and In Vivo Oxidative Stress. *Foods* **2019**, *9*, 5. [CrossRef]
9. Radovanović, B.C.; Anđelković, S.; Radovanović, A.B.; Anđelković, M.Z. Antioxidant and antimicrobial activity of polyphenol extracts from wild berry fruits grown in southeast Serbia. *Trop. J. Pharm. Res.* **2013**, *12*, 813–819. [CrossRef]
10. Veličković, J.M.; Kostić, D.A.; Stojanović, G.S.; Mitić, S.S.; Mitić, M.N.; Ranđelović, S.S.; Đorđević, A.S. Phenolic composition, antioxidant and antimicrobial activity of the extracts from *Prunus spinosa* L. fruit. *Hem. Ind.* **2014**, *68*, 297–303. [CrossRef]
11. Kumarasamy, Y.; Cox, P.J.; Jaspars, M.; Nahar, L.; Sarker, S.D. Comparative studies on biological activities of *Prunus padus* and *P. spinosa*. *Fitoterapia* **2004**, *75*, 77–80. [CrossRef]
12. Marakoğlu, T.; Arslan, D.; Özcan, M.; Hacıseferoğulları, H. Proximate composition and technological properties of fresh blackthorn (*Prunus spinosa* L. subsp dasyphylla (Schur.)) fruits. *J. Food Eng.* **2005**, *68*, 137–142. [CrossRef]
13. Ozzengin, B.; Zannou, O.; Koca, I. Quality attributes and antioxidant activity of three wild plums from Prunus spinosa and Prunus domestica species. *Meas. Food* **2023**, *10*, 100079. [CrossRef]
14. Fang, J. Classification of fruits based on anthocyanin types and relevance to their health effects. *Nutrition* **2015**, *31*, 1301–1306. [CrossRef] [PubMed]
15. Hidalgo, G.I.; Almajano, M.P. Red Fruits: Extraction of Antioxidants, Phenolic Content, and Radical Scavenging Determination: A Review. *Antioxidants* **2017**, *6*, 7. [CrossRef]
16. Alarcón, R.; Pardo-de-Santayana, M.; Priestley, C.; Morales, R.; Heinrich, M. Medicinal and local food plants in the south of Alava (Basque Country, Spain). *J. Ethnopharmacol.* **2015**, *176*, 207–224. [CrossRef]
17. Gunes, F. Medicinal plants used in the Uzunkopru district of Edirne, Turkey. *Acta Soc. Bot. Pol.* **2017**, *86*. [CrossRef]
18. Tardío, J.; Sánchez-Mata, M.d.C.; Morales, R.; Molina, M.; García-Herrera, P.; Morales, P.; Díez-Marqués, C.; Fernández-Ruiz, V.; Cámara, M.; Pardo-de-Santayana, M. Ethnobotanical and food composition monographs of selected Mediterranean wild edible plants. In *Mediterranean Wild Edible Plants*; Springer: Cham, Switzerland, 2016; pp. 273–470.
19. Karakas, N.; Okur, M.E.; Ozturk, I.; Ayla, S.; Karadag, A.E.; Polat, D.Ç. Antioxidant activity of blackthorn (*Prunus spinosa* L.) fruit extract and cytotoxic effects on various cancer cell lines. *Medeni. Med. J.* **2019**, *34*, 297.
20. Vokou, D.; Katradi, K.; Kokkini, S. Ethnobotanical survey of Zagori (Epirus, Greece), a renowned centre of folk medicine in the past. *J. Ethnopharmacol.* **1993**, *39*, 187–196. [CrossRef]

21. Condello, M.; Meschini, S. Role of Natural Antioxidant Products in Colorectal Cancer Disease: A Focus on a Natural Compound Derived from *Prunus spinosa*, Trigno Ecotype. *Cells* **2021**, *10*, 3326. [CrossRef]
22. Popović, B.; Blagojević, B.; Pavlović, R.Ž.; Mićić, N.; Bijelić, S.; Bogdanović, B.; Mišan, A.; Duarte, C.M.; Serra, A.T. Comparison between polyphenol profile and bioactive response in blackthorn (*Prunus spinosa* L.) genotypes from north Serbia-from raw data to PCA analysis. *Food Chem.* **2020**, *302*, 125373. [CrossRef]
23. Magiera, A.; Czerwinska, M.E.; Owczarek, A.; Marchelak, A.; Granica, S.; Olszewska, M.A. Polyphenol-Enriched Extracts of *Prunus spinosa* Fruits: Anti-Inflammatory and Antioxidant Effects in Human Immune Cells Ex Vivo in Relation to Phytochemical Profile. *Molecules* **2022**, *27*, 1691. [CrossRef] [PubMed]
24. Avasilcai, L.; Teliban, G.; Morariu, D.I.; Stoleru, V.; Bibire, N.; Vieriu, M.; Panainte, A.D.; Munteanu, N. Parameters of chemical composition of *Phaseolus coccineus* L. pods grown in protected areas. *Methods* **2017**, *68*, 2955–2958. [CrossRef]
25. Backes, E.; Leichtweis, M.G.; Pereira, C.; Carocho, M.; Barreira, J.C.; Genena, A.K.; Baraldi, I.J.; Barreiro, M.F.; Barros, L.; Ferreira, I.C. *Ficus carica* L. and *Prunus spinosa* L. extracts as new anthocyanin-based food colorants: A thorough study in confectionery products. *Food Chem.* **2020**, *333*, 127457. [CrossRef] [PubMed]
26. Gündüz, G.T. Antimicrobial activity of sloe berry purees on *Salmonella* spp. *Food Control* **2013**, *32*, 354–358. [CrossRef]
27. Leichtweis, M.G.; Pereira, C.; Prieto, M.; Barreiro, M.F.; Baraldi, I.J.; Barros, L.; Ferreira, I.C. Ultrasound as a rapid and low-cost extraction procedure to obtain anthocyanin-based colorants from *Prunus spinosa* L. fruit epicarp: Comparative study with conventional heat-based extraction. *Molecules* **2019**, *24*, 573. [CrossRef]
28. Kavaz Yuksel, A. The Effects of Blackthorn (*Prunus Spinosa* L.) Addition on Certain Quality Characteristics of Ice Cream. *J. Food Qual.* **2015**, *38*, 413–421. [CrossRef]
29. Ürkek, B.; Şengül, M.; Akgül, H.İ.; Kotan, T.E. Antioxidant activity, physiochemical and sensory characteristics of ice cream incorporated with sloe berry (*Prunus spinosa* L.). *Int. J. Food Eng.* **2019**, *15*. [CrossRef]
30. Ruiz-Rodríguez, B.M.; De Ancos, B.; Sánchez-Moreno, C.; Fernández-Ruiz, V.; de Cortes Sánchez-Mata, M.; Cámara, M.; Tardío, J. Wild blackthorn (*Prunus spinosa* L.) and hawthorn (*Crataegus monogyna Jacq.*) fruits as valuable sources of antioxidants. *Fruits* **2014**, *69*, 61–73. [CrossRef]
31. Celik, F.; Gundogdu, M.; Alp, S.; Muradoglu, F.; Ercişli, S.; Gecer, M.K.; Canan, I. Determination of phenolic compounds, antioxidant capacity and organic acids contents of *Prunus domestica* L., *Prunus cerasifera* Ehrh. and *Prunus spinosa* L. fruits by HPLC. *Acta Chromatogr.* **2017**, *29*, 507–510. [CrossRef]
32. Băbălău-Fuss, V.; Senila, L.; Becze, A.; Al-Zaben, O.B.; Dirja, M.; Tofană, M. Fatty acids composition from rosa canina and prunus spinosa plant fruit oil. *Stud. Univ. Babes-Bolyai Chem.* **2021**, *66*, 41–48. [CrossRef]
33. Atik, I.; Karasu, S.; Sevik, R. Physicochemical and bioactive properties of cold press wild plum (*Prunus spinosa*) and sour cherry (*Prunus cerasus*) kernel oils: Fatty acid, sterol and phenolic profile. *Riv. Ital. Sostanze Grasse* **2022**, *991*, 13–20.
34. Stanković, M.; Maksimović, S.; Tadić, V.; Arsić, I. The oil content of wild fruits from different plant species obtained by conventional Soxhlet extraction technique. *Acta Fac. Med. Naissensis* **2018**, *35*, 193–200.
35. Opriş, O.; Soran, M.-L.; Lung, I.; Stegarescu, A.; Guţoiu, S.; Podea, R.; Podea, P. Optimization of extraction conditions of polyphenols, antioxidant capacity and sun protection factor from *Prunus spinosa* fruits. Application in sunscreen formulation. *J. Iran. Chem. Soc.* **2021**, *18*, 2625–2636. [CrossRef]
36. Cefali, L.C.; Ataide, J.A.; Fernandes, A.R.; Sousa, I.M.O.; Goncalves, F.; Eberlin, S.; Davila, J.L.; Jozala, A.F.; Chaud, M.V.; Sanchez-Lopez, E.; et al. Flavonoid-Enriched Plant-Extract-Loaded Emulsion: A Novel Phytocosmetic Sunscreen Formulation with Antioxidant Properties. *Antioxidants* **2019**, *8*, 443. [CrossRef] [PubMed]
37. Mandic, S.; Savanovic, D.; Velmir, A.; Kalaba, V.; Savanovic, J.; Jokanovic, V. Effect of incorporating blackthorn fruit (*Prunus spinosa* L.) extract in natural casing on quality of Kranjska sausage. *Sci. J. Meat Technol.* **2018**, *59*, 80–90. [CrossRef]
38. Natić, M.; Pavlović, A.; Bosco, F.L.; Stanisavljević, N.; Zagorac, D.D.; Akšić, M.F.; Papetti, A. Nutraceutical properties and phytochemical characterization of wild Serbian fruits. *Eur. Food Res. Technol.* **2019**, *245*, 469–478. [CrossRef]
39. Gil, M.I.; Tomas-Barberan, F.A.; Hess-Pierce, B.; Kader, A.A. Antioxidant capacities, phenolic compounds, carotenoids, and vitamin C contents of nectarine, peach, and plum cultivars from California. *J. Agric. Food Chem.* **2002**, *50*, 4976–4982. [CrossRef]
40. Wang, H.; Cao, G.; Prior, R.L. Total antioxidant capacity of fruits. *J. Agric. Food Chem.* **1996**, *44*, 701–705. [CrossRef]
41. Lungu, I.; Huzum, B.; Humulescu, I.A.; Cioancă, O.; Morariu, D.; Şerban, I.-L.; Hăncianu, M. Flavonoids as promising therapeutic and dietary agents. *Med.-Surg. J.* **2020**, *124*, 151–156.
42. Ozkan, G. Phenolic compounds, organic acids, vitamin C and antioxidant capacity in *Prunus spinosa* L. *Comptes Rendus De L'academie Bulg. Des Sci.* **2019**, *72*, 267–273.
43. Luna-Vázquez, F.J.; Ibarra-Alvarado, C.; Rojas-Molina, A.; Rojas-Molina, J.I.; Bah, M. Prunus. In *Fruit and Vegetable Phytochemicals: Chemistry and Human Health*, 2nd ed.; Wiley: New York, NY, USA, 2017; pp. 1215–1226.
44. Yang, C.S.; Landau, J.M.; Huang, M.T.; Newmark, H.L. Inhibition of carcinogenesis by dietary polyphenolic compounds. *Annu. Rev. Nutr.* **2001**, *21*, 381–406. [CrossRef] [PubMed]
45. Lall, R.K.; Syed, D.N.; Adhami, V.M.; Khan, M.I.; Mukhtar, H. Dietary polyphenols in prevention and treatment of prostate cancer. *Int. J. Mol. Sci.* **2015**, *16*, 3350–3376. [CrossRef] [PubMed]
46. Santos, I.S.; Ponte, B.M.; Boonme, P.; Silva, A.M.; Souto, E.B. Nanoencapsulation of polyphenols for protective effect against colon–rectal cancer. *Biotechnol. Adv.* **2013**, *31*, 514–523. [CrossRef]

47. Verma, S.; Singh, A.; Mishra, A. Gallic acid: Molecular rival of cancer. *Environ. Toxicol. Pharmacol.* **2013**, *35*, 473–485. [CrossRef] [PubMed]
48. Dai, J.; Mumper, R.J. Plant phenolics: Extraction, analysis and their antioxidant and anticancer properties. *Molecules* **2010**, *15*, 7313–7352. [CrossRef] [PubMed]
49. Tripoli, E.; La Guardia, M.; Giammanco, S.; Di Majo, D.; Giammanco, M. Citrus flavonoids: Molecular structure, biological activity and nutritional properties: A review. *Food Chem.* **2007**, *104*, 466–479. [CrossRef]
50. Marčetić, M.; Samardžić, S.; Ilić, T.; Božić, D.D.; Vidović, B. Phenolic Composition, Antioxidant, Anti-Enzymatic, Antimicrobial and Prebiotic Properties of *Prunus spinosa* L. Fruits. *Foods* **2022**, *11*, 3289. [CrossRef]
51. Chun, O.K.; Kim, D.O.; Moon, H.Y.; Kang, H.G.; Lee, C.Y. Contribution of individual polyphenolics to total antioxidant capacity of plums. *J. Agric. Food Chem.* **2003**, *51*, 7240–7245. [CrossRef]
52. Dragovic-Uzelac, V.; Levaj, B.; Mrkic, V.; Bursac, D.; Boras, M. The content of polyphenols and carotenoids in three apricot cultivars depending on stage of maturity and geographical region. *Food Chem.* **2007**, *102*, 966–975. [CrossRef]
53. Fanning, K.J.; Topp, B.; Russell, D.; Stanley, R.; Netzel, M. Japanese plums (*Prunus salicina Lindl.*) and phytochemicals–breeding, horticultural practice, postharvest storage, processing and bioactivity. *J. Sci. Food Agric.* **2014**, *94*, 2137–2147. [CrossRef]
54. Ruiz, D.; Egea, J.; Gil, M.I.; Tomas-Barberan, F.A. Characterization and quantitation of phenolic compounds in new apricot (*Prunus armeniaca* L.) varieties. *J. Agric. Food Chem.* **2005**, *53*, 9544–9552. [CrossRef] [PubMed]
55. Tomas-Barberan, F.A.; Gil, M.I.; Cremin, P.; Waterhouse, A.L.; Hess-Pierce, B.; Kader, A.A. HPLC-DAD-ESIMS analysis of phenolic compounds in nectarines, peaches, and plums. *J. Agric. Food Chem.* **2001**, *49*, 4748–4760. [CrossRef] [PubMed]
56. Najgebauer-Lejko, D.; Liszka, K.; Tabaszewska, M.; Domagala, J. Probiotic Yoghurts with Sea Buckthorn, Elderberry, and Sloe Fruit Purees. *Molecules* **2021**, *26*, 2345. [CrossRef] [PubMed]
57. Mechchate, H.; Es-Safi, I.; Haddad, H.; Bekkari, H.; Grafov, A.; Bousta, D. Combination of Catechin, Epicatechin, and Rutin: Optimization of a novel complete antidiabetic formulation using a mixture design approach. *J. Nutr. Biochem.* **2021**, *88*, 108520. [CrossRef] [PubMed]
58. Binang, K.; Takuwa, D.T. Development of reverse phase-high performance liquid chromatography (RP-HPLC) method for determination of selected antihypertensive active flavonoids (rutin, myricetin, quercetin, and kaempferol) in medicinal plants found in Botswana. *Phys. Sci. Rev.* **2021**. [CrossRef]
59. Khoo, N.K.; White, C.R.; Pozzo-Miller, L.; Zhou, F.; Constance, C.; Inoue, T.; Patel, R.P.; Parks, D.A. Dietary flavonoid quercetin stimulates vasorelaxation in aortic vessels. *Free Radic. Biol. Med.* **2010**, *49*, 339–347. [CrossRef]
60. Li, P.G.; Sun, L.; Han, X.; Ling, S.; Gan, W.T.; Xu, J.W. Quercetin induces rapid eNOS phosphorylation and vasodilation by an Akt-independent and PKA-dependent mechanism. *Pharmacology* **2012**, *89*, 220–228. [CrossRef]
61. Chung, K.-T.; Wei, C.-I.; Johnson, M.G. Are tannins a double-edged sword in biology and health? *Trends Food Sci. Technol.* **1998**, *9*, 168–175. [CrossRef]
62. Koleckar, V.; Kubikova, K.; Rehakova, Z.; Kuca, K.; Jun, D.; Jahodar, L.; Opletal, L. Condensed and hydrolysable tannins as antioxidants influencing the health. *Mini Rev. Med. Chem.* **2008**, *8*, 436–447. [CrossRef]
63. Okuda, T.; Ito, H. Tannins of constant structure in medicinal and food plants—Hydrolyzable tannins and polyphenols related to tannins. *Molecules* **2011**, *16*, 2191–2217. [CrossRef]
64. He, J.; Giusti, M.M. High-purity isolation of anthocyanins mixtures from fruits and vegetables–A novel solid-phase extraction method using mixed mode cation-exchange chromatography. *J. Chromatogr. A* **2011**, *1218*, 7914–7922. [CrossRef]
65. Seeram, N.P.; Momin, R.A.; Nair, M.G.; Bourquin, L.D. Cyclooxygenase inhibitory and antioxidant cyanidin glycosides in cherries and berries. *Phytomedicine* **2001**, *8*, 362–369. [CrossRef]
66. Khoo, H.E.; Azlan, A.; Tang, S.T.; Lim, S.M. Anthocyanidins and anthocyanins: Colored pigments as food, pharmaceutical ingredients, and the potential health benefits. *Food Nutr. Res.* **2017**, *61*, 1361779. [CrossRef] [PubMed]
67. Wallace, T.C. Anthocyanins in cardiovascular disease. *Adv. Nutr.* **2011**, *2*, 1–7. [CrossRef] [PubMed]
68. Katanić Stanković, J.S.; Mićanović, N.; Grozdanić, N.; Kostić, A.Ž.; Gašić, U.; Stanojković, T.; Popović-Djordjević, J.B. Polyphenolic Profile, Antioxidant and Antidiabetic Potential of Medlar (*Mespilus germanica* L.), Blackthorn (*Prunus spinosa* L.) and Common Hawthorn (*Crataegus monogyna Jacq.*) Fruit Extracts from Serbia. *Horticulturae* **2022**, *8*, 1053. [CrossRef]
69. Sabatini, L.; Fraternale, D.; Di Giacomo, B.; Mari, M.; Albertini, M.C.; Gordillo, B.; Rocchi, M.B.L.; Sisti, D.; Coppari, S.; Semprucci, F. Chemical composition, antioxidant, antimicrobial and anti-inflammatory activity of *Prunus spinosa* L. fruit ethanol extract. *J. Funct. Foods* **2020**, *67*, 103885. [CrossRef]
70. Brglez Mojzer, E.; Knez Hrncic, M.; Skerget, M.; Knez, Z.; Bren, U. Polyphenols: Extraction Methods, Antioxidative Action, Bioavailability and Anticarcinogenic Effects. *Molecules* **2016**, *21*, 901. [CrossRef]
71. Veličković, I.; Žižak, Ž.; Rajčević, N.; Ivanov, M.; Soković, M.; Marin, P.D.; Grujić, S. *Prunus spinosa* L. leaf extracts: Polyphenol profile and bioactivities. *Not. Bot. Horti Agrobot. Cluj-Napoca* **2021**, *49*, 12137. [CrossRef]
72. Wang, M.; Zhang, Z.; Sun, H.; He, S.; Liu, S.; Zhang, T.; Wang, L.; Ma, G. Research progress of anthocyanin prebiotic activity: A review. *Phytomedicine* **2022**, *102*, 154145. [CrossRef]
73. Pinacho, R.; Cavero, R.Y.; Astiasarán, I.; Ansorena, D.; Calvo, M.I. Phenolic compounds of blackthorn (*Prunus spinosa* L.) and influence of in vitro digestion on their antioxidant capacity. *J. Funct. Foods* **2015**, *19*, 49–62. [CrossRef]
74. Rajbhar, K.; Dawda, H.; Mukundan, U. Polyphenols: Methods of extraction. *Sci. Revs. Chem. Commun.* **2015**, *5*, 1–6.

75. Wissam, Z.; Ghada, B.; Wassim, A.; Warid, K. Effective extraction of polyphenols and proanthocyanidins from pomegranate's peel. *Int. J. Pharm. Pharm. Sci.* **2012**, *4*, 675–682.
76. Dent, M.; Dragović-Uzelac, V.; Penić, M.; Bosiljkov, T.; Levaj, B. The effect of extraction solvents, temperature and time on the composition and mass fraction of polyphenols in Dalmatian wild sage (*Salvia officinalis* L.) extracts. *Food Technol. Biotechnol.* **2013**, *51*, 84–91.
77. Turkmen, N.; Velioglu, Y.S.; Sari, F.; Polat, G. Effect of extraction conditions on measured total polyphenol contents and antioxidant and antibacterial activities of black tea. *Molecules* **2007**, *12*, 484–496. [CrossRef]
78. Suwal, S.; Marciniak, A. Technologies for the Extraction, Separation and Purification of polyphenols–A Review. *Nepal J. Biotechnol.* **2018**, *6*, 74–91. [CrossRef]
79. Tešić, Ž.L.; Gašić, U.M.; Milojković-Opsenica, D.M. Polyphenolic profile of the fruits grown in Serbia. In *Advances in Plant Phenolics: From Chemistry to Human Health*; ACS Publications: Washington, DC, USA, 2018; pp. 47–66.
80. D'Alessandro, L.G.; Dimitrov, K.; Vauchel, P.; Nikov, I. Kinetics of ultrasound assisted extraction of anthocyanins from *Aronia melanocarpa* (black chokeberry) wastes. *Chem. Eng. Res. Des.* **2014**, *92*, 1818–1826. [CrossRef]
81. Barba, F.J.; Zhu, Z.; Koubaa, M.; Sant'Ana, A.S.; Orlien, V. Green alternative methods for the extraction of antioxidant bioactive compounds from winery wastes and by-products: A review. *Trends Food Sci. Technol.* **2016**, *49*, 96–109. [CrossRef]
82. Lucchesi, M.E.; Chemat, F.; Smadja, J. Solvent-free microwave extraction of essential oil from aromatic herbs: Comparison with conventional hydro-distillation. *J. Chromatogr. A* **2004**, *1043*, 323–327. [CrossRef] [PubMed]
83. Veličković, I.; Žižak, Ž.; Rajčević, N.; Ivanov, M.; Soković, M.; Marin, P.D.; Grujić, S. Examination of the polyphenol content and bioactivities of *Prunus spinosa* L. fruit extracts. *Arch. Biol. Sci.* **2020**, *72*, 105–115. [CrossRef]
84. Marchelak, A.; Owczarek, A.; Matczak, M.; Pawlak, A.; Kolodziejczyk-Czepas, J.; Nowak, P.; Olszewska, M.A. Bioactivity Potential of *Prunus spinosa* L. Flower Extracts: Phytochemical Profiling, Cellular Safety, Pro-inflammatory Enzymes Inhibition and Protective Effects Against Oxidative Stress In Vitro. *Front. Pharmacol.* **2017**, *8*, 680. [CrossRef]
85. Ferreira, L.; Pereira, R.R.; Carvalho-Guimaraes, F.B.; Remigio, M.; Barbosa, W.L.R.; Ribeiro-Costa, R.M.; Silva-Junior, J.O.C. Microencapsulation by Spray Drying and Antioxidant Activity of Phenolic Compounds from Tucuma Coproduct (*Astrocaryum vulgare* Mart.) Almonds. *Polymers* **2022**, *14*, 2905. [CrossRef]
86. Mitroi, C.L.; Gherman, A.; Gociu, M.; Bujancă, G.; Cocan, E.N.; Rădulescu, L.; Megyesi, C.I.; Velciov, A. The antioxidant activity of blackthorn fruits (*Prunus Spinosa* L. *) review. J. Agroaliment. Process. Technol.* **2022**, *28*, 288–291.
87. Lapidot, T.; Harel, S.; Akiri, B.; Granit, R.; Kanner, J. pH-dependent forms of red wine anthocyanins as antioxidants. *J. Agric. Food Chem.* **1999**, *47*, 67–70. [CrossRef] [PubMed]
88. Wang, S.Y.; Lin, H.S. Antioxidant activity in fruits and leaves of blackberry, raspberry, and strawberry varies with cultivar and developmental stage. *J. Agric. Food Chem.* **2000**, *48*, 140–146. [CrossRef] [PubMed]
89. Fraternale, D.; Giamperi, L.; Bucchini, A.; Sestili, P.; Paolillo, M.; Ricci, D. *Prunus spinosa* fresh fruit juice: Antioxidant activity in cell-free and cellular systems. *Nat. Prod. Commun.* **2009**, *4*, 1665–1670.
90. Olesińska, K.; Sugier, D.; Sęczyk, Ł. The influence of selected preservation methods and storage time on the content of antioxidants in blackthorn (*Prunus spinosa* L.) fruits. *Agron. Sci.* **2019**, *74*, 53–62. [CrossRef]
91. Smullen, J.; Koutsou, G.A.; Foster, H.A.; Zumbe, A.; Storey, D.M. The antibacterial activity of plant extracts containing polyphenols against *Streptococcus mutans*. *Caries Res.* **2007**, *41*, 342–349. [CrossRef]
92. Temiz, M.A.; Okumus, E.; Yaman, T.; Keles, O.F. Mixture of leaf and flower extract of *Prunus spinosa* L. alleviates hyperglycemia and oxidative stress in streptozotocin-induced diabetic rats. *S. Afr. J. Bot.* **2021**, *141*, 145–151. [CrossRef]
93. Temiz, M.A.; Temur, A. The effect of olive leaf extract on digestive enzyme inhibition and insulin production in streptozotocin-induced diabetic rats. *Ank. Üniversitesi Vet. Fakültesi Derg.* **2019**, *66*, 163–169.
94. Crnić, I.; Frančić, T.; Dragičević, P.; Balta, V.; Dragović-Uzelac, V.; Đikić, D.; Landeka Jurčević, I. Blackthorn Flower Extract Impact on Glycaemic Homeostasis in Normoglycaemic and Alloxan-Induced Hyperglycaemic C57BL/6 Mice. *Food Technol. Biotechnol.* **2021**, *59*, 349–359. [CrossRef]
95. Cosmulescu, S.; Trandafir, I.; Nour, V. Phenolic acids and flavonoids profiles of extracts from edible wild fruits and their antioxidant properties. *Int. J. Food Prop.* **2017**, *20*, 3124–3134. [CrossRef]
96. Varga, E.; Domokos, E.; Fogarasi, E.; Steanesu, R.; Fulop, I.; Croitoru, M.D.; Laczko-Zold, E. Polyphenolic compounds analysis and antioxidant activity in fruits of *Prunus spinosa* L. *Acta Pharm. Hung.* **2017**, *87*, 19–25. [PubMed]
97. Velickovic, J.; Ilic, S.; Mitic, S.; Mitic, M.; Kostic, D. Comparative analysis of phenolic and mineral composition of hawthorn and blackthorn from southeast Serbia. *Oxid. Commun.* **2016**, *39*, 2280–2290.
98. Egea, I.; Sánchez-Bel, P.; Romojaro, F.; Pretel, M.T. Six edible wild fruits as potential antioxidant additives or nutritional supplements. *Plant Foods Hum. Nutr.* **2010**, *65*, 121–129. [CrossRef]
99. Coppari, S.; Colomba, M.; Fraternale, D.; Brinkmann, V.; Romeo, M.; Rocchi, M.B.L.; Di Giacomo, B.; Mari, M.; Guidi, L.; Ramakrishna, S.; et al. Antioxidant and Anti-Inflammaging Ability of Prune (*Prunus Spinosa* L.) Extract Result in Improved Wound Healing Efficacy. *Antioxidants* **2021**, *10*, 374. [CrossRef] [PubMed]
100. Šamec, D.; Durgo, K.; Grúz, J.; Kremer, D.; Kosalec, I.; Piljac-Žegarac, J.; Salopek-Sondi, B. Genetic and phytochemical variability of six Teucrium arduini L. populations and their antioxidant/prooxidant behaviour examined by biochemical, macromolecule-and cell-based approaches. *Food Chem.* **2015**, *186*, 298–305. [CrossRef] [PubMed]

101. Kello, M.; Kulikova, L.; Vaskova, J.; Nagyova, A.; Mojzis, J. Fruit peel polyphenolic extract-induced apoptosis in human breast cancer cells is associated with ROS production and modulation of p38MAPK/Erk1/2 and the Akt signaling pathway. *Nutr. Cancer* **2017**, *69*, 920–931. [CrossRef] [PubMed]
102. Murati, T.; Miletic, M.; Kolaric, J.; Lovric, V.; Kovacevic, D.B.; Putnik, P.; Jurcevic, I.L.; Dikic, D.; Dragovic-Uzelac, V.; Kmetic, I. Toxic activity of *Prunus spinosa* L. flower extract in hepatocarcinoma cells. *Arch. Hig. Rada Toksikol.* **2019**, *70*, 303–309.
103. Condello, M.; Pellegrini, E.; Spugnini, E.P.; Baldi, A.; Amadio, B.; Vincenzi, B.; Occhionero, G.; Delfine, S.; Mastrodonato, F.; Meschini, S. Anticancer activity of "Trigno M", extract of *Prunus spinosa* drupes, against in vitro 3D and in vivo colon cancer models. *Biomed. Pharmacother.* **2019**, *118*, 109281. [CrossRef]
104. Novembre, E.; Dini, L.; Bernardini, R.; Resti, M.; Vierucci, A. Unusual reactions to food additives. *Pediatr. Med. Chir.* **1992**, *14*, 39–42. [PubMed]
105. Balta, I.; Sevastre, B.; Miresan, V.; Taulescu, M.; Raducu, C.; Longodor, A.L.; Marchis, Z.; Maris, C.S.; Coroian, A. Protective effect of blackthorn fruits (*Prunus spinosa*) against tartrazine toxicity development in albino Wistar rats. *BMC Chem.* **2019**, *13*, 104. [CrossRef] [PubMed]
106. Hassan, G. Effects of some synthetic coloring additives on DNA damage and chromosomal aberrations of rats. *Arab. J. Biotechnol.* **2010**, *13*, 13–24.
107. Balta, V.; Đikić, D.; Crnić, I.; Odeh, D.; Orsolic, N.; Kmetič, I.; Murati, T.; Dragović Uzelac, V.; Landeka Jurčević, I. Effects of four-week intake of blackthorn flower extract on mice tissue antioxidant status and phenolic content. *Pol. J. Food Nutr. Sci.* **2021**, *70*, 361–375. [CrossRef]
108. Đikić, D.; Balta, V.; Pedisić, S.; Zorić, Z.; Padovan, J.; Butorac, D.; Milić, A.; Jurić, D.; Jurčević, I.L. Polyphenol bioavailability and modulatory potential on brain antioxidative markers in C57BL/6 mouse. *Period. Biol.* **2022**, *124*, 41–54. [CrossRef]
109. Milutinović, M.; Dimitrijević-Branković, S.; Rajilić-Stojanović, M. Plant extracts rich in polyphenols as potent modulators in the growth of probiotic and pathogenic intestinal microorganisms. *Front. Nutr.* **2021**, *8*, 688843. [CrossRef] [PubMed]
110. Blagojević, B.; Četojević-Simin, D.; Djurić, S.; Lazzara, G.; Milioto, S.; Agić, D.; Vasile, B.S.; Popović, B.M. Anthocyanins and phenolic acids from *Prunus spinosa* L. encapsulation in halloysite and maltodextrin based carriers. *Appl. Clay Sci.* **2022**, *222*, 106489. [CrossRef]

MDPI AG
Grosspeteranlage 5
4052 Basel
Switzerland
Tel.: +41 61 683 77 34

Molecules Editorial Office
E-mail: molecules@mdpi.com
www.mdpi.com/journal/molecules

www.ingramcontent.com/pod-product-compliance
Lightning Source LLC
Chambersburg PA
CBHW061514210926
51850CB00008B/1917
9783725858859